**Contributions
to Current Research
in Geophysics (CCRG)**

7

Geothermics and Geothermal Energy

Editors: Ladislaus Rybach

Institute of Geophysics
ETH Zurich, Switzerland

Lajos Stegena

Eötvös University
Budapest, Hungary

Reprinted from PAGEOPH

1979   Birkhäuser Verlag, Basel und Stuttgart

Reprinted from Pure and Applied Geophysics
(PAGEOPH),
Volume 117 (1978), Nos 1/2

CIP-Kurztitelaufnahme der Deutschen Bibliothek

Geothermics and geothermal energy: reprinted from
PAGEOPH/ed.: Ladislaus Rybach; Lajos Stegena.
– Basel, Stuttgart: Birkhäuser, 1979.
  (Contributions to current research in geophysics;
  7)
  ISBN 3-7643-1062-6
NE: Rybach, Ladislaus [Hrsg.]

© Birkhäuser Verlag Basel, 1979

ISBN 3-7643-1062-6

Contents

Editors' Note 1

General Geothermics

J. W. ELDER: Magma Traps: Part I, Theory 3
J. W. ELDER: Magma Traps: Part II, Application 15
ARTHUR H. LACHENBRUCH: Heat Flow in the Basin and Range Province and Thermal
   Effects of Tectonic Extension 34
CRAIG S. WEAVER and DAVID P. HILL: Earthquake Swarms and Local Crustal
   Spreading Along Major Strike-slip Faults in California 51
V. M. HAMZA: Variation of Continental Mantle Heat Flow with Age: Possibility of
   Discriminating Between Thermal Models of the Lithosphere 65
LADISLAUS RYBACH: The Relationship Between Seismic Velocity and Radioactive
   Heat Production in Crustal Rocks: An Exponential Law 75
GÜNTER BUNTEBARTH: The Degree of Metamorphism of Organic Matter in Sedimen-
   tary Rocks as a Paleo-Geothermometer, Applied to the Upper Rhine Graben . . . 83

Regional Heat Flow

VLADIMÍR ČERMÁK and ECKART HURTIG: The Preliminary Heat Flow Map of Europe
   and Some of its Tectonic and Geophysical Implications 92
R. I. KUTAS, E. A. LUBIMOVA AND YA. B. SMIRNOV: Heat Flow Map of the European
   Part of the U.S.S.R. 104
JACEK MAJOROWICZ: Mantle Heat Flow and Geotherms for Major Tectonic Units in
   Central Europe 109
CRISAN DEMETRESCU: On the Geothermal Regime of Some Tectonic Units in Roma-
   nia 124
M. LODDO and F. MONGELLI: Heat Flow in Italy 135
YORAM ECKSTEIN: Review of Heat Flow Data from the Eastern Mediterranean
   Region 150

Geothermal Potential

L. J. PATRICK MUFFLER and ROBERT L. CHRISTIANSEN: Geothermal Resource Assess-
   ment of the United States 160
T. J. LEWIS, A. S. JUDGE and J. G. SOUTHER: Possible Geothermal Resources in the
   Coast Plutonic Complex of Southern British Columbia, Canada 172
V. M. HAMZA, S. M. ESTON and R. L. C. ARAUJO: Geothermal Energy Prospects in
   Brazil: A Preliminary Analysis 180
K. G. ERIKSSON, K. AHLBOM, O. LANDSTRÖM, S. Å. LARSON, G. LIND and D. MALM-
   QVIST: Investigation for Geothermal Energy in Sweden 196
NIELS BALLING and SVEND SAXOV: Low Enthalpy Geothermal Energy Resources in
   Denmark 205
PAUL MORGAN and CHANDLER A. SWANBERG: Heat Flow and the Geothermal Poten-
   tial of Egypt 213

Exploration, Characterisation and Exploitation of Geothermal Resources

CHANDLER A. SWANBERG and PAUL MORGAN: The Linear Relation Between Temperatures Based on the Silica Content of Groundwater and Regional Heat Flow: A New Heat Flow Map of the United States 227

INGVAR B. FRIDLEIFSSON: Applied Volcanology in Geothermal Exploration in Iceland 242

F. D'AMORE, J. C. SABROUX and P. ZETTWOOG: Determination of Characteristics of Steam Reservoirs by Radon-222 Measurements in Geothermal Fluids 253

EMANUEL MAZOR: Noble Gases in a Section across the Vapor Dominated Geothermal Field of Larderello, Italy 262

ALFRED H. TRUESDELL and NANCY L. NEHRING: Gases and Water Isotopes in a Geochemical Section across the Larderello, Italy, Geothermal Field 276

MORTON C. SMITH: Heat Extraction from Hot, Dry, Crustal Rock 290

ALAIN C. GRINGARTEN: Reservoir Lifetime and Heat Recovery Factor in Geothermal Aquifers used for Urban Heating 297

Geothermal Effects of Hydrothermal Circulation

J. W. PRITCHETT and S. K. GARG: Flow in an Aquifer Charged with Hot Water from a Fault Zone 309

C. H. SONDERGELD and D. L. TURCOTTE: Flow Visualization Studies of Two-Phase Thermal Convection in a Porous Layer 321

JUN IRIYAMA and YASUE ŌKI: Thermal Structure and Energy of the Hakone Volcano, Japan 331

Subject Index 339

Editors' Note

The rapidly growing interest in geothermal energy as an alternative energy source has raised many basic questions in geothermics. Especially, the contribution of the various disciplines of earth sciences to the understanding of development and distribution of geothermal resources, along with various aspects of their utilisation, became evident. At the 16th IUGG General Assembly 1975 in Grenoble the International Heat Flow Commission (IHFC) decided to hold a Symposium 'Geothermics and Geothermal Energy', with the aim of clarifying some of these questions.

The joint General Assembly of the International Associations of Seismology and Physics of the Earth's Interior (IASPEI) and Volcanology and Chemistry of the Earth's Interior (IAVCEI), held in Durham/England from 9 to 19 August 1977, provided an international forum for presenting and discussing results of current research in general and applied geothermics at its *Joint Symposium 'Geothermics and Geothermal Energy'* (11–12 August 1977). The Symposium was co-sponsored by the IHFC; its convenors were the IHFC members L. Stegena and L. Rybach.

The Opening Lecture of the Symposium was delivered by IHFC President E. Lubimova (USSR): 'On the heat loss of the earth'. Invited papers were given by V. Čermák (ČSSR) and E. Hurtig (GDR), P. Muffler and R. L. Christiansen (USA) and M. C. Smith (USA). Session Chairmen were L. Stegena (Hungary), A. Lachenbruch (USA), F. Evison (New Zealand), G. Pálmason (Iceland) and L. Rybach (Switzerland). The great number of the presented contributions as well as the broad international participation in this Symposium has clearly demonstrated the rapidly growing importance of geothermics.

This volume contains the papers presented at the Symposium 'Geothermics and Geothermal Energy' with a few modifications. The following papers were given in Durham but are not included in this issue (for various reasons, e.g. no manuscripts submitted or published elsewhere):

E. E. DAVIS, C. R. B. LISTER and U. S. WADE (USA): Detailed heat flow measurements in faulted and unfaulted young oceanic crust.

S. W. RICHARDSON and P. C. ENGLAND (UK): Age dependence of continental heat flow.

M. H. DODSON (UK): A model for terrestrial convection.

F. HORVÁTH, L. BODRI and L. STEGENA (Hungary): The heat anomaly of the Pannonian Basin and its tectonophysical background.

W. A. ELDERS (USA): Physical effects of water/rock reactions in evolving geothermal systems.

P. L. ERNST (FRG): Exploitation of geothermal energy in Germany.

M. L. GUPTA (India): Heat flow distribution and possibilities of exploiting geothermal energy in India.

M. TONELLI (Italy): Airborne and space geothermics.

H. M. IYER and T. HITCHCOCK (USA): Use of seismic noise and P-wave delays in geothermal exploration.

R. ALVAREZ (Mexico): Telluric detection of geothermal areas.

J. R. BLOOMER (UK): Thermal conductivity of Mesozoic mudstones from Southern England.

D. S. CHAPMAN, F. H. BROWN, K. L. COOK, W. P. NASH, W. T. PARRY, W. R. SILL, R. B. SMITH, S. H. WARD and J. A. WHELAN (USA): Roosevelt Hot Springs-Utah: a Basin and Range geothermal system.

F. F. EVISON (New Zealand): The downflow regime in natural hydro-thermal systems.

On the other hand, the present volume contains several papers which were scheduled for presentation at the Durham Meeting but could not be given there for various reasons.

In these Proceedings the contributions of the Symposium have been arranged in the following order: (i) General Geothermics, (ii) Regional Heat Flow, (iii) Geothermal Potential, (iv) Exploration, Characterisation and Exploitation of Geothermal Resources, (v) Geothermal Effects of Hydrothermal Circulation. In editing the Proceedings an attempt was made to use *SI units* throughout. This goal has nearly been achieved; at least both units are given, particularly in the figures. The kind cooperation of numerous colleagues who acted as reviewers is gratefully acknowledged.

Sincere thanks are due to Dr. R. E. Long, Organizing Secretary of the IASPEI/IAVCEI General Assembly (Department of Geological Sciences, Durham University) for providing an ideal infrastructure for the Symposium. The participants departed with the feeling of not only having discussed data and results but also having gained deeper insight into the manifold interrelations of geothermics and geothermal energy.

Budapest and Zurich, July 1978

L. STEGENA
L. RYBACH

Pageoph, Vol. 117 (1978/79), Birkhäuser Verlag, Basel

Magma Traps: Part I, Theory

By J. W. Elder[1])

Abstract – The role of the buoyancy barrier of the stratified crust in controlling the ascent of magmas is represented by a model which operates close to lithostatic equilibrium and in which fractional crystallization of the magma or partial melting of the ambient rock can occur. The density structure of the crust has a powerful effect in trapping magma and thereby in controlling the occurrence of high level geothermal systems. The model is tested in Part II (this volume).

Key words: Ascent of magma; Lithostatic equilibrium; Oceanic and continental crust; Density of magma.

1. Introduction

For the heat flowing from the interior of the earth to be most readily available for use by high-level geological systems or human exploitation it needs to be retained temporarily in a high-level reservoir. Perhaps the simplest systems of this kind are deep sedimentary basins which are heated by normal near surface conductive heat flow. A good example is the Pannonian basin of eastern Europe. The behaviour of these rather passive systems are well understood. Such is not the case for active systems related to contemporary volcanism. Strong geothermal activity, as found in modern hydrothermal systems, is undoubtedly related to ambient volcanism. Yet it is curious that often only minor parasitic hydrothermal systems are directly associated with active volcanic systems. This leads us to consider the very simple idea that in such systems the bulk of the energy has been dissipated at the surface and to raise the question of under what circumstances a magmatic system can retain part or all of its energy at depth. We are lead to the notion of geothermal or magma traps.

Where a volume of magma is trapped below the surface in a zone sufficiently permeable and with access to surface and connate waters, a hydrothermal system can develop. I have presented elsewhere an outline of the processes involved (ELDER, 1977).

A variety of trapping mechanisms have already been identified. Trapping can occur at quite deep levels simply because the rising magma volume freezes – and certain granitic systems, because of the chemical buffering of water, are especially

---

[1]) Geology Department, Manchester University, Manchester M13 9PL, England.

prone to this (HARRIS *et al.*, 1970). Trapping can occur at high levels in many ways, a notable example being in Recent subglacial volcanism of Iceland because of the enhanced porosity and permeability of the deposits (for example: FRIDLEIFSSON, 1976).

In this study, however, I wish to draw attention to what is perhaps the simplest trapping mechanism of all – namely the static effects of the ambient stratified crust and mantle on the magma column. The key idea is not new, and it has been used by several authors (for example: HOLMES, 1965; RAMBERG, 1963 – esp. pp. 53ff; MODRINIAK and STUDT, 1959). As far as I am aware, its testing against data from the central West Greenland Upper Cretaceous – Lower Tertiary sedimentary and volcanic area, described in Part II (this volume), is the most detailed study to date.

The study is incomplete in so far as purely static constraints are imposed on the model. The next step would be a study of reservoir dynamics.

2. Lithostatic controls of volcanism

Consider those aspects of lithothermal systems which are mechanically controlled by variations of the ambient crust. The emphasis is not so much on the magma but on the gross features of its container. For the want of a word I refer to lithostatics. In this and subsequent sections we calibrate these ideas by discussion of particular systems.

If the solid earth were a homogeneous body, production of a sufficiently voluminous partially molten region at depth would always lead to a surface discharge, since the magma density being less that of its solid parent, the magma column, even if it reached the surface would have less weight than a similar body of solid rock. But all magmas do not reach the surface. This is simply because the crust is stratified, with material near the surface not only less dense than that at depth but is also less dense than some magmas. For these magmas the uppermost layers of the earth present a buoyancy barrier which can be penetrated or punctured only under special circumstances.

Manometric model

The key idea used in the quantitative discussion is the notion of lithostatic equilibrium. Isostatic readjustments are geologically very rapid, with a time scale of order 10^4 yr (HASKELL, 1937). Lithostatic models of crustal and upper mantle systems with time scales greater than 10^4 yr provide therefore good approximations to the overall distribution of mass. All the models below treat the system under discussion as if it were a simple manometer constructed from two vertical limbs – for example, one limb may be the unchanged crust the other limb the thinned crust. The essence of the calculations is that since the manometer is close to static equilibrium, equating the two pressures at the base of each limb is sufficient to determine the situation.

The structure of a volcanic system will also be treated as a manometer operating close to static equilibrium. Clearly the lava pile will continue to grow locally until it reaches static equilibrium. Only this terminal or equilibrium state is used here for the identification of the volcanic manometer.

Rearrangement of the crust and upper mantle occurs on a time scale of order 100 Myr. Hence volcanic systems can be regarded in the gross aspects of their behaviour as quasi-steady systems embedded in an ambient system which is slowly changing. Although the volcanic system remains close to equilibrium this does not imply that it also necessarily changes slowly – as we shall see below.

Crustal layers

Consider a layered crust as sketched in Fig. 1 made up of a stack of layers of thickness h_i and density ρ_i where, counting the layers downwards 0 to $n + 1$: layer 0

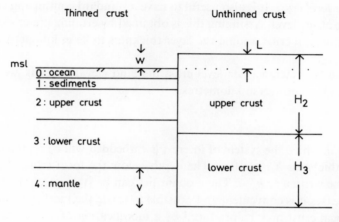

Figure 1
Schematic representation of the crust.

represents the ocean; layer $n + 1$ represents the deepest layer of interest, part of the mantle. If a layer is absent, we simply set its corresponding thickness to zero. Here only the following are considered, with $n = 3$:

0. the ocean of density 1.03 g/cm³, taken as 1 g/cm³ in all calculations;
1. unconsolidated sediments of density 2.5 g/cm³;
2. upper crust of consolidated sediments and gneissic basement of density 2.7 g/cm³;
3. lower crust to density 3.0 g/cm³, gravitationally indistinguishable from basalt;
4. mantle of density 3.3 g/cm³.

Reference crust. It is particularly useful to have a standard or reference crust. This can be used as the basis of a variety of models. Although the choice is somewhat arbitrary at least it is only arbitrary once. Here I make a choice by requiring that the ocean and continental parts of the reference crust are: (i) mutually in static equilibrium and (ii) are separately close to what is known of the structure of stable cratonic areas and stable abyssal oceanic areas. Thus, if we choose the layer thicknesses, H_j for $j = 0$–3, in km, as:

oceanic crust: 5, 2, 0, 5;
cratonic crust: 0, 0, 25, 10.

then the land elevation of the cratonic crust is 1.03 km. This figure is a little higher than would be suggested by the global average land elevation of 0.8 km (see for example: WOOLLARD, 1969 and ELDER, 1976). A cratonic lower crust of 8 km gives a land elevation of 0.85 km. Changing the cratonic upper crust to 20 km, gives a land elevation of 0.12 km. This latter case is for example the sort of situation suitable for representation of young or small continental type crustal areas as in New Zealand.

Reference sea-level crust. It is also useful to have a standard continental crust with its upper surface at sea level. Assuming this is obtained by eroding upper crust from the standard continental crust we find the layer thickness to be in kilometres: 0, 0, 19.3, 10.

Relative to the reference sea-level crust the land elevation, L is given by $H_2 = 19.3 + 5.5L$, all quantities in kilometres.

Local crust

The crust in which the system of interest is embedded will be represented by a set of layers of thickness h_j, $j = 0$–n. The depth below the local crust of an object of interest will be written as h_{n+1}. These quantities can be stated separately.

In many cases it is convenient and possible to relate the local crust to a reference crust. The local crust may be produced by a modification of a reference crust. For example, in the discussion below of tertiary volcanism in the 'rift' zone of central West Greenland the local crust is considered as being produced by thinning the old continental crust and is represented by:

$$h_1 = \xi h_{11}, \qquad h_2 = (1 - \xi)H_2, \qquad h_3 = (1 - \xi)H_3 + \xi H_{31}, \qquad (1)$$

with $0 \le \xi \le 1$. Here a single parameter ξ allows representation of a range of crusts from $\xi = 0$, unmodified continental crust to $\xi = 1$, oceanic crust. Note that if the depth of an object of interest is given in terms of H_{n+1}, the quantity h_{n+1} will also be a function of ξ.

Isostasy

Apart from a few isolated zones such as oceanic trenches the upper layers of the earth are close to lithostatic equilibrium. Where departures from equilibrium occur, the local crust returns to equilibrium after the disturbing stresses are removed with a time scale of order 10^4 yr. Systems of much greater time scale embedded in the crust can be treated as if the ambient crust was permanently in lithostatic equilibrium. Hence we have:

$$\sum_0^n (H_j - h_j) = v + w, \qquad \sum_0^n \rho_j(H_j - h_j) = \rho_{n+1}v \qquad (2)$$

where w is the difference of levels of the tops of the two crusts and v is the difference of levels of their bases. In the simplest case of given H_j, h_j these relations determine v, w. In the case of a continental reference crust the relation of the interfaces to sea level is given by reference to the land elevation, L.

It is more usual for one or more of the h_j to be the unknown. Consider the commonest situation where this is h_0, the water depth. For convenience write:

$$h_a = \sum_0^n H_j,$$

$$p_a = \sum_0^n \rho_j H_j,$$

$$h_t = \sum_0^n h_j \equiv h_0 + h_t',$$

$$p_t = \sum_0^n \rho_j h_j \equiv \rho_0 h_0 + p_t', \qquad (3)$$

$$\tilde{v} = (p_a - p_t' - \rho_0 h_0)/\rho_{n+1}$$

$$\tilde{u} = h_a - h_t' - w - \tilde{v}$$

$$\tilde{w} = h_a - h_t' - \tilde{v}.$$

(a) *Continental reference crust.* Two cases arise. Notice that $\tilde{v} > 0$.

(i) $h_0 \equiv 0$. Isostasy requires:

$$w + v + h_t' = h_a, \qquad \rho_{n+1}v + p_t' = p_a \qquad (4)$$

whence $v = \tilde{v}, w = \tilde{w}$. The local surface is $(L - w)$ above sea level. This is the solution provided $\tilde{w} \leq L$. Otherwise $h_0 > 0$.

(ii) $h_0 > 0$; $w \equiv L$, given. Isostasy requires:

$$w + v + h_0 + h_t' = h_a, \qquad \rho_{n+1}v + \rho_0 h_0 + p_t' = p_a, \qquad (5)$$

whence by first eliminating v we have

$$v = \tilde{v}, \qquad h_0 = \rho_{n+1}\tilde{u}/(\rho_{n+1} - \rho_0).$$

In either case the elevation of all the interfaces is then determined.

(b) *Oceanic reference* ($L \equiv 0$). Two cases arise here also and can be treated in a similar fashion. Notice that now $\tilde{v} < 0$.

(i) $h_0 \equiv 0$. We find:

$$v = -\tilde{v}, \qquad w = -\tilde{w} \tag{6}$$

The local land elevation is w. This is the solution provided $\tilde{w} \leq 0$.

(ii) $h_0 > 0$, $w \equiv 0$. We find:

$$v = -\tilde{v}, \qquad h_0 = \rho_{n+1}\tilde{u}/(\rho_{n+1} - \rho_0). \tag{7}$$

3. The magma column

Magma fractionation

The lava pile thickness is determined by conditions in the magma column and in the ambient ground.

As the magma rises from depth its character changes. In models of the type discussed here the information of interest is $\rho(z)$, the density of the magmatic fluid as a function of depth z below the top of the lava pile. Clearly $\rho(z)$ is a piecewise continuous function, but its detailed form is virtually unknown. In the simplest case I represent $\rho(z)$ as a single step function:

$$
\begin{aligned}
z \leq z_*, &\qquad \rho = \rho_b, \\
z > z_*, &\qquad \rho = \rho_a,
\end{aligned}
\tag{8}
$$

with $\rho_b < \rho_a$ and z_* defined, for a given magmatic source, by the 'characteristic pressure' p_* at $z = z_*$, as shown in Fig. 2. It is important to appreciate that ρ_a, ρ_b and p_* are not independent parameters of the models described here. They only occur in combination to evaluate the pressures within the magma column. Thus the pressure, $p(z)$ as a function of depth, z in the magma column, where g is the acceleration of gravity is given by:

$$
\begin{aligned}
z < z_* &: p = g\rho_b z, \\
z = z_* &: p = p_* = g\rho_b z_*, \\
z > z_* &: p = p_* + g\rho_a(z - z_*).
\end{aligned}
\tag{9}
$$

The choice of a characteristic pressure is both convenient and reasonable. For a given magmatic source, a statement of ρ_a and ρ_b is essentially a statement of composition, and the temperature is then defined by the dynamical conditions, for example isenthalpic for rapid injection of magma, so that pressure is the dominant free thermodynamic variable. It is important to realize however that p_* is a phenomenological variable solely related to a step function approximation to $\rho(z)$. Nevertheless it can be thought of, rather crudely, as representing the pressure level at which the greatest

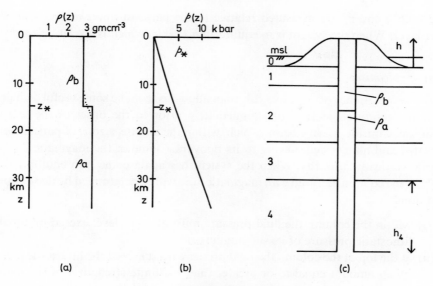

Figure 2
Schematic illustration of the conditions in an open magma column. (a) Density $\rho(z)$; and (b) pressure $p(z)$ as a function of depth below the upper surface of the column. The dotted line shows a possible actual density profile, together with the step function approximation drawn as a solid line. (c) Schema for a hydrostatic model for determining the thickness of a lava pile.

rate of change of composition with depth occurs. p_* is regarded solely as a property of the source magma, and not of its surroundings.

An independent determination of ρ_a, ρ_b and p_* is difficult. Various petrogenetic models of the derivation of magmas have been proposed (WYLLIE, 1971). Most of these schemes have sequential differentiation. For example in basaltic systems there is proposed middle level differentiation at 5–10 kbar and high level differentiation at typically 0–3 kbar. Possible values for the magma densities are: $\rho_a = 3.1$ g cm^{-3} for the hyperbasic magma; and $\rho_b = 2.8$ g cm^{-3} for the high level magma. Evidence from oceanic island volcanism suggests that these upper levels are well within the oceanic crust (MALAHOFF, 1969), so that for example in the case of Hawaii with a 9 km pile on a 5 km oceanic crust the pressure p_* is rather less than 5 kbar.

Consider the system of a magma column penetrating a layered crust and producing a volcanic pile of ultimate elevation h. The lithostatic pressure, P at the source level

$$P = \sum_{0}^{n+1} \rho_i h_i, \tag{10}$$

where h_{n+1} is the mantle depth of the source. The pressure, p in the magma column at the source level

$$p = \rho_b(h + z_*) + \rho_a\left(\sum_{0}^{n+1} h_i - z_*\right). \tag{11}$$

Here both h and z_* are measured relative to the palaeosurface, or the base of the water layer. When the system is in equilibrium $P = p$, which is a relation for h.

Onset of volcanism

The very existence of extrusive volcanics allows us to make some useful deductions about the condition of the crust immediately prior to the onset of the extrusive volcanism. Consider a crust below which, within the mantle, a zone of partial melting is created and maintained. Owing to its buoyancy some of the magmatic fluid will migrate vertically upwards. When the system has again come into equilibrium the height of the column perfused with magmatic fluid will be determined by two dominant conditions:

  (i) within the column the fluid pressure must at every level exceed or equal the lithostatic pressure of its surroundings;
  (ii) at the top of the column the fluid pressure must exceed the lithostatic pressure by an amount equal to or greater than the finite strength σ of the material above, otherwise its advance will cease.

In addition, the crust and mantle are stratified in a manner such that if $P(z)$ is the lithostatic pressure as a function of depth, the gradient dP/dz generally increases with depth. Thus, for example, as shown in Fig. 3, from a source represented by

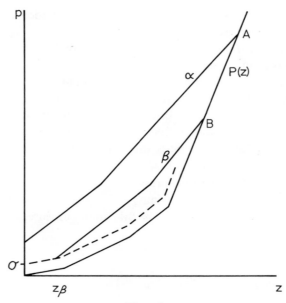

Figure 3

Diagrammatic representation of the pressure, p, as a function of depth, z, in a stratified crust: lithostatic pressure, $P(z)$; fluid pressure for two different magma columns α, β for sources at A, B. Finite strength σ.

point A, the pressure distribution in the magma column is given by line α, so that a discharge is possible. But for a source represented by point B, and the pressure distribution in the magma column represented by line β, a discharge is not possible and the column is terminated at depth z_β.

The magnitude of the finite strength, σ is not well known. A nominal value for crustal rocks of 1 kbar is frequently used and there is an estimate for basaltic lavas of about 0.1 kbar (ELDER, 1976). The role of σ is largely in controlling conditions at and before the onset of eruptions. For values of order 0.1 kbar the effects are minor; for values of order 1 kbar most of the models do not even allow eruptions. For the want of anything better I used $\sigma = 0.3$ kbar in the numerical models.

It is worth noting that where wet sediments are encountered by an intruding magma there may be a type of phreatic eruption generated by the high gas pressure of the vaporized water (see for example, ELDER 1977). This effect is equivalent to having a lower finite strength – indeed, often a negative finite strength.

Consider, as sketched in Fig. 4, that the stratified crust and mantle is given as a

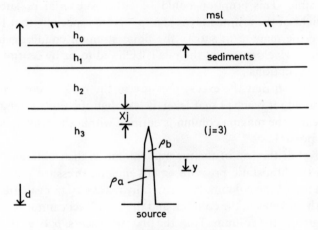

Figure 4

Schematic arrangement for consideration of the conditions in the magma column which allow it to rise. A layered crust with a magma column reaching from a source in the mantle into layer j. The magma column has an upper portion of density ρ_b and a lower portion of density ρ_a.

sequence of layers of thickness h_i and density ρ_i. To be determined is the disposition of the magma column both in vertical extent and the level z_* between the deep denser magma of density ρ_a and the shallow lighter magma of density ρ_b, if any.

(i) The source depth below sea level:

$$d = \sum_0^{n+1} h_i. \tag{12}$$

The lithostatic pressure at the level of the source is given by equation (10) which includes the contributions from the sea layer if present. Note that in both relations (10) and (12) we use h_{n+1} for the depth of the source below the base of the thinned crust. The quantity, d the source depth is taken as fixed throughout the development so that eqn. (12) is actually a relation for h_{n+1}.

(ii) The depth y below the top of the sediments at which the characteristic pressure p_* is reached is:

$$y = d - h_0 - (P - p_*)/\rho_a. \tag{13}$$

(iii) When $y < 0$ the dense hyperbasic magma is above the top of the sediments.

(iv) When $y > 0$ the dense magma level is below the palaeosurface. If there is a zone of light magma, the excess pressure p' at the top of the sediments is given by:

$$p' = p_* - \rho_b y. \tag{14}$$

(v) When first $p' > \sigma_1$, that is, the excess pressure in the magma column at the top of the sediments exceeds the finite strength of the sediments, eruption of the light magma is possible. This eruption could be either subaerial or subaqueous. The consequent height of the erupted lavas is p'/ρ_b. Note especially that I do not subsequently require the magma to satisfy the finite strength condition in its own pile since the ultimate development of the pile is believed to be by central eruptions and associated flank eruptions.

(vi) If $p' < \sigma_1$, so that the excess pressure in the magma column at the top of sediments, assuming it reached that level, is less than the finite strength of the sediments, then clearly the magma column is entirely within the crust, and in particular no eruption is possible.

(vii) The level of the top of the magma column is then obtained by finding the level at which the lithostatic pressure equals magma pressure + finite strength of the rock at the top of the column. It is generally necessary to evaluate the two possibilities: (a) both magmas in the column, and then if a level cannot be found; (b) only the denser magma in the column. Thus the pressure excess, p_c over the finite strength at depth x_j in layer j in the magma column is:

$$p_c = p' + \rho_b x_j. \tag{15}$$

The lithostatic load is:

$$P_1 = \sum_0^{j-1} \rho_i h_i + \rho_j \left(x_j - \sum_0^{j-1} h_i \right). \tag{16}$$

We search successively with $j = 1, 2, 3, \ldots$ till we find $p_c = P_1$ and a corresponding value of x_j. If this fails the column contains only the denser magma and we replace p_c eqn (15) and repeat the search with:

$$p_c = p_* - \rho_a y + \rho_a x_j. \tag{17}$$

The behaviour of the magma column is illustrated in Fig. 5 using the data of Table 1, with $p_* = 5$ kbar and $\sigma = 0.3$ kbar. (1) For a given source, there will be a thickness of crust for which no discharge is possible. The magma column will rise above the source as the crust is thinned, that is as ξ increases from zero, but there is a range of ξ for which the magma column remains below the land surface. (2) The striking aspect of this model is the sharpness with which, at a critical crustal thinning,

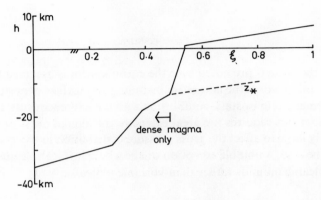

Figure 5

Level of the top of the magma column, h as a function of crustal thinning parameter ξ. Negative values of h refer to depth below the original surface. Values in kilometres. The level z_* is also shown.

Table 1

Model data

ρ_0	1.00 g/cm³	Ocean water density, nominal value
ρ_1	2.5 g/cm³	Unconsolidated sediment density
ρ_2	2.7 g/cm³	Upper crust, or consolidated sediment density
ρ_3	3.0 g/cm³	Lower crust density
ρ_4	3.3 g/cm³	Mantle density
L	0.6 km (0.4 km)	Land elevation of ice free continental interior. Initial estimate and revised estimate.
H_2	25 km	Upper crust thickness, unthinned
H_3	10 km	Lower crust thickness, unthinned
H_{31}	5 km	Oceanic crust thickness (completely thinned lower crust)
ρ_b	2.8 g/cm³	High level magma density
ρ_a	3.1 g/cm³	Deep level magma density
σ	0.3 kbar	Fracturing strength of rock
p_*	5 kbar (3 kbar)	Characteristic pressure, initial estimate and revised estimate.
ξ	0.55	Crustal thinning parameter, model design value

Notes:

(a) $H_0 = 0$, dry continental interior; $H_1 = 0$, no unconsolidated sediments on continental interior remote from the embayment.

(b) The actual density of ocean water is 1.03 g/cm³ but the nominal value of 1.00 g/cm³ has been used in all calculations here.

eruptive volcanism commences. The sharp start is controlled by the non-uniform density stratification of the crust and its finite strength.

Thus we have a very diagnostic observation. There is a sharply defined crustal thickness. In the case of a crust gradually being thinned there would be negligible surface manifestation until the critical condition was reached, whereupon a convulsive start would be made of an eruptive phase which would continue till the lava pile or plateau was built or the source turned off.

4. Final remark

In most of the models presented here the entire system is assumed to be in static equilibrium. In other words the new crust, including any surface deposits and internal mass rearrangement, is in isostatic equilibrium with the reference crust. It is important to appreciate that this requires the areal extent of the altered parts of the new crust to be sufficiently large to affect the gross pressure distribution in the system. This will not always be the case. A notable exception will be where surface volcanism is confined to localized volcanic mounds rather than volcanic plateaux.

REFERENCES

ELDER, J. W., *The bowels of the earth* (Oxford University Press 1976), 222 pp.

ELDER, J. W. (1977), *Model of hydrothermal ore genesis*, J. Geo. Soc. London, *Volcanic Processes in Ore Genesis*, Special Issue 7, 4–13.

FRIDLEIFSSON, I. B., *Lithology and structure of geothermal reservoir rocks in Iceland, 371–376*, in *Proc. Second U.N. Symp., Development and Use of Geothermal Resources, San Francisco* (U.S. Govt. Printing Office, Washington, DC, USA 1976).

HARRIS, P. G., KENNEDY, W. Q. and SCARFE, C. M. (1970), *Volcanism versus plutonism – the effect of chemical composition, 187–200*, in Newall, G. and Rast, N. (eds.), *Geological Journal*, Special Issue 2, 380 pp.

HASKELL, N. A. (1937), *The viscosity of the asthenosphere*, Amer. J. Sci. 33, 22–28.

HOLMES, A., *Principles of Physical Geology* (Nelson, London 1965), 1288 pp.

MALAHOFF, A. (1969), *Gravity anomalies over volcanic regions, 364–379*, in Hart, P. J. (ed.) *The Earth's Crust and Upper Mantle*, Geophysical monograph 13, Amer. Geophys. Union.

MODRINIAK, N. and STUDT, F. E. (1959), *Geological structure and volcanism of the Taupo – Tarawera district*, N.Z. J. Geol. Geophys. 2, 654–684.

RAMBERG, H. (1963), *Experimental study of gravity tectonics by means of centrifuged models*, Bull. Geol. Soc. Uppsala 42, 1–97.

WOOLLARD, G. P. (1969), *Standardization of gravity measurements, and regional variations in gravity, 283–292, 320–341*, in Hart, P. J. (ed.), *The Earth's Crust and Upper Mantle*, Geophysical monograph 13, Amer. Geophys. Union.

WYLLIE, P. J., *The Dynamic Earth* (Wiley, New York 1971), 416 pp.

(Received 1st November 1977, revised 21st February 1978)

Pageoph, Vol. 117 (1978/79), Birkhäuser Verlag, Basel

Magma Traps: Part II, Application

By J. W. Elder[1])

Abstract – A model of the lithostatic control of the ascent of magma, described in Part I (this volume), is tested against data from the Upper Cretaceous–Lower Tertiary sedimentary and volcanic region of central West Greenland: the thickness of sedimentary rock; the thickness of the pillow breccias; the total thickness of the lava pile; the depth of the post volcanic paleosurface. The local development is largely determined by a single parameter, the proportion of crustal thinning, and requires a magma source at 75 km depth with differentiation at 11 km depth. The model is applied in outline to the development of continental and orogenic volcanism in New Zealand.

Key words: West Greenland; New Zealand; Crustal thinning; Crustal melting; Lava pile; Bouguer anomalies.

1. Introduction

The role of the buoyancy barrier of the stratified crust in controlling the ascent of magmas has been represented by a model, described in Part I (this volume), which presumes that the system operates close to lithostatic equilibrium. The model is tested against data from the central West Greenland Upper Cretaceous–Lower Tertiary sedimentary and volcanic area. There the idea works well and encourages its application to other fields. A sketch of an application to trapping within a region of acidic volcanism is given.

The notation, definitions and quantities are those of Part I.

2. Sub-continental volcanism

Direct evidence for the above hypothesised control of the rise of the magma column by the buoyancy barrier of the crust is shown in a study of the development of the crustal structure in central West Greenland during Cretaceous–Lower Tertiary time (ELDER, 1978).

As seen today, the volcanic region of central West Greenland, apart from the relatively small oceanic basin of Baffin Bay, lies on and within a large block of ancient

---

[1]) Geology Department, Manchester University, Manchester M13 9PL, England.

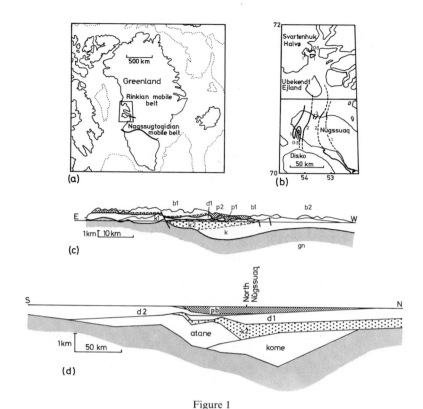

Figure 1
Localities and stratigraphy of central West Greenland region.

(a) Greenland and its surroundings. The dotted contour shows ocean floor 2000 m below sea level. The box encloses the central West Greenland region.

(b) The locality of the Cretaceous–Tertiary sediments and Tertiary volcanism. The sediment trough, shown as contours of depth, in kilometres, below mean sea level of the base of the 3.2 km s^{-1} sedimentary layer, determined by refraction and reflection seismic measurements.

(c) Schematic representation of the stratigraphy in an 80 km section along northern Nûgssuaq, shown as profiles projected on the east–west vertical surface 70°45′N. Viewed looking south, vertical exaggeration 2:1.
Formations:

    b2 upper basalts, feldspar-phyric
    b1 lower basalts, olivine rich
    p2 pillow breccia, olivine basalt
    p1 highly altered pillow breccia, olivine basalt
    d2 Upper Danian, marine sandstone (small isolated 100 m thick patches in this section – not shown)
    d1 Lower Danian, marine shale and sandstone
    k2 Upper Cretaceous (Coniacian to Maastrichtian), shale and sandstone of mixed origin.
    k1 Middle Cretaceous (Barremian to Turonian), non-marine shale and sandstone.
    k Middle to Lower Cretaceous (presumed) and older (?)
    gn gneiss basement, identified seismically, or as seen in coastal outcrop

Two topographic profiles also shown: (i) on 70°45′N; (ii) along the high ridge behind the coast. (Drawn from GGU geological maps, 1:100 000, 70 V.1 and 70 V.2)

(d) Gross sedimentary stratigraphy, shown as a schematic reconstructed axial section, at the time of onset of subaerial basalts. Drawn by projecting known sections on the line NE Disko to East Svartenhuk Halvø with the known or supposed subaerial basalt level taken as a level surface – the top surface of the diagram. Vertical exaggeration 20:1. The formations are: the gneissic basement; Kome, Lower Cretaceous (Barremian-Aptian, 118–106 Myr ago); Atane, Cretaceous (Upper Turonian-Conician, 94–82 Myr ago); k2, marine and terrestrial formations, Upper Cretaceous; d1, Lower Danian; d2, Upper Danian; pb, basaltic pillow breccia. (Adapted from a diagram prepared by E. J. Schiener for the book *Geology of Greenland*, GGU 1976, personal communication. See HENDERSON *et al.*, 1976.)

crust of continental scale – see Fig. 1(a). The basement of the embayment is composed
of granodioritic gneisses of generally amphibolite facies of the Rinkian mobile belt
which was active 1680–1870 Myr ago and lies just to the north of the slightly older
Nagssugtoqidian mobile belt (ESCHER and PULVERTAFT, 1976). As this block migrated
northward since before 200 Myr ago (ROY, 1973) there has been an intra-continental
zone of extensive rifting (HENDERSON, 1973) and the development of small temporary
ocean covered basins but in which the crust remained largely continental except for a
progressive thinning which reached a localized peak of activity in the central West
Greenland area about 60 Myr ago. Subsequent tectonic activity has been minor so
that much of the crustal structure developed at the time has been preserved both
on the land and off-shore.

The crustal trough

Sediments have been deposited extensively over the intra-continental zone during
at least the past 120 Myr (oldest sediments are Barremian age, 118 Myr ago) to
thicknesses of several kilometres. Off-shore, both in the bathyl region of Baffin Bay
and on the intra-continental slopes, there are thick deposits over broad areas and
especially within minor marginal basins (see for example KEEN et al., 1974). The
off-shore sediments are of uncertain age and have been delineated only rather
crudely. On land, however, in the central West Greenland area there is an extensive
and well-developed succession of terrestrial and marine sediments.

The form of the crustal trough, otherwise referred to as the embayment, is shown
in Fig. 1(b). The stratigraphy on land is known in great detail (HENDERSON et al.,
1976). A transverse schematic section as seen today is shown in Fig. 1(c). This
represents the stratigraphy across the deepest part of the basin. This section exposes
all the major formations of the area. An axial schematic, as reconstructed for about
60 Myr ago and much simplified, is shown in Fig. 1(d). This shows most clearly the
sequence of subsidence which produced the basin. In more recent time much of this
material has been uplifted by up to at least 2 km.

A thick sequence of terrestrial sediments deposited in a subsiding basin was
followed by marine transgression and subsequent deposition of shallow marine
sandstones and shales. The initial volcanism was at least in part subaqueous, pillow
breccias of thickness 0.5 km being typical. The rate of development of the lava pile
exceeded that of the subsidence since the bulk of the lava pile is subaerial.

The lava plateau

The most prominent feature of the area and that which is of interest here is the
lava plateau. This structure and its rocks as seen on land have been studied by many
workers (CLARKE and PEDERSEN, 1976). The areal extent on land is well exposed and
recent magnetic surveys, together with dredging, have been made to delineate the

outline off-shore. Two such compilations are included in Fig. 2. The area shown in outline is about 5×10^4 km.

These flood basalts have been named as the West Greenland Basalt Group (HALD and PEDERSEN, 1975) with formations:

Lower basalt: Vaigat Formation, tholeiitic olivine-rich and picrite basalts;
Upper basalt: Maligât Formation, plagioclase–porphyritic tholeiitic basalts together with a thin Hareøen Formation of transitional olivine–porphyritic basalts.

Duration estimates from dating measurements. Volcanism seems to have started in the Danian. For example, JÜRGENSEN and MIKKELSEN (1974) have identified coccoliths in Danian andesitic tuffs at Marrait kitdlit (south Nûgssuaq) and at Kangilia (north Nûgssuaq), an age range of 59.5–63 Myr. These tuffs can also be correlated with early native iron bearing lavas suggesting that they were deposited contemporaneously with the lower picrite lavas and breccias of the Vaigat Formation (see CLARKE and PEDERSEN, 1976, p. 373).

An estimate of the duration of the episode of lava production can be made from the observation that of the lavas so far measured for direction of magnetisation all but a few are the same – reversed. The notable exception is the well-established zone of normal magnetisation low in the Vaigat Formation (ATHAVALE and SHARMA. 1975). In view of the rather scattered and unsystematic sampling of the lava pile, particularly poor in its upper levels, this observation is by no means certain but let us accept it for the moment. Then, provided some other evidence gives us an approximate age of the lavas, we can look for a suitable known interval of reversal. Whereas the earth's magnetic field has had normal and reversed intervals averaging about 0.3 Myr in the recent past, before 45 Myr ago the intervals were generally longer (HEIRTZLER *et al.*, 1968). For example, during 55–65 Myr ago the normal interval averaged 0.6 Myr, the reversed interval 1.2 Myr. In Danian time there was an exceptionally long period of reversal from 60.5–62.7 Myr ago. This event fits with the 'Danian' reversed volcanics. Thus, even allowing for the undoubted errors in the dating of the magnetic time scale, this suggests that these lavas were extruded in this period of reversal – a total time of about 2.2 Myr. The interval of 2.2 Myr is a lower bound for this interval of lava production.

The bulk of the lavas could have been produced in a few million years, perhaps about 3 Myr. The time of development of the lava pile was short compared to that of the embayment as a whole, so that the production of the lava pile can be treated as a system isolated in time although spatially embedded in the embayment.

The pillow breccia thickness. The deposits of pillow breccias, indicated in Fig. 2, are within the area of the sediment trough. This sedimentary and breccia trough can be regarded as an elongated depressed block, probably composed of numerous smaller blocks, which was subsiding during the sedimentation. It is important to realise,

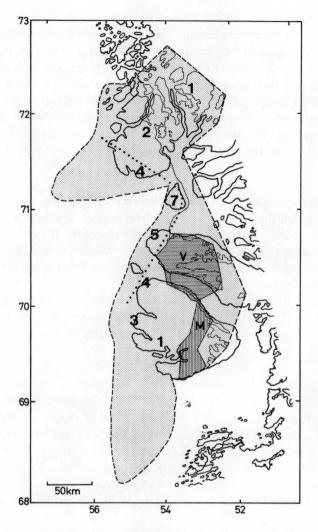

Figure 2

Gross outline of the central West Greenland volcanic pile now. Eastward land exposure indicated by fine lines, strippled area drawn 'headland to headland.' Off-shore outline given by DENHAM (1974) and in the extreme south by PARK *et al.* (1971). Some of the material in the north off-shore may be ice-rafted deposits, identified north and west of Svartenhuk Halvø by BAKER and FRIEDMAN (1973). The line of crustal thinning is shown dotted.

Also shown basaltic breccia occurrence, in the southern part of the volcanic area, from 69°–71°N. The outlines have been drawn 'headland to headland' assuming continuity across the Vaigat, using the 1:500 000 Sondre Strømfjord–Nûgssuaq, Geological Map, GGU 1973. V, Vaigat formation breccia (sub-aqueous lower basalt); M, Maligât formation breccia (sub-aqueous upper basalt).

The numerical labels indicate lava pile thickness, in km, based on a compilation by ELDER (1978) from data given by: MÜNTHER (1973), DREVER (1958), HALD (1973), HALD and PEDERSEN (1975), various others and the author.

however, that the breccias are not a regional chronostratigraphical unit. They are locally early lithostratigraphic elements. In a wider context they are aqueous facies: of the Vaigat Formation, in Nûgssuaq; and the Maligât Formation, in south Disko. Thus the presumably deeper parts of the trough, in Nûgssuaq were filled with sub-aqueous basalts and then covered with sub-aerial basalts before the onset of sub-aqueous volcanism in south Disko. In the model presented below this is interpreted as due to a continuing subsidence within the trough controlled by the amount of local crustal thinning. The thickness of these breccias reaches 0.7 km with typical thicknesses of 0.5 km. This thickness is an important quantity in calibrating the models below.

The lava pile thickness. The thickness of the lava plateau provides one of the dominant quantities in the identification of the manometer system. The lava pile is extensively exposed on land so that reasonably reliable estimates of the thickness can be made – these are indicated in Fig. 2. The pile is roughly lens shaped, more or less flat bottomed with a domed upper surface reaching a maximum thickness in excess of 7 km. The average thickness is about 5 km.

We have a pile of original average thickness about 5 km spread over an area of about 5×10^4 km. Thus the original pile had a volume of 2.5×10^5 km^3, somewhat bigger than Hawaii with 1.4×10^5 km^3 and comparable in volume with that of other lava plateaux.

Amount of subsidence. It is possible to make an estimate of the amount of lowering of the palaeosurface, at the base of the pillow breccias. Sub-aqueous breccias form part of the Maligât Formation, but were proceeded in Nûgssuaq and north Disko by 1.5 km and 2 km respectively of the Vaigat Formation. The Maligât Formation breccias are typically up to 0.5 km thick. Clearly the situation will be complicated by any differential vertical displacements. If these are ignored, the above figures suggest a subsidence of the palaeosurface of at least 2 to 2.5 km, with 3 km quite possible. A typical figure of 2.5 km is used in the models below.

Post-volcanic uplift. The most casual inspection on land of the sedimentary basin shows that there was a substantial overall subsidence before and during the volcanism. And yet it is clear that subsequently the land had risen up. For example, in north Nûgssuaq we see pillow breccias at 1500 m above sea level. This, together with the measurement of breccia uplift together with the estimate of the elevation of the original palaeosurface $L = 0.6$ km, suggests a net vertical displacement of about 2 km.

The elevation of the land in and adjacent to the embayment after the volcanism is clearly related to regional processes. An uplift of 2 km could arise from the combined effect of a continued crustal thinning of about 0.5 in the intra-continental zone leading to the formation of Baffin Bay together with a mean erosion of about 1.5 km of land (ELDER, 1978).

The crust as a whole – the gravity data, overall features

Gravity data, converted to a relative Bouguer anomaly map, is shown in Fig. 3. In the eastern part of the area shown there is an extensive region of roughly uniform gravity, this region being bordered by the -60 milligal contour, further east of which gravity is somewhat lower with a typical level of -70 milligal. In the western part of the area shown, there is another extensive region of gentle variation, this being bordered by the $+60$ milligal contour, with a typical level immediately outside the volcanic region of $+70$ milligal. The total range of the change in gravity across the structure reaches extreme values of about 170 milligal, but is more typically about 120–140 milligal.

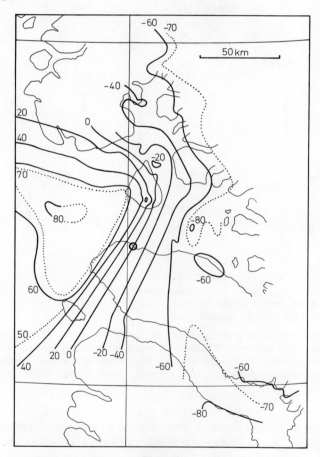

Figure 3

Gravity anomaly map. Contour labels in milligals. Measurements converted to Bouguer values *relative* to the value at the Tuperssuartâ base, north Nûgssuaq, shown as a heavy circle, based on land and sea ice stations. Note: The contours do *not* show absolute Bouguer values. To obtain the absolute Bouguer values *add* 20 milligal to the values shown.

   Between these two regions of relatively uniform gravity there is a pronounced
westerly increase across an area which occurs in a strip 40 km wide at its narrowest,
trending very roughly north–south through the area. This indicates a dramatic crustal
change in which the western crust has been substantially modified.

   The gravitational effects of lateral variation of the uppermost clastics and volcanics
are disregarded in the gravity interpretation. The gravity data is already corrected
to sea level as if upper crust extended to sea level, so that in the gravity model
$h_0 = h_1 = H_0 = H_1 = 0$. We have the situation sketch in Fig. 4(a). Define $\xi = \xi(s)$

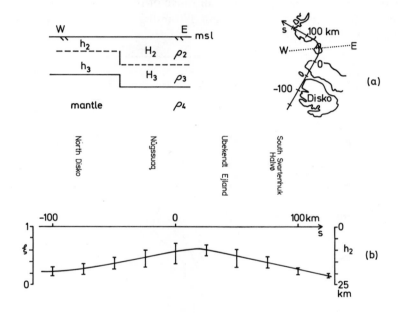

Figure 4
Gravity interpretation.
(a) Simple gravity model, vertical section drawn normal to the line of crustal thinning. Map showing line
of crustal thinning, distance s and a possible 'profile' line.

(b) Proportion of crustal thinning, as a function of position, s along the line of crustal thinning, measured
relative to the Tuperssuartâ base, north Nûgssuaq. Distances in kilometres. Also shown on the right-hand
side is the thickness of upper crust to the west, h_2. The vertical bars indicate the range of ξ for the
fit to the profile at the plotted s, drawn as $\pm$ twice the root mean square difference between the model and
the actual profiles.

as the proportion of crustal thinning where h_2 and h_3 are given by equation (I.1) and
$0 < \xi < 1$ where s is the distance, measured from a reference point – here the
Tuperssuartâ base, north Nûgssuaq – along the trend of the gravity data, taken to
be the line zero milligal shown in Fig. 3. Hence for a chosen value of the thicknesses
of the pre-existing crust the gravity data can be used to obtain the proportion of
crustal thinning, $\xi(s)$ as a function of position along the trend.

The variation of the proportion of crustal thinning, $\zeta(s)$ is shown in Fig. 4(b). There is strong thinning in the region from Disko to Svartenhuk Halvø and extreme thinning is found between north Nûgssuaq and south Ubekendt Ejland where the thinning exceeds 0.7 and the upper crust is a mere 8 km thick. The present data gives the location of greatest thinning at $s = 20$ km, a few kilometres south of Ubekendt Ejland. Also shown are the ranges of values of ζ. The range crudely indicates the local variability of the crustal thinning, whereas the solid line gives a smoothed value. The amount of crustal thinning is rather variable on the local scale.

In later discussion it is useful to have a typical value of ζ for use in the initial design of various models. The value chosen is $\bar{\zeta} = 0.55$. This is an average value over the central part of the region of thinning where the full stratigraphic sequence is measurable.

Volcanism within the embayment

I consider the crustal restructuring as determined by a single process operating on two scales:

(i) an intra-continental zone of progressive crustal thinning, determined by processes on a continental scale which, in particular, is responsible for the post-volcanic uplift;

(ii) a local region of rapid short-lived crustal thinning, superimposed on the slower but long duration continental scale thinning, and determined principally by local conditions beneath the embayment.

We ignore the details at the margins of the two parts and treat them as two distinct layered blocks. The requirement of static equilibrium then imposes constraints on the thicknesses and densities of the various layers.

We have considered the development of a basin as a consequence of thinning the crust below the basin – see equations (I.1) to (I.7). We have considered the development of a magma column and the lava pile in a given basin – see equations (I.12) to (I.17). In this combined model as sketched in Fig. 5 both processes are at work. As the crust thins and sinks down relative to the remaining undisturbed crust, the magma column experiences a changed ambient pressure distribution which modifies its behaviour.

It should be noted, in this model, that the depth of water is controlled by a competition between the crustal thinning, which leads to a lowering of the crust and the sedimentation which leads to an effective raising of the palaeosurface. Thus the water depth is an important factor in the identification of the model.

Since the initial development of the lava pile, immediately after onset of the extrusive volcanism, is rapid compared to all the other processes, in particular the sedimentation, in the model once onset has been reached further sedimentation is cut off. In reality some sedimentation will still occur, but the total extra thickness

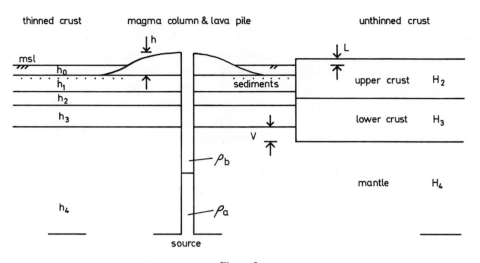

Figure 5
Schema and nomenclature of the combined model of crustal thinning and evolution of the lava pile.

will be a very small proportion of the lava pile thickness. It is worth remarking that the very small amount of intercalated sediment in the subaqueous breccias and lavas is clear evidence of the rapidity of the initial extrusion phase. The effect of turning off the sedimentation has the effect in the model of allowing the magmatic system to rise slightly higher than it would otherwise do, since there is less lighter material to act as a buoyancy barrier, and to give a slightly greater water depth owing to the somewhat greater crustal load.

The model has been adjusted to provide a fit to the following data:

(i) The rate of production of sediment, a parameter of the model, must be such that at onset of volcanism the local terrestrial sediment thickness is that found by field measurement in the range 0–3 km and typically 2.0 km.

(ii) The depth of water at onset must be such that the initial volcanics are locally subaqueous, and remain subaqueous until a breccia pile of measured thickness develops, in the range 0–0.7 km and typically 0.5 km.

(iii) The total thickness of the lava pile must reach the estimated values in the range 0–8 km, and typically 5 km.

(iv) The depth to the palaeosurface at the end of the volcanic phase must be in the estimated range 0–3 km and typically 2.5 km.

Although the model contains several free parameters they are severely constrained to rather narrow limits. Many possibilities can be immediately discarded as having for example: no discharge; zero water depth; no basalt. The method of finding acceptable combinations has been iterative, by varying successively one parameter at a time to find the best range for that parameter, and then successively repeating the scan of the parameters to obtain the best fit. The criterion for the goodness of fit was the value of the root mean square of the proportional differences (model value — field value)/(field value) of the above four quantities.

Take as given the data of Table 1, of Part I, and the ultimate upper crust thinned to the design value $\xi = \bar{\xi} = 0.55$. The model is then defined by choosing: sediment thickness parameter h_{11}; mantle depth, H_4; pressure scale, p_*; and ambient land elevation, L. We start with extreme values in the ranges: h_{11}, 0–10 km; H_4, 0–100 km; $p_* = 0$–10 kbar; $L = 0$–1 km and search for some intermediate values for which there is an absolute minimum in the root mean square of the fit. A minimum root mean square of the fit of 3.1 percent is found with $C = 7.9$ km; $H_4 = 41$ km; $p_* = 3.0$ kbar; $L = 0.43$ km. The sensitivity of the fit can be demonstrated by varying these four parameters separately near their optimum values. The root mean square of the fit is better than or equal to 10 percent for values varied singly in the range:

$$h_{11} = 6.8\text{–}8.5 \text{ km}; \qquad H_4 = 37\text{–}42 \text{ km}; \qquad p_* = 2.4\text{–}3.4 \text{ kbar}; \qquad L = 0.38\text{–}0.50 \text{ km}.$$

The minima for h_{11} and L are sharp, but that for H_4 and p_* are very flat bottomed. The very deep flat root mean square minima for variation of H_4 and p_* indicate that values outside the stated ranges produce quite unacceptable models.

The effect of choosing one of these parameters different from its optimum value is as follows:

(i) H_4: Low values give a thin pile and thick sediments and initial eruptions in deep water. High values give a thick pile but eruptions throughout on dry land.

(ii) p_*: The effects are less pronounced than those of H_4. Low values give a thinner pile and initial eruptions in deeper water. High values produce the opposite effects.

(iii) h_{11}: Low values which correspond to thin sediments, produce a thicker pile and a deeper final water depth. High values which correspond to thick sediments produce a thinner pile.

(iv) L: Low values produce larger water depths, thin sediments and a thicker pile. High values produce the opposite effects.

The model of best fit requires a source of the hyperbasic magma at a depth of 75 km bsml, which differentiates at a depth below the discharge surface of 11 km.

The behaviour of the model with the parameters of best fit is illustrated in Fig. 6. Let us consider what happens as the crust is progressively thinned.

(1) $\xi < 0.15$. During this interval terrestrial sediments accumulate on top of the subsiding old crust.

(2) $\xi \approx 0.15$. In spite of the accumulation of terrestrial sediments the crust has subsided sufficiently to allow the transgression of ocean water over the land surface of the developing embayment.

(3) $0.15 < \xi < 0.25$. Subaqueous marine sediments are now deposited. In reality the entire embayment was not inundated at a single instant rather a progressive marine transgression occurred. The evidence suggests that this transgression was prominently from the north and west but that the subsidence was jerky rather than uniform (see also HENDERSON *et al.*, 1976). The model here gives only a gross average representation.

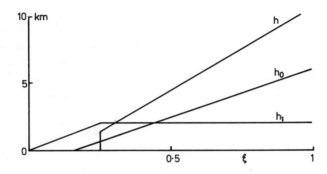

Figure 6
Results of the best combined model. Water depth, or depth of paleosurface, h_0; sediment thickness, h_1; lava pile thickness, h. Values in kilometres as a function of the proportion of crustal thinning, ξ.

(4) $\xi \approx 0.25$. Throughout the preceding interval the magma column has been rising higher. It now erupts through the top of the sediments into shallow ocean water to produce the subaqueous pillow breccias. The use in the model of a finite strength of the pre-existing rock leads to the sharp occurrence of the initial eruptions.

(5) $\xi > 0.25$. The lava pile now builds steadily in response to the continuing crustal thinning. Early in the life of the pile its thickness exceeds the depth of oceanic water and a regression of oceanic water occurs. As with the earlier transgression the regression in reality is not sudden. Indeed there is good evidence for a minor succession of local regressions and transgressions, particularly noted on north Disko (PEDERSEN, 1975).

Once the regression has occurred it is no longer appropriate to refer to the water level, rather the quantity h_0 is now the depth of the palaeosurface on which the lavas were originally deposited.

The lava pile continues to grow until the local crustal thinning ceases.

In essence, these models treat the system as a kind of manometer imbedded in the crust and upper mantle. A number of simple but distinctive and diagnostic features allow us in effect to read the manometer at various instants during the development of the system. These readings are more than sufficient to calibrate the fossil manometer and thereby obtain an understanding of the essential simplicity of the overall behaviour of the system.

The correlation of crustal thickness and lava thickness

We have assumed that the source region was distributed beneath the volcanic zone at a uniform depth and deduced that there will be a relation between crustal thickness and lava thickness. If this is correct, it implies that the source region was created and maintained solely by large-scale processes arising in a region much greater in area than the volcanic area itself. There is indeed a strong relation between the crustal thickness deduced from the gravity data and estimates of lava pile thickness as shown in Fig. 7.

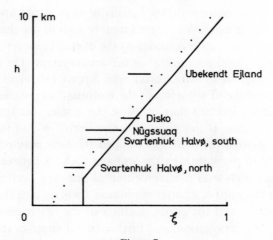

Figure 7

Correlation of lava pile thickness, h, with proportion of crustal thinning, ξ. The short bars indicate the range of ξ corresponding to the rather broad regions over which the thicknesses have been measured. The solid line is obtained from the model.

If the field data used to calibrate the model were precise, it would be possible to obtain an estimate of the finite strength σ. The dotted line shown is for $\sigma = 0.1$ kbar and the heavy line for $\sigma = 0.3$ kbar. In general there is a tendency to overestimate the local pile height as measured in the field by an uncertain amount so that $\sigma = 0.3$ kbar is probably about right.

This correlation suggests that the development of the embayment and especially its associated volcanism arose from a single localized event at a uniform depth in the mantle.

As a final remark, it is worth noting that the different ages of the pillow breccias, mentioned above, suggests that the crustal thinning proceeded more or less uniformly from zero to its local final value. The crust of south Disko, potentially less thinable, needed to continue its thinning for longer before the onset of volcanism as compared to that of Nûgssuaq. Thus the younger breccias of the Maligât Formation occur in a region less thinned than those of the older breccias of the Vaigat Formation.

3. *Application to acidic volcanism*

Let us now consider the case of acidic volcanism as found today, for example, in the North Island of New Zealand. A scenario of the sequence of events has been given by KEAR (1959) and more recently by BALLANCE (1976). Briefly the sequence of events has been: (i) in Northland: 20–15 Myr ago, subaqueous basalts; 18–6 Myr ago, andesites followed by dacites; 6–3 Myr ago, rhyolites; 3–0 Myr, basalts, (ii) in Rotorua-Taupo area: 3–0.75 Myr ago, rhyolites and ignimbrites; 0.75–0 Myr ago, continued rhyolites plus andesites, (iii) in Mt. Egmont area 3–0 Myr ago, high potash andesite. The details are quite complex especially as it would appear as if at least two systems have been in operation. Here I merely wish to use this type of situation as a background for a first step in considering the overall creation of geothermal and magma traps in a high level continental or orogenic system. The kernel of the idea has already been sketched by MODRINIAK and STUDT (1959) in providing an explanation for the sequence of volcanics in the Rotorua-Taupo area.

In the system of section two part or all of the matter was transportable to the top of the magma column. If this column is terminated within the crust below the solid surface, even though the matter is trapped the heat may not be. This heat is potentially available to partially melt the ambient rocks. I believe that the evidence is overwhelming that high-level partial melting of this type occurs (see for example STEINER, 1958). For the want of a better hypothesis I will assume that basaltic magma is the ubiquitous product of the partial melting of the upper mantle and probably subsequent high-level fractionation, and further that it supplies the heat to produce high-level acidic magmas by partial melting of the crust. In the model sketched below I also assume that the melting of the crust is confined to the level at the top of the basaltic magma column. This assumption is one of many that are possible – melting of the crust may well occur throughout the vertical extent of the basaltic magma column. Most crustal melting will occur where the bulk of the basaltic magma is trapped and the numerous pod-like gabbro bodies suggest that this is high in the basaltic system. Nevertheless these questions cannot be resolved until a study of reservoir dynamics has been made and will be put aside for this paper. Thus what follows is a theoretical sketch which merely suggests the sort of possibilities that arise from the control of such a system by a stratified crust.

Simple crustal accumulation model

Suppose an active portion of oceanic crust is transgressed by a thickening wedge of cratonic material (or what is the same in effect, a portion of active oceanic crust passing under a wedge of cratonic material). The wedge can be represented, for example, by:

$$h_1 = H_1 + \xi h_{11}, \qquad h_2 = \xi h_{21}, \qquad h_3 = H_3 + \xi(h_{31} - H_3). \tag{1}$$

Here ξ is a descriptive parameter, a measure of distance into the wedge, such that $\xi = 0$ corresponds to original oceanic crust and $\xi = 1$ to a cratonic crust upon which is a sedimentary deposit of thickness $(H_1 + h_{11})$. These sediments may be of terriginous or volcanic origin. The quantity ξ is simply a measure of the thickness of the crust. In a static model it is irrelevant how this arises. It could come from a direct local crustal thickening, or from a wedge of crust passing over the source region.

What would be appropriate values of these parameters? With New Zealand and the Taupo depression in mind, for example, and $h_{21} = 20$ km, $h_{31} = 10$ km then a land elevation of about 0.5 km and sedimentary layer of 5 km or so requires $h_{11} \approx 5$ km with $\xi = 0.75$ ($C = 5$ km, $\xi = 0.75$ gives $L = 0.49$ km, $h_1 = 5.57$ km and a total crustal thickness of 29.5 km).

A more flexible approach to the accumulation of the crust is to maintain the same total mass for the fully built crust, but otherwise to allow the proportions to be set explicitly. Thus let us take,

$$h_1 = \xi\lambda h_{11} \qquad h_2 = \xi(1 - \lambda)h_{21} \qquad h_3 = H_3 + \xi(h_{31} - H_3). \qquad (2)$$

Here λ represents the relative proportions of the thicknesses of the accumulating sediments and upper crust. If we choose h_{11}, h_{21} so that the crust has the same mass/unit area as before for $\xi = 1$ and all λ, then with $h_{21} = 20$ km, as before, we need $h_{11} = (\rho_2/\rho_1)h_{21}$, namely 21.6 km, with standard values.

Purely for simplicity of presentation the magma density of the crustal melt will be written $(\rho - \Delta\rho)$ where ρ is the parent crustal rock density and given by stating $\Delta\rho/\rho$.

With the second crustal model the behaviour of the whole system will now be evaluated. Some typical results are shown in Fig. 8 for $\lambda = 0.1$, $\Delta\rho/\rho = 0.2$. First consider the case with $H_4 = 40$ km.

(i) For $\xi < 0.93$ the paleosurface is below sea level.

(ii) For $\xi < 0.88$ basaltic magma penetrates the crust to produce suboceanic

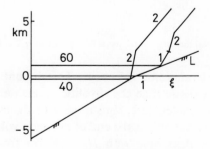

Figure 8

Crustal melting model showing elevation of the paleosurface, L, and elevation of the top of the lava pile as a function of ξ, for two cases of source depth, H_4: 40 km; 60 km. Labels 1, 2 refer to upper and lower crustal melting. $\Delta\rho/\rho = 0.2$, $\lambda = 0.1$.

volcanism reaching about 0.3 km below sea level. The basaltic pile will be thickest for $\xi = 0$ and of zero thickness for $\xi = 0.83$.

(iii) For $0.88 < \xi < 0.90$, and thereafter there is a rapid build up of acidic lavas, the top of the basalt column is in the upper crust and volcanics from upper crustal melts are extruded – here presumed to be rhyolitic.

(iv) For $\xi > 0.9$ the top of the basalt column is in the lower crust and lower crustal melts are extruded – here presumed to be andesitic.

A much deeper source, here with $H_4 = 60$ km, gives entirely subaerial lavas and a somewhat extended interval of rhyolites. With $\lambda = 0.1$, that is rather small, the crust has only a thin cover of sediments and near $\xi = 1$, is rather like that in the western part of the North Island, New Zealand. If we took the volcanoes: Egmont 2.5 km; Ruapehu, 2.8 km as representative this suggests $\xi \approx 0.95$, with the palaeosurface barely above sea level and a total crustal thickness of 29–30 km.

Notice that for smaller values of $\Delta\rho/\rho$ the thickness and elevation of the acid volcanics is smaller – they have less buoyancy. On the other hand larger values of source depth H_4 produce a thicker basaltic pile, but a thinner acidic pile – simply because the basaltic volcanism continues to larger values of ξ.

Similar data, for the case $\lambda = 0.5$, $\Delta\rho/\rho = 0.2$ is shown in Fig. 9. The overall behaviour is similar to that with $\lambda = 0.1$, but here with a rather large value of λ the sedimentary cover is rather thick and rhyolitic volcanism occurs over an extended range of ξ. Further there is a pronounced discontinuity in the pile elevation when

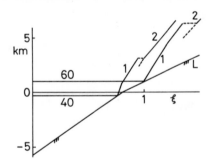

Figure 9
Crustal melting model, as in Fig. 8, but with $\lambda = 0.5$.

switching from rhyolitic to andesitic volcanism. This discontinuity arises because the andesitic magma is more dense than the rhyolitic magma and with equal geometry the andesite column must be less tall. Thus if we take the model as it stands, if ξ is increasing, there is an interval of ξ at the end of rhyolitic volcanism before andesites can be extruded – in the data here with $H_4 = 40$ km from $\xi = 0.94$–0.99; with $H_4 = 60$ km from $\xi = 1.35$–1.48.

These systems are relatively insensitive, as far as elevation is concerned, to density changes produced in the crust since the effect is proportional to $\Delta\rho/\rho_4$ where $\Delta\rho$ is the density change and ρ_4 is the mantle density. As illustration consider as standard a crust with a 25 km layer of density 2.7 g cm^3 from the lower 5 km of which acid volcanics of thickness 5 km and density 2.5 g cm^{-3} are produced and deposited on the surface. Consider two cases: (i) the crustal layer is unaltered – the new elevation is 1.21 km above that of the standard crust; (ii) the bottom 5 km is altered to density 3.0 g cm^{-3} – the new elevation is 0.76 km above that of the standard crust. The difference between the two cases of 0.45 km is 5 $\Delta\rho/\rho_4$ km where $\Delta\rho = 0.3$ g cm^{-3} and $\rho_4 = 3.3$ g cm^{-3}.

Acid magma densities. We can turn the argument on its head by posing the question as follows. Assuming that the acid volcanics are produced, at least in part, from partially melted crust what are the limits on the density of the magma to allow extrusive volcanism?

The extreme case would be material derived from the base of the lower crust. Using our standard cratonic crust, if ρ is the magma density then the acid lava pile height:

$$h = (\rho_2/\rho - 1)h_2 + (\rho_3/\rho - 1)h_3. \tag{3}$$

For $\rho = 2.6, 2.7, 2.8$ g cm^{-3} then $h = 2.5, 1.1, -0.2$ km. In the case of the andesitic volcanoes of New Zealand reaching 2–3 km above the paleosurface, clearly we need $\rho \lesssim 2.6$ g cm^{-3}.

A similar argument applied to rhyolitic discharges originating at the extreme of the base of the upper crust gives a pile height:

$$h = (\rho_1/\rho - 1)h_1 \tag{4}$$

In the case of the rhyolitic domes of New Zealand rarely reaching 1 km above the paleosurface we need $(\rho_1 - \rho)/\rho_1 \lesssim 0.04$. This value is far too small! Clearly some other factor must be responsible for inhibiting the development of the extruded rhyolites. Plainly these systems do not reach hydrostatic equilibrium. This could arise owing to: (i) extreme slowness of development of a very viscous material; (ii) and consequent dominance of erosion; (iii) but more frequently because of the wide dispersal of the material as ignimbrite eruptions.

We can retain the use of a static model by the device of imposing a critical condition on the most viscous magmas. Various possibilities suggest themselves. For example, rather than restrict say the andesitic source to the head of the basalt column, if and only if it is located in the lower crust, we could allow in addition partial melting of lower crust whenever rhyolitic holdup occurs. The rhyolites may leave their source region merely as buoyant blobs and not under the control of a plumbing system. This possibility is consistent with the occurrence of rhyolites in ring structures, typically 40 km in diameter within which there are numerous domes and a coherent pattern of volcanism.

4. Final remarks

A simple way of looking at systems of the type considered here is to regard them as systems in which the essential process is mass rearrangement. This occurs in two fairly distinct forms.

(i) *One shot rearrangement.* Matter is simply moved to a different place. An example is extrusive basaltic volcanism for which it is presumed that a liquid derived from partially melted mantle is moved upward through the crust possibly forming a deposit at some intermediate level owing to partial fractional crystallisation.

(ii) *Multi-shot rearrangement.* Matter may be reprocessed several times. As an extreme we can envisage an extended period of andesitic/rhyolitic volcanism together with local erosion and sedimentation. The net effect on the mass of the local crust may be quite small. It is important to notice in lithostatic models of this case that the end result will be independent of the details of the local density distribution provided the total mass is unchanged. The only way of identifying the detailed crustal structure is through the particular sequence of events.

One of the paradoxes of the known kinematics of the global surface is the occurrence of vigorous volcanism above the cold recharge regions of the mantle (so-called subduction zones). This apparent paradox arises because of a preoccupation with the temperature needed to produce partial melting. There is a further vital factor in producing surface volcanism, namely the necessity of the magmatic system, even if it exists, to penetrate the buoyancy barrier of the crust. Thus in a cold recharge region the local isotherms will be depressed so that the melting point will be met rather deeper. But this is just what we want. Greater source depth leads to greater head and possibility of surface volcanism.

All the versions of the model presented here presume a given source region and that the production is not limited by the ability of the source to produce magma. Production is limited by the buoyancy barrier of the crust. The model assumes that geothermal traps occur at the level of interfaces within the magma column whether produced by fractional crystallisation or partial melting of the ambient rock. Finally it is necessary to remind ourselves that having a trap does not necessarily mean that anything will be trapped – only if the trappable matter comes that way.

Acknowledgements

The measurements described in section two were made with the support of the Geological Survey of Greenland and grants from the Royal Society of London.

REFERENCES

ATHAVALE, R. N. and SHARMA, P. V. (1975), *Paleomagnetic results on early Tertiary Lava flows from West Greenland and their bearing on the evolution of the Baffin Bay–Labrador Sea Region*, Can. J. Earth Sci. *12*, 1–18.

BALLANCE, P. F. (1976), *Evolution of the upper Cenozoic magmatic arc and plate boundary in northern New Zealand*, Earth Planet. Sci. Lett. *28*, 356–370.

BAKER, S. R. and FRIEDMAN, G. M. (1973), *Sedimentation in an arctic marine environment: Baffin Bay between Greenland and the Canadian Archipeligo*, Pap. Geol. Surv. Can. *71* (23), 471–498.

CLARKE, D. B. and PEDERSEN, A. K. (1976), *The tertiary volcanic province of West Greenland, 364–385* in Escher, A. and Watt, W. S. (eds.), *The Geology of Greenland* (Geological Survey of Greenland Copenhagen 1976), 603 pp.

DENHAM, L. R. (1974), *Offshore geology of northern West Greenland (69°–75°N)*, Rapp. Grønlands Geol. Unders. *63*, 24 pp.

DREVER, H. I. (1958), *Geological results of four expeditions to Ubekendt Ejland, West Greenland*, Arctic *11*, 198–210.

ELDER, J. W. (1978), *Development of the Cretaceous–Tertiary crustal structure, central West Greenland*. Geological Survey of Greenland Bulletin (in preparation).

ESCHER, A. and PULVERTAFT, T. C. R. (1976), *Rinkian mobile belt of West Greenland, 140–149*, in Escher, A. and Watt, W. S. (eds.), *The Geology of Greenland* (Geological Survey of Greenland, Copenhagen 1976), 603 pp.

HALD, N. (1973), *Preliminary results of the mapping of the Tertiary basalts in western Nûgssuaq*, Rapp. Grønlands Geol. Unders. *53*, 11–19.

HALD, N. and PEDERSEN, A. K. (1975), *Lithostratigraphy of the early Tertiary volcanic rocks of central West Greenland*, Rapp. Grønlands Geol. Unders. *69*, 17–23.

HEIRTZLER, J. R., DICKSON, G. O., HERRON, E. M., PITMAN, W. C. III and LE PICHON, X. (1968), *Marine magnetic anomalies, geomagnetic field reversals, and motions of the ocean floors and continents*, J. Geophys. Res. *73*, 2119–2136.

HENDERSON, G. (1973), *The geological setting of the West Greenland basin in the Baffin Bay region*, Pap. Geol. Surv. Can. *71*(23), 521–544.

HENDERSON, G., ROSENKRANTZ, A. and SCHIENER, E. J. S. (1976), *Cretaceous–Tertiary sedimentary rocks of West Greenland, 340–362*, in Escher, A. and Watt, W. S. (eds.), *The Geology of Greenland* (Geological Survey of Greenland, Copenhagen 1976), 603 pp.

JÜRGENSEN, T. and MIKKELSEN, N. (1974), *Coccoliths from volcanic sediments (Danian) in Nûgssuaq, West Greenland*, Bull. Geol. Soc. Denmark *23*, 225–229.

KEAR, D. (1959), *Stratigraphy of New Zealand's Cenozoic volcanism northwest of the volcanic belt*, N.Z. J. Geol. Geophys. *2*, 578–589.

KEEN, C. E., KEEN, M. J., ROSS, D. I., and LACK, M. (1974), *Baffin Bay: small ocean basin formed by sea floor spreading*, Bull. Amer. Ass. Petrol. Geol. *58*, 1089–1108.

MODRINIAK, N. and STUDT, F. E. (1959), *Geological structure and volcanism of the Taupo–Tarawera district*, N.Z. J. Geol. Geophys. *2*, 654–684.

MÜNTHER, V. (1973), *Results from a geological reconnaissance around Svartenhuk Halvø, West Greenland*, Rapp. Grønlands Geol. Unders *50*, 26 pp.

PARK, I., CLARKE, D. B., JOHNSON, J. and KEEN, M. J. (1971), *Seaward extension of the west Greenland Tertiary volcanic province*, Earth Planet. Sci. Lett. *10*, 235–238.

PEDERSEN, A. K. (1975), *New mapping in North-Western Disko 1972*, Rapp. Grønlands Geol. Unders. *69*, 25–32.

ROY, J. L. (1973), *Latitude maps of the eastern North American–Western European paleoblock*, Pap. Geol. Surv. Can. *71*(23), 3–22.

STEINER, A. (1958), *Petrogenetic implications of the 1954 Ngauruhoe lava and its xenoliths*, N.Z. J. Geol. Geophys. *1*, 325–363.

(Received 1st November 1977, revised 21st February 1978)

Pageoph, Vol. 117 (1978/79), Birkhäuser Verlag, Basel

Heat Flow in the Basin and Range Province and Thermal Effects of Tectonic Extension

By Arthur H. Lachenbruch[1])

Abstract – In regions of tectonic extension, vertical convective transport of heat in the lithosphere is inevitable. The resulting departure of lithosphere temperature and thickness from conduction-model estimates depends upon the mechanical mode of extension and upon how rapidly extension is (and has been) taking place. Present knowledge of these processes is insufficient to provide adequate constraints on thermal models. The high and variable regional heat flow and the intense local heat discharge at volcanic centers in the Basin and Range province of the United States could be accounted for by regional and local variations in extensional strain rate without invoking anomalous conductive heat flow from the asthenosphere. Anomalous surface heat flow typical of the province could be generated by distributed extension at average rates of about $\frac{1}{2}$ to $1\%/\text{m.y.}$, similar to rates estimated from structural evidence. To account for higher heat flow in subregions like the Battle Mountain High, these rates would be increased by a factor of about 3, and locally at active bimodal volcanic centers, by an order of magnitude more.

Key words: Convective heat transfer; Thermal models of lithosphere; Extensional strain rate; Reduced heat flow; Basaltic magmatism.

1. Introduction

For more than a decade it has been recognized that within the ocean basins and on the continents, heat flow is generally lower in those regions that have been relatively undisturbed by tectonic and magmatic events for longer periods of time (Polyak and Smirnov, 1966; Hamza and Verma, 1969; McKenzie, 1967; Chapman and Pollack, 1975). Loosely stated, heat-flow correlates negatively with the time elapsed since the last major 'thermo-tectonic event.' With the refinement of plate-tectonic models, this correlation has taken on explicit physical meaning for the major ocean basins; the last thermo-tectonic event is generally the creation of sea floor at a spreading axis, and the time is determined from the subsequent amount of spreading. Specific thermo-mechanical models of the axial process, controlled by many types of observations, provide the initial condition necessary to describe the time-evolution of surface heat flow as the conductive decay of a thermal transient (e.g., McKenzie, 1967; Turcotte and Oxburgh, 1967; Sclater and Francheteau, 1970; Forsyth and Press, 1971). A useful extension of this concept to the continents has been presented

---

[1]) U.S. Geological Survey, Menlo Park, California 94025, USA.

recently by CROUGH and THOMPSON (1976). However for the continents the problem is more difficult; there may be several types and intensities of 'thermo-tectonic events' and we have no satisfactory general thermo-mechanical models of any of them (some thermo-tectonic events, e.g., subduction at a continental margin, can reduce the surface heat flow). Consequently the initial thermal condition, and even the time origin from which to reckon the subsequent cooling of the continents are largely matters of speculation.

In this context we investigate a region that is evidently in the throes of one type of continental 'thermo-tectonic event,' the Basin and Range province of the western United States. The region is characterized by late Cenozoic basaltic magmatism, distributed tectonic extension and high and variable heat flow. We consider how these three characteristics might be related to one another in terms of simple thermo-mechanical models, and what such models might imply about the present thermal regime of the lithosphere in this province. (For a more detailed discussion see LACHENBRUCH and SASS, 1978).

2. Simple thermo-mechanical models of an extending lithosphere

To understand the thermal regime (and the mechanics) of the lithosphere in an active province like the Basin and Range, it is necessary to consider how the anomalous heat flow observed at the surface is ultimately transported from the asthenosphere: *Is the transport of anomalous heat through the lower lithosphere primarily by conduction or by convection?* The evidence for widespread basaltic magmatism and tectonic extension since early Miocene time suggests that convection must at least play a contributing role, for continuity requires that mass must be moving upward in an extending lithosphere. The importance of convective transport in an extending lithosphere depends upon the mechanical mode of extension, and, of course, upon the rate at which the area of the province is (and has been) increasing, i.e. upon the extensional strain rate.

Three simple thermo-mechanical models of an extending lithosphere are illustrated in Fig. 1. For each, we assume that the horizontal strain rate s is uniform throughout the lithosphere and that it has not changed for a sufficiently long period that a thermal and mechanical steady state obtains. The depth R to the base of the lithosphere is given by the intersection of the geotherm and the basalt dry solidus $\theta_m(z)$ which we take to be

$$\theta_m \ [\text{°C}] = 1050 + 3z \ [\text{km}]. \tag{1}$$

The large amounts of magma (presumed to be basalt) required by two of the modes (Figs. 1b and c) probably could not be generated shallower than the base of a lithosphere defined in this way (WYLLIE, 1971). Conductive flux from the asthenosphere into the base of the lithosphere is denoted by q_a. The reduced heat flow at the surface

q_r is the regional heat flow q reduced by the radioactive contribution from the lithosphere A_0D (assumed for convenience to be the same as the contribution from the crust) where A_0 is heat production at the surface, and D is a characteristic depth (see LACHENBRUCH and SASS, 1978).

In the models of Figs. 1a and b, the lithosphere is stretched like pulled taffy. Under such circumstances we should expect the lithosphere to thin until it could carry off heat at the same rate as the asthenosphere could supply it. Thereafter

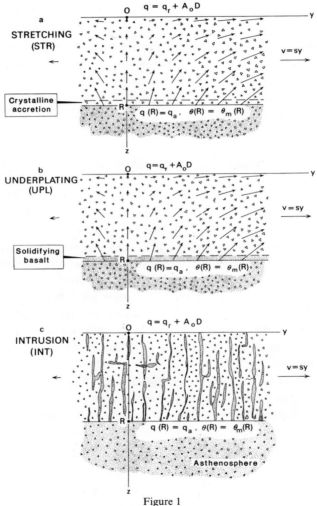

Figure 1

Simple thermo-mechanical models of a steadily extending lithosphere of thickness R with base at solidus temperature, $\theta_m(R)$. (a) STRETCHING: Homogeneous solid-state stretching with accretion of crystalline material at the base. (b) UNDERPLATING: Homogeneous solid-state stretching with accretion of solidifying basalt at the base. (c) INTRUSION: Extension by distributed dike intrusion. Arrows in (a) and (b) are velocity vectors (v, w), and s is horizontal strain rate.

accretion at the base would compensate for the tendency to thin and the solidus isotherm θ_m would remain at a stationary depth R. In the '*stretching*' mode (Fig. 1a) the asthenosphere material that accretes onto the lithosphere (to compensate for thinning) is all crystalline. By contrast, in the *underplating* mode (Fig. 1b), the accreting material is all liquid basalt which gives up its latent heat as it crystallizes at the base of the lithosphere. In these two modes convective transport in the lithosphere is in the solid state (see velocity vectors, Fig. 1 and equation (3b)). In the *intrusion* mode, Fig. 1c, vertical convective transport is by fluid basalt in distributed intermittent dike-like intrusions which, over long time periods, generate an average stationary thermal state. The intrusion accommodates lithosphere extension without vertical solid-state convection. Models similar to this one have been used by BODVARSSON (1954) and PALMASON (1973) to describe conditions in Iceland.

Assuming the radioactivity to be distributed exponentially with depth and denoting thermal conductivity by K and volumetric specific heat by ρc, we can write the differential equation governing the stretching and underplating modes (Figs. 1a and b) as follows:

$$K\frac{d^2\theta}{dz^2} + w\rho c\,\frac{d\theta}{dz} = -A_0\,e^{-z/D}. \tag{2}$$

The terms from left to right represent respectively conduction, convection, and radioactive heat production. The average upward component of velocity at any depth is w, and we denote by v the horizontal velocity component; both are referred to an arbitrary origin $y = 0$ on the surface $z = 0$. Thus for an incompressible lithosphere, the continuity condition yields

$$\frac{\partial w}{\partial z} = \frac{\partial v}{\partial y} = s, \tag{3a}$$

$$w = sz, \qquad v = sy. \tag{3b}$$

For the dike-intrusion mode (Fig. 1c) the differential equation of heat transfer is

$$K\frac{d^2\theta}{dz^2} = A_0\,e^{-z/D} + s\rho c\left[\theta - \left(\theta_m + \frac{L}{c}\right)\right] - w\rho c\,\frac{d\theta_m}{dz}. \tag{4}$$

The second term on the right (including the brackets) represents heat given up by solidification and cooling of the magma in place; the last term represents heat given up by the magma during its rise. Equations (3) apply also to this mode, with w representing the average upward rate of flow of magma per horizontal unit area.

Boundary conditions for all three modes can be written

$$\theta = 0, \qquad z = 0 \tag{5a}$$

$$K\frac{d\theta}{dz} = q_a + \left(\frac{L}{c}\,R\rho cs\right), \qquad z = R. \tag{5b}$$

where the term in parenthesis is set equal to zero for stretching or intrusion, and for the underplating mode it is retained. Latent heat is denoted by L.

Simple analytical solutions to these equations for all three modes are presented elsewhere. The lithosphere thickness (i.e., depth to dry-basalt melting) is obtained from them by adjusting R to satisfy the condition (see equation (1))

$$\theta(R) = \theta_m(R). \tag{6}$$

In a later section we shall drop this condition and apply the results only to the crust.

The time necessary to approach the steady-state condition described by the above equations can be specified approximately in terms of the time constant τ for the extending layer (see Fig. 5) and conditions at the lower boundary. As a rough rule of thumb, equilibrium will be approached in about 1 time constant for extension at constant basal temperature, and in 3 time constants or so for extension at constant basal flux, even if these boundary values undergo large changes at the time extension is initiated. (Other conceivable boundary conditions could result in response times either shorter or longer than those indicated by this range.) As the present episode of Basin-and-Range extension has probably been in progress for only 15–20 m.y. it is unlikely that the steady-state models will closely approximate conditions in a lithosphere more than about 50 km thick (τ (50 km) $\simeq$ 20 m.y., Fig. 5). However, conditions at depth are unknown, and the simplicity of steady-state models justifies using them as a limiting case to obtain insight.

With reasonable assumptions for radioactive heat production ($A_0 D = 0.5$ HFU), volumetric specific heat ($\rho c = 0.6$ cal cm^{-3} °C), thermal conductivity ($K = 6$ mcal cm^{-1} sec^{-1} °C^{-1}), and latent heat parameter ($L/c = 350$°C), the solution for each model yields relations among reduced heat flow (q_r), extension rate (s), lithosphere thickness (R), and conducted flux from the asthenosphere (q_a). Specification of any two will yield the others and the lithosphere geotherm, $\theta(z)$, as well (for analytical results and additional discussion, see LACHENBRUCH and SASS, 1978). These simple modes are easily combined to obtain more realistic geologic models by matching boundary conditions so that, for example, one mode might operate in the crust and a second in the mantle lithosphere (see e.g., Fig. 5, inset).

3. Relation between steady-state heat flow and extension rate

With the idealized lithosphere specified by the models of Fig. 1, we return to the question of whether the anomalous surface heat flow might come from the asthenosphere by conduction or by convection. In the conduction case ($s = 0$), regional variations in q_r would mimic variations in q_a; except for neglected radioactivity of the upper mantle we would have $q_r \simeq q_a$. In the special convective case illustrated in Fig. 2, q_a is assumed to be uniform and regional variations in q_r result from regional variations in extension rate s or equivalently, in the associated convective flux from the asthenosphere.

Figure 2 shows that intrusion is the most efficient extension mode in the sense that it produces the greatest surface heat flow for a given asthenosphere flux and extension rate; *stretching* is the least efficient, as it does not involve the release of magmatic heat of crystallization. The magmatic modes, *intrusion* and *underplating*, taken in combination can account for dike- and sill-like intrusion of the lithosphere; evidence for widespread basaltic magmatism suggests that these extension modes may be important in the Basin and Range province.

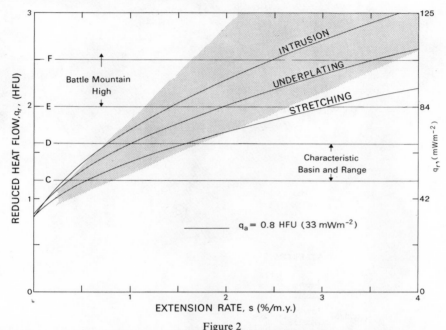

Figure 2

Variation in steady-state reduced heat flow q_r with extension rate s for uniform conductive flux from the asthenosphere, $q_a = 0.8$ HFU (33 mW/m^2).

The generalized map of heat flow and physiographic provinces in the western United States (Fig. 3) shows that with the exception of the Eureka Low (EL, Fig. 3, an anomaly believed to be of hydrologic origin, LACHENBRUCH and SASS, 1977) heat flow is generally high in the Basin and Range province. There, and in other less studied cross-hatched regions of the map, the reduced heat flow (q_r) is characteristically in the range 1.2–1.6 HFU (50–67 mWm^{-2}); typically the surface heat flow is larger than q_r by about 0.5 HFU (20 mWm^{-2}) (LACHENBRUCH and SASS, 1977). In the Battle Mountain High subprovince (BMH, Fig. 3), the reduced heat flow is generally greater than 2 HFU (84 mWm^{-2}), and it averages about 2.5 HFU (104 mWm^{-2}). These ranges are shown as bands on the ordinate scale of Fig. 2.

Thus for the simple steady-state model of Fig. 2, characteristic Basin and Range heat flow could be generated by the magmatic modes (intrusion and underplating)

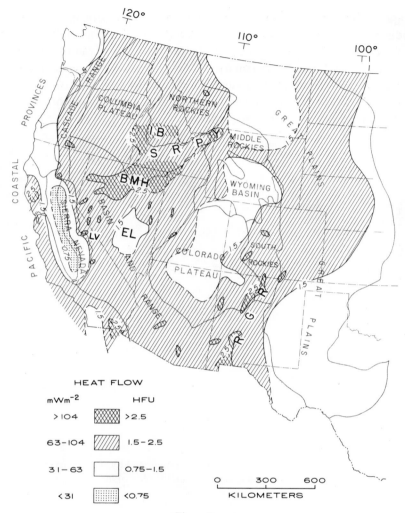

Figure 3

Generalized map of heat flow and physiographic provinces in the western United States (after LACHENBRUCH and SASS, 1978). Abbreviations are BMH for Battle Mountain High, SRP for eastern and central Snake River Plain, IB for Idaho Batholith, Y for Yellowstone geothermal area, RGR for Rio Grande Rift, EL for Eureka Low, and LV for Long Valley volcanic center.

with extension rates less than 1%/m.y., and the average Battle Mountain High condition ($q_r \simeq 2.5$ HFU) with rates of only about 3%/m.y. ($\sim 10^{-15}$ sec^{-1}). As the extension rate s approaches zero, the convective contribution of the lithosphere vanishes and asthenosphere flux q_a approaches reduced heat flow q_r; for the models of Fig. 2 this is 0.8 HFU (33 mWm^{-2}), chosen to coincide with the value characteristic of the stable eastern U.S. (ROY et al., 1968).

Solid curves in Fig. 4 show the 'lithosphere thickness' (or depth to the basaltic

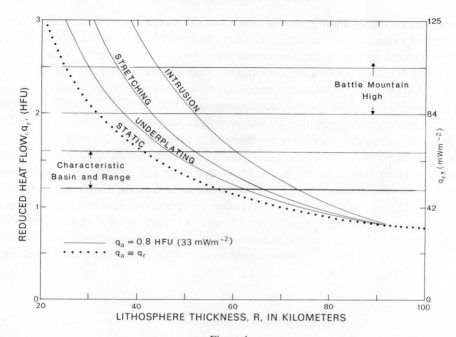

Figure 4

Solid curves show relation between steady-state reduced heat flow and lithosphere thickness for conditions presented in Fig. 2. Dotted curve represents the purely conductive case, $s = 0$.

source), R, corresponding to conditions in Fig. 2. The dotted curve ('static,' Fig. 4) shows the corresponding lithosphere thickness for the case ($s = 0$) in which heat transport through the lithosphere is exclusively by conduction. The figure illustrates the model-dependence of estimates of the lithosphere thickness based on observations of heat flow; all four curves represent steady-state conditions (not strictly applicable to the thicker lithospheres). Convective transport associated with lithosphere extension decreases thermal gradients in the lower lithosphere and increases estimates of lithosphere thickness (solid curves, Fig. 4).

Several estimates of the rates of extension of the northern portion of the Basin and Range province (the Great Basin) have been made by summing fault displacement occurring since early Miocene time (for a summary, see THOMPSON and BURKE, 1974). Although they vary widely, most estimates fall in the range of about 0.5 to 1%/m.y. It is seen from the simple model of Fig. 2 (band between C and D) that extension at such rates in the magmatic modes would be adequate to account for reduced heat flow characteristics of the Basin and Range province. Conversely if distributed extension did indeed occur at such rates and the magmatic modes are applicable, the observed characteristic anomalous flux cannot have been transported from the asthenosphere by conduction (see Fig. 10a, LACHENBRUCH and SASS, 1978).

Although distributed extension requires convective heat transport in the lithosphere, the converse is not generally true. Only dike-intrusion and stretching require horizontal extension; substantial heat-flow anomalies could be caused in a non-extending lithosphere by the formation sills which make room for themselves by displacing the lower lithosphere downward. Thus high heat flow can also be caused by convection in the lithosphere unrelated to extension, and, of course, by high asthenosphere flux q_a for the pure conduction case. The simple model of Fig. 2 would be quite inadequate to account for the very high heat flow in the Battle Mountain High if extension rates there were comparable to those estimated for the Great Basin as a whole. However, there is some independent evidence that the extension rate might be substantially greater in the Battle Mountain High and similar hot subprovinces (LACHENBRUCH and SASS, 1978). If this turns out to be the case, there may be no need to invoke high asthenosphere flux, or convective processes unrelated to extension, to account for the high and variable heat flow in the Basin and Range province. For a rather broad range of steady extension models and choices of asthenosphere flux q_a, the regional heat flow falls in the shaded region of Fig. 2, defined by the dimensional relation $q_r[\text{HFU}] = 0.8 + 1$ to $2 \times s$ [%/m.y.].

4. Extension and bimodal volcanic centers

There is evidence that like the anomalous regional heat flow, the anomalous local heat discharge at volcanic centers in the Basin and Range province also is a consequence of extension. In these features, upper crustal silicic magma chambers a few tens of kilometers across seem to survive for periods of a few million years. Their large heat loss is evidently supplied from below by associated basaltic intrusion (SMITH and SHAW, 1975); this association of basaltic and silicic magmas has led to the designation 'bimodal' for these features (CHRISTIANSEN and LIPMAN, 1972). It is likely that rapid local extension controls the passive rise of this basalt through the lithosphere and consequently that local extension controls the location of the volcanic centers (LACHENBRUCH et al., 1976; LACHENBRUCH and SASS, 1978). Calculations of heat and mass budgets for volcanic centers suggest that their area might increase some 10% or 20% for each m.y. of activity, and there is independent seismic evidence that such rapid local spreading does indeed occur (WEAVER and HILL, 1978/79). Thus bimodal silicic volcanic centers might be produced by a process related to the intrusion mode with local extension rates an order of magnitude greater than those required to account for the high regional heat flow.

5. Extension, basalt production, and the lithosphere geotherms

The effects of extension on lithosphere temperature is illustrated in Fig. 5 for the model shown schematically in the inset. This is not a 'preferred model' but an illustrative example obtained by modifying the analytical results for the intrusion and stretching modes and matching boundary conditions at their interface. Extension in the lithosphere mantle is accommodated by solid-state stretching. Basalt generated in the asthenosphere rises through the lower lithosphere in discontinuous blebs (perhaps in the manner described by WEERTMAN, 1971) and intrudes the crust where it accommodates extension by forming distributed dike-like bodies. (If we replaced the upper 10 km of the crust by a layer of normal faulting (a form of stretching), the results would not be drastically different (LACHENBRUCH and SASS, 1978).) The solid curves labeled C, D, E, and F represent geotherms in the extending lithosphere for reduced heat flows of 1.2, 1.6, 2.0 and 2.5 HFU, respectively; the letters on the dashed curves identify the corresponding geotherms for the case ($s = 0$) in which the transport of anomalous heat in the lithosphere is exclusively by conduction. Numbers on the solid curves are the required extension rates in %/m.y. For the average Battle Mountain High condition (curves F, Fig. 5) accounting for convection in the extending lithosphere almost doubles the estimate of lithosphere thickness and decreases estimates of temperature in the lower crust more than 300°C. For the characteristic

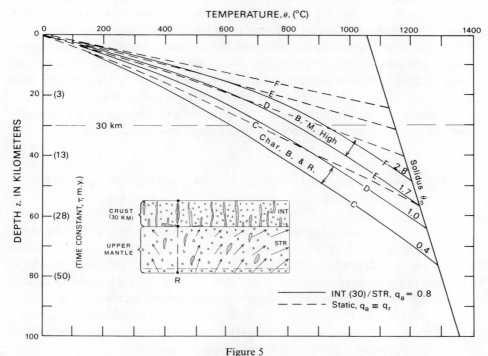

Figure 5

Solid curves show steady-state geotherms for a lithosphere in which the 30-km crust extends in the intrusive mode and the upper mantle extends in the stretching mode (see inset and text). Dashed curves are corresponding geotherms for static case $s = 0$. Numbers on solid curves are extension rates in %/m.y. Curves labeled C, D, E, and F are for reduced heat flows of 1.2, 1.6, 2.0, and 2.5 HFU, respectively (1 HFU $\simeq$ 42 mW/m²). The time constant τ for a layer of thickness z is $\rho c z^2/4K$.

Basin and Range condition (curves C to D, Fig. 5) the estimate of average lithosphere thickness is increased from less than 50 km in the static conduction model to 70 km in the extension model.

Magma (presumed to be basalt) must be supplied by the asthenosphere at a rate equal to the product of s and the cumulative thickness of lithosphere layers extending in magmatic modes, 30 km in the example of Fig. 5. Thus for the characteristic Basin and Range condition (average of curves C and D, Fig. 5) s is 0.7%/m.y. and basalt must enter the base of the lithosphere at the average rate of 0.2 km/m.y.; for the Battle Mountain High, curve F, the value is 0.8 km/m.y. These can be compared to the 1–2 km layer of basalt extruded in perhaps 2–3 m.y. during the prolific Columbia Plateau events (BAKSI and WATKINS, 1973; SWANSON et al., 1975) implying average (extruded) basalt production at the rate of 0.3–1.0 km/m.y. For an average extension rate of 0.7 %/m.y. the 800-km wide Great Basin would be extending at 6 mm/yr and the asthenosphere beneath it would (in this model) have to supply basalt at the same total rate as that beneath an oceanic spreading center with a half rate of about 1.8 cm/yr (assuming a 5-km basaltic oceanic crust, see LACHENBRUCH and SASS, 1978).

Other models are easily obtained with different combinations of the simple modes and different choices of q_a, but the example serves to illustrate the potentially important role of lithosphere extension in determining the thermal regime and thickness of the lithosphere, and boundary conditions on heat and mass flux in the asthenosphere. The high and variable regional heat flow of the Basin and Range province and the thermal budget of its volcanic centers can be accounted for by regional and local variations in the rates of lithosphere extension without calling on anomalous conductive flux from the asthenosphere; the required extension rates are generally consistent with fragmentary evidence from structural and seismic studies.

6. The thermal regime of an extending crust with transport in the upper mantle unspecified

Although the foregoing formulation of the problem of an extending lithosphere yields useful insight from simple analytical results, it is hazardous for several reasons. The lithosphere is defined in terms of two distinct conditions neither of which can be realistic in detail. Mechanically it is specified as a region in which the horizontal strain rate s is uniform; beneath the lithosphere s presumably increases if the asthenosphere is pulling the lithosphere apart, or decreases if the extending lithosphere is dragging the asthenosphere apart. Thermodynamically its base is defined by a simple melting curve (equation 1) which must represent an enormously complex process, sensitive to unknown details of chemical composition (see e.g., YODER, 1976; WYLLIE, 1971). We have assumed a stationary state, which for some cases may require uniform conditions for times as great as 50 m.y. or more. The radioactivity was assumed to vary with depth as $A_0 \exp(-z/D)$, $D = 10$ km, and hence the radioactive contribution of the upper mantle was largely neglected. Additionally the thermal properties were assumed to be uniform throughout the lithosphere.

Useful results of a more conservative nature can be obtained by applying the extension models of Fig. 1 only to the crust (of specified thickness R') and leaving conditions on heat and mass transport in the upper mantle unspecified. The assumptions of uniform extension rate and thermal properties, and the assumed distribution of radioactivity are then more reasonable as they apply only to the crust. Furthermore that stationary state is more likely to obtain, as it requires only that extension rate and boundary conditions change little over times large relative to a crustal time constant. (For the 30-km crust of the Basin and Range province, the time constant is about 7 m.y., Fig. 5.) Under these conditions, for any observed reduced heat flow (and surface radioactivity), the crustal regime is determined completely by specifying any one of the quantities: extension rate s, temperature θ_n, or heat flow q_n, where the subscript n refers to conditions just beneath the base of the crust, a convention borrowed from seismology. Relations among these three parameters for each extension mode are shown in Figs. 6a, b, and c by the solid curves for average conditions in the Battle Mountain High ($q_r = 2.5$ HFU), and by the broken curves for typical conditions in the Basin and Range province as a whole ($q_r = 1.4$ HFU). As before, we assume an average thermal conductivity of 6 mcal/cm^{-1} sec^{-1} °C^{-1} ($\simeq 2.5$ Wm^{-1} °K^{-1}) and a surface heat production A_0, of 5 HGU (2.1 μWm^{-3}); doubling the assumed heat production would increase the basal temperature by 80°C or so.

For the simple static conduction model ($s = 0$), the average temperature at the base of the crust would be 1320°C for the Battle Mountain High and 780°C for typical Basin and Range (Fig. 6a) and the reduced heat flow q_r would equal the mantle heat flow q_n (Fig. 6b) as is frequently assumed. The crustal geotherm for BMH for this case is labeled 1 in Fig. 6d, and the number 1 in Figs. 6a, b, and c designates the corresponding coordinates ($s = 0, \theta_n = 1320$°C, $q_n = q_r = 2.5$ HFU). Thus for the static case the lower crust would be nearly or quite all melted as its temperature would lie above the basalt dry solidus (BDS, Fig. 6d). To bring the average basal temperature down to BDS (1140°C at 30 km), would require extension at the rate of 3.7%/m.y. in the underplating and stretching modes (a condition denoted respectively by numbers 2 and 3, Figs. 6a, b, c, and d), or a rate of 1.25%/m.y. for extension in the intrusion mode (denoted by '4', Figs. 6a, b, c, and d). Combined modes usually result in intermediate values of extension rate (LACHENBRUCH and SASS, 1978). For a given extension rate (in this case, 3.7%/m.y.) underplating and stretching yield the same crustal geotherm (2, 3, Fig. 6d), but underplating requires much less conducted flux q_n from the mantle (compare points 2 and 3, Fig. 6c) because of the heat released by the crystallizing sill at the base of the crust. For a given basal temperature (in this case 1140°C) intrusion requires greater mantle flux q_n than either of the other modes (compare point 4 to 3 and 2, Fig. 6c), but it achieves the high reduced heat flow (in this case $q_r = 2.5$ HFU) with a much smaller spreading rate

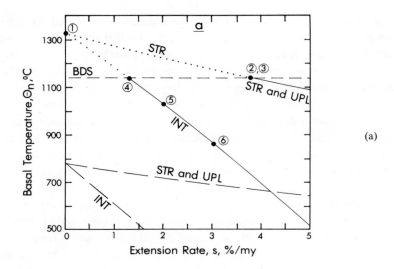

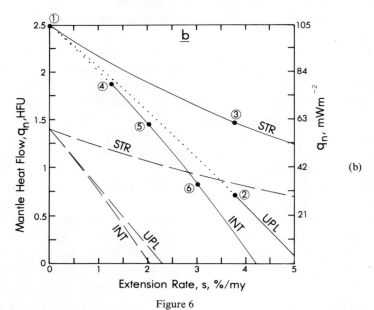

Figure 6

(a) Steady-state relations between basal temperature θ_n and extensional strain rate s in a 30-km crust with a reduced heat flow typical of the Battle Mountain High ($q_r = 2.5$ HFU, solid curves) and that typical of the Basin and Range province in general ($q_r = 1.4$ HFU, broken curves) for each extension mode (STR, UPL, and INT, Fig. 1). Intrusion and underplating are not valid on dotted portion of curves (see text). (b) Relation between mantle heat flow q_n and extensional strain rate for same conditions as Fig. 6a.

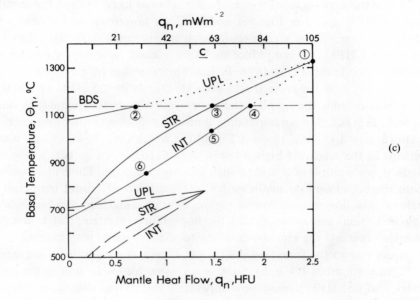

(c)

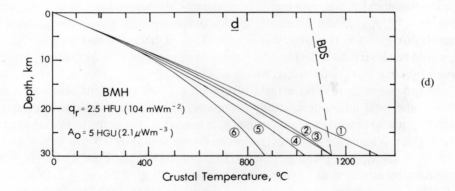

(d)

Figure 6(*cont.*)

(c) Relation between basal temperature θ_n, and mantle heat flow for same conditions as Fig. 6a. (d) Crustal geotherms for BMH ($q_r = 2.5$ HFU) under the following conditions: Curve 1, static mode, $s = 0$; curve 2, underplating, $s = 3.7\%$/m.y.; curve 3, stretching, $s = 3.7\%$/m.y.; curves 4, 5, and 6 are for intrusion with $s = 1.25$, 2 and 3%/m.y., respectively. Corresponding numbers in parts a, b, and c identify coordinates of parameters for each geotherm in part d. BDS is basalt dry solidus.

(compare point 4 to 3 and 2, Fig. 6a). Geotherms for BMH extending in the intrusion mode at rates of 2%/m.y. and 3%/m.y. are shown by curves 5 and 6, Fig. 6d; for the latter rate the average basal temperature is reduced to 860°C and the mantle flux to only 0.8 HFU (point 6, Figs. 6a and b). Thus horizontal strain rates s less than 10^{-15} sec^{-1} ($=3.2$%/m.y.) can result in an enhancement of the heat flow across the crust by 1.7 HFU and can reduce the average basal temperature by almost 500°C. Indeed in the limiting case ($q_n = 0$, Fig. 6b) extension by intrusion at the rate of 4.2%/m.y. in a 30-km crust would generate the entire 2.5 HFU of BMH with no contribution from the mantle; the subcrustal temperature would be only 660°C ($q_n = 0$, Fig. 6c). The corresponding limiting condition for the underplating mode occurs at $s = 5.1$%/m.y. ($q_n = 0$, Fig. 6b) in which case the subcrustal temperature remains at the relatively high value of 1080°C ($q_n = 0$, Fig. 6c). In the stretching mode, q_r is a multiple of q_n and no such limiting case exists. These illustrations underscore the broad options available for interpretation of crustal temperatures from regional heat-flow observations in regions of extensional tectonics even when steady-state conditions are assumed, and the importance of attempting to determine the extension rate and the kinematics of crustal extension and magmatism.

According to the foregoing discussion, if crustal temperatures are below BDS, 1.25%/m.y. (number 4, Fig. 6) might be a reasonable lower limit to the average s in BMH, and 1.85 HFU a reasonable upper limit to q_n if quasi-steady extension models can be applied there. If significant underplating occurred or if magma supplies were inadequate to maintain the intrusion mode, s would have to be substantially greater, and q_n substantially smaller.

If $q_r = 1.4$ HFU (dashed curves, Figs. 6a, b, c) represents a reasonable average for the Basin and Range province, and post early Miocene extension has been at the rate of 0.5%/m.y. in the magmatic modes (intrusion and underplating), then the mantle flux q_n would typically be about 1.1 HFU; if extension had been at 1%/m.y., q_n would be about 0.8 HFU (Fig. 6b). Estimates of subcrustal temperature θ_n for these conditions can be read from Fig. 6a.

In general, if there is no magma formation (i.e., no net latent heat absorption) in the subcrustal lithosphere, steady extension of the upper mantle will cause q_n to exceed q_a just as crustal extension causes q_r to exceed q_n. From the foregoing example and Fig. 2 it is seen that steady extension by intrusion at only 0.5%/m.y. is consistent with $q_r = 1.4$, $q_n = 1.1$ and $q_a = 0.8$ HFU. Similarly from Figs. 2 and 6b, steady extension by intrusion at 2.5%/m.y. is consistent with $q_r = 2.5$, $q_n = 1.15$ and $q_a = 0.8$ HFU; in the latter case, the mantle lithosphere is thinner (Fig. 4) and most of the convective enhancement of surface heat flow occurs in the crust.

In applications of the underplating model to the crustal domain, basaltic sills would form intermittently at the base of the crust giving up their heat of crystallization and sensible heat at the average ambient temperature. In this application, underplating generates a quasi-stationary state like the intrusion model, and neither are valid where crustal temperatures lie above the basalt dry solidus (dotted portions of curves, Fig. 6).

For stationary crustal extension in the magmatic modes, the crustal thickness remains constant and the crustal fraction of basalt, $b(t)$ increases with time t according to $[1 - b(t)] = [1 - b(0)] \exp(-st)$. Thus if the basaltic fraction were initially $\frac{1}{3}$, it would grow to $\frac{1}{2}$ after 30 m.y. of extension at 1%/m.y. or after 10 m.y. at 3%/m.y. Hence the present composition of the crust imposes some constraints on the history of extension in the magmatic modes. (Contributions of more felsic mantle-derived magmas would, of course, increase these time estimates.) In the stretching mode, the crust thins according $\exp(-st)$; it would thin from 40 km to 30 km in 30 m.y. for $s = 1\%$/m.y. or in 10 m.y. for $s = 3\%$/m.y. Hence the present thickness of the crust imposes some constraints on the history of extension in the stretching mode.

7. Concluding statement

Faulting, inelastic deformation, and movements of magma and water generally result in relative vertical motion of mass in tectonically active parts of the lithosphere. These movements can result in the convective transport of large amounts of heat. As the kinematics of these processes are not well understood on the continents there are many options for interpreting heat flow in terms of the thermal regime of active parts of the continental lithosphere. This has been illustrated for the Basin and Range province of the western United States which is similar to many other continental regions of extensional (normal) faulting and basaltic magmatism. Although the processes are known to be intermittent and non-uniformly distributed, we have attempted to describe them with idealized one-dimensional steady-state models. Even under these assumptions our ignorance of the rate and mechanical mode of lithosphere extension leaves broad options for estimates of lithosphere temperature and thickness when the heat flow is known. Under certain circumstances extension of regions 10^3 km wide at rates on the order of only 1 cm/yr could double the conductive heat loss at the surface. Estimates of upper mantle temperatures could be reduced by hundreds of °C and estimates of lithosphere thickness could be increased by tens of kilometers compared to predictions from simple heat-conduction models. The simple extension models suggest that the anomalous regional heat flow in the Basin and Range province could be accounted for (without anomalous conductive flux from the asthenosphere) by sustained distributed extension at strain rates of perhaps $\frac{1}{2}$ to 1%/m.y., similar to rates estimated by other means. Higher heat flow in the Battle Mountain High subprovince might represent extension rates greater by a factor of about three; for the intense local heat discharge at active bimodal volcanic centers, rates greater by an order of magnitude more are indicated. More refined estimates of extensional strain rate are needed to constrain thermo-mechanical models of the lithosphere; they could lead to useful information on processes controlling magmatism and crustal evolution and on boundary conditions for heat and mass flux in the asthenosphere.

8. Acknowledgements

I am grateful to my colleague, John Sass, for helpful discussions and assistance with computations. I thank L. J. P. Muffler and James C. Savage for useful comments on the manuscript.

REFERENCES

BAKSI, A. K. and WATKINS, N. D. (1973), *Volcanic production rates: Comparison of oceanic ridges, islands, and the Columbia Plateau basalts*, Science *180*, 493–496.

BODVARSSON, G. (1954), *Terrestrial heat balance in Iceland*, Timarit, 69–76.

CHAPMAN, D. S. and POLLACK, H. N. (1975), *Global heat flow: A new look*, Earth Planet. Sci. Lett. *28*, 23–32.

CHRISTIANSEN, R. L. and LIPMAN, P. W. (1972), *Cenozoic volcanism and plate-tectonic evolution of the western United States—Part II, Late Cenozoic*, Royal Soc. London Philos. Trans., Ser. A, *271*, 249–284.

CROUGH, S. T. and THOMPSON, G. A. (1976), *Thermal model of continental lithosphere*, J. Geophys. Res. *81*, 4857–4862.

FORSYTH, D. W. and PRESS, F. (1971), *Geophysical tests of petrological models of the spreading lithosphere*, J. Geophys. Res. *76*, 7963–7979.

HAMZA, V. M. and VERMA, R. K. (1969), *The relationship of heat flow with age of basement rocks*, Bull. Volcanol. *33*, 123–152.

LACHENBRUCH, A. H. and SASS, J. H. (1977), *Heat flow in the United States and the thermal regime of the crust*, in *The Earth's Crust* (ed. J. G. Heacock) (Am. Geophys. Union Geophys. Mon. 20) 626–675.

LACHENBRUCH, A. H. and SASS, J. H., *Models of an extending lithosphere and heat flow in the Basin and Range province*, in *Geol. Soc. America Memoir 152* (Geol. Soc. America 1978) in press.

LACHENBRUCH, A. H., SASS, J. H., MUNROE, R. J. and MOSES, T. H., JR. (1976), *Geothermal setting and simple heat conduction models for the Long Valley caldera*, J. Geophys. Res. *81*, 769–784.

MCKENZIE, D. P. (1967), *Some remarks on heat flow and gravity anomalies*, J. Geophys. Res. *72*, 6261–6273.

PALMASON, G. (1973), *Kinematics and heat flow in a volcanic rift zone, with application to Iceland*, Geophys. J. R. astr. Soc. *33*, 451–481.

POLYAK, B. G. and SMIRNOV, YA. B. (1966), *Heat flow on continents*, Doklady Akad. Nauk SSSR, *168*, 26–29.

ROY, R. F., BLACKWELL, D. D. and BIRCH, F. (1968), *Heat generation of plutonic rocks and continental heat flow provinces*, Earth Planet. Sci. Lett. *5*, 1–12.

SCLATER, J. G. and FRANCHETEAU, J. (1970), *The implications of terrestrial heat-flow observations on current tectonic and geochemical models of the crust and upper mantle of the earth*, Geophys. J. R. astr. Soc. *20*, 493–509.

SMITH, R. L. and SHAW, H. R., *Igneous-related geothermal systems*, in *Assessment of Geothermal Resources of the United States—1975* (ed. D. E. White and D. L. Williams) (U.S. Geol. Survey Circ. 726, 1975), pp. 58–83.

SWANSON, D. A., WRIGHT, T. L. and HELZ, R. T. (1975), *Linear vent systems and estimated rates of magma production and eruption for the Yakima Basalt on the Columbia Plateau*, Am. Jour. Sci. *275*, 877–905.

THOMPSON, G. A. and BURKE, D. B. (1974), *Regional geophysics of the Basin and Range province*, Annual Rev. Earth Planet. Sci. *2*, 213–238.

TURCOTTE, D. L. and OXBURGH, E. R. (1967), *Finite-amplitude convective cells and continental drift*, J. Fluid Mech. *28*, 29–42.

WEAVER, C. S. and HILL, D. P. (1978/79), *Earthquake swarms and local crustal spreading along major strike-slip faults in California*, Pure appl. Geophys. *117*, 51–64.

WEERTMAN, J. (1971), *Theory of water-filled crevasses in glaciers applied to vertical magma chambers beneath oceanic ridges*, Jour. Geophys. Res. *76*, 1171–1183.

WYLLIE, P. J., *A discussion of water in the crust*, in *The Structure and Physical Properties of the Earth's Crust* (ed. J. G. Heacock) (Am Geophys. Union Geophys. Mon. 14, 1971) pp. 257–260.

YODER, H. S., JR., *Generation of Basaltic Magma* (Natl. Acad. Sciences, Washington, D.C., 1976) 265 pp.

(Received 11th November 1977, revised 21st March 1978)

Pageoph, Vol. 117 (1978/79), Birkhäuser Verlag, Basel

Earthquake Swarms and Local Crustal Spreading Along Major Strike-slip Faults in California

By CRAIG S. WEAVER and DAVID P. HILL[1])

Abstract – Earthquake swarms in California are often localized to areas within dextral offsets in the linear trend in active fault strands, suggesting a relation between earthquake swarms and local crustal spreading. Local crustal spreading is required by the geometry of dextral offsets when, as in the San Andreas system, faults have dominantly strike-slip motion with right-lateral displacement. Three clear examples of this relation occur in the Imperial Valley, Coso Hot Springs, and the Danville region, all in California. The first two of these areas are known for their Holocene volcanism and geothermal potential, which is consistent with crustal spreading and magmatic intrusion. The third example, however, shows no evidence for volcanism or geothermal activity at the surface.

Key words: Strike-slip faults; Crustal spreading centers; Fault plane solutions; Stress directions; Brawley and Coso Hot Springs geothermal areas (USA).

1. Introduction

The step-like pattern formed by the boundary between spreading plates in ocean basins is clearly marked by linear trends of earthquake epicenters. As SYKES (1967) has shown, the focal mechanisms of these earthquakes reflect the spreading motion between adjacent plates; predominantly normal mechanisms are associated with earthquakes along the ridges (spreading centers) and strike-slip mechanisms with earthquakes along the transform faults joining offset ridge segments. SYKES (1967) also found that earthquakes along the spreading centers tend to occur in swarm-like sequences, whereas earthquakes along the transform faults typically occur as main shock-aftershock sequences (see also Fig. 6, KLEIN, 1976). Several authors have noted the tendency for earthquake swarms to be associated with areas of recent volcanism in both oceanic and continental environments (SYKES, 1967; WARD, 1972; HILL *et al.*, 1975; COMBS and ROTSTEIN, 1976).

In this paper we point out what appears to be a genetic relation between earthquake swarms and three small-scale analogs of oceanic spreading centers along major continental strike-slip fault zones in California. Locations of these centers and fault zones are shown in Fig. 1. Two of these local spreading centers, Brawley

---

[1]) U.S. Geological Survey, Menlo Park, CA 94025, USA.

in the Imperial Valley and Danville just east of San Francisco Bay, lie along the San Andreas fault zone. The third, Coso Hot Springs, lies at the south end of the Owens Valley fault zone. Both Brawley and Coso Hot Springs are recognized geothermal areas. Evidence has yet to be found, however, that the swarm activity in the Danville area is related to anomalous thermal activity or magmatic intrusion in the underlying crust.

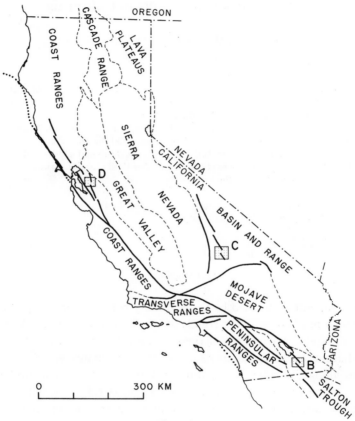

Figure 1

Map showing location (heavy squares) of three local crustal spreading areas in California. B is Brawley, C is Coso Hot Springs, and D is Danville. Major faults are indicated by heavy lines (dotted at sea). Tectonic provinces are outlined by dashed lines.

In each case, the evidence for local crustal spreading is based on the facts that (1) the earthquake swarm activity is localized within a right-stepping offset between active strands of a right-lateral strike-slip fault, and (2) this geometry (or the conjugate with a left-stepping offset between active strands of a left-lateral strike-slip fault) requires some mode of opening or extension in the area between offset fault strands. The analogy between this geometry and that of ocean floor spreading is clear: the offset strike-slip fault strands correspond to transform faults and the intervening region of local crustal extension corresponds to an isolated spreading center (see Fig. 2).

Dense seismographic networks recently installed in the Imperial Valley, Coso Range, and Danville areas have made possible accurate hypocenter locations and fault plane solutions for relatively small earthquakes ($M_L \lesssim 4.5$) in each area, and in particular these networks have allowed a detailed study of earthquake swarms.

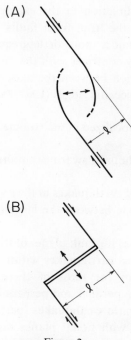

Figure 2

Simple models of crustal spreading centers. The small arrows indicate direction of strike-slip movement and direction of crustal spreading. *l* is the dimension of the crustal spreading center. A. Local spreading center. B. Mid-ocean ridge spreading center.

We use these seismic results to map regional contemporary faulting trends and stress orientations (as determined from focal mechanisms) and to infer some properties of the opening mode in these local extension centers. Finally, we extend an earlier geometrical model for earthquake swarms proposed by HILL (1977) and show that our seismic data are consistent with this model.

2. Common structural and seismological relations

Analysis of the data from these dense seismic networks reveals six basic structural and seismological relations common to each of the areas.

1. The dimension of the offset, l, between parallel fault strands (see Fig. 2) is of the same order as the maximum depth of earthquake occurrence (or thickness of the seismogenic zone), h. In each case l is 5 to 10 km, and h is 7 to 10 km. The value of the ratio $(l/h) \approx 1$ for these local crustal spreading centers contrasts markedly with values of $(l/h) \gg 1$ typical of major oceanic spreading centers.

2. One aspect of this large difference in scale appears to be a difference in the opening geometry between the local crustal spreading centers and major oceanic ridge systems. The opening direction in the examples of local crustal spreading considered here is oblique to the 'transform faults' (see 6 below and Fig. 2) and evidently distributed throughout a more or less equidimensional volume between the offset faults. This geometry contrasts with the more familiar orthogonal pattern that forms a stable configuration between the axes of linear spreading centers and transform faults along major oceanic ridges (LACHENBRUCH, 1976; OLDENBURG and BRUNE, 1975).

3. Earthquake swarms are localized in the volume of crust between the offset fault strands.

4. The duration of the earthquake swarms is limited, ranging from a few hours to less than a week.

5. Maximum magnitudes of earthquakes in the swarms are in the range 4.5 to 4.7, and the difference in magnitude between the largest and next-largest events is less than 0.5 magnitude units.

6. The direction of the T axes for fault plane solutions from earthquakes along the offset fault strands and swarms earthquakes within the offset volume is nearly invariant. Fault plane solutions for most of these earthquakes show strike-slip mechanisms with nodal planes parallel and perpendicular to the strike of the offset fault strands. A few of the swarm earthquakes (particularly in the Coso area) have normal fault plane solutions with nodal planes oriented roughly 45° (clockwise) from the strike of the offset fault strands. This relation suggests that the direction of the least principal stress, σ_3, is not significantly perturbed in the vicinity of the offset from its regional orientation.

3. Examples

Imperial Valley. Perhaps the clearest example illustrating a relation between the occurrence of earthquake swarms and local crustal extension is provided by the dextral offset between the Imperial and Brawley faults just south of the town of Brawley in the Imperial Valley (see Fig. 3). Oblique opening of the Gulf of California resulting from crustal spreading of the East Pacific Rise is generally supposed to extend as far north as the Imperial Valley in a series of right-stepping transform fault segments separated by small spreading centers (LOMNITZ et al., 1970; ELDERS et al., 1972). It was not until after installation of a 20-station seismograph array in the

Imperial Valley in 1973, however, that sufficient resolution in hypocenter locations was available to define the relation between the recurring earthquake swarm activity in the Brawley area and the right-stepping offset between the seismically active Imperial and Brawley faults (HILL *et al.*, 1975).

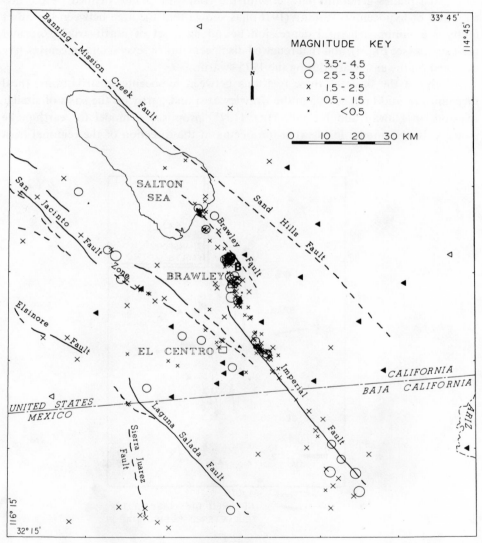

Figure 3

Location of earthquakes in the Imperial Valley region with respect to major faults and geothermal anomalies. The smallest and largest earthquakes plotted have magnitudes of 0.5 and 4.5, respectively. Triangles are locations of seismographic stations. The capital letter B indicates the Brawley geothermal area (adapted from HILL *et al.*, 1975).

A major earthquake swarm that occurred in this area in January 1975 was studied in detail by JOHNSON and HADLEY (1976) using the master event technique to achieve precise hypocenter locations. They found that the earthquakes tended to fall on sets of conjugate planes oriented parallel and perpendicular to the trends of the Imperial and Brawley faults. Most of the focal mechanisms were strike-slip and had nodal planes generally parallel with the conjugate planes defined by the distribution of hypocenters. SHARP (1976) has shown that the area between the offset faults is a minor structural depression bound by a set of north-striking normal fault scarps (see Fig. 4), and that renewed displacements of several tens of centimeters occurred on the east scarp during the 1975 swarm.

Partly on the basis of these relations between hypocenter distributions, focal mechanisms, and fault scarps in the Brawley area and partly on the basis of similar relations in Hawaii and Iceland, HILL (1977) proposed a model for earthquake swarms. In this model local extension occurs in the direction of the regional least

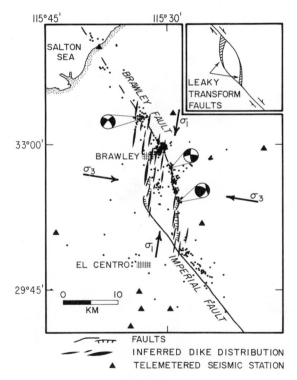

Figure 4

Map of the central part of the Imperial Valley, showing epicenters for earthquake swarms in June 1973 (HILL *et al.*, 1975) and January 1975 (JOHNSON and HADLEY, 1976). Hachures indicate down-dropped side of normal fault scarps (SHARP, 1976). Focal spheres illustrate typical fault plane solutions, with darkened quadrants representing compressions, white quadrants dilatations. Insert suggests the overall pattern of inferred dike distribution in terms of leaky transform faults. (Figure adapted from HILL, 1977.)

principal stress, σ_3, through magma injection into dikes with their long dimension in the direction of the regional greatest principal stress, σ_1 (see Fig. 4), and earthquake swarms occur as a sequence of shear failures along conjugate planes oblique to the principal stress directions (see Fig. 1 in HILL, 1977). JOHNSON (1977) has found additional evidence supporting this model in the Brawley area from detailed studies of the spatial distribution of hypocenters and focal mechanisms in more recent swarm sequences.

Coso Range. The Coso Range in California lies immediately east of the southern Sierra Nevada (see Fig. 1) in the Basin and Range physiographic province. An episode of rhyolitic volcanism began in late Pleistocene time in the central part of this range and continued until recent times (DUFFIELD and BACON, 1977). To assess the possible geothermal potential of the Coso Range, the U.S. Geological Survey deployed a 16-station seismic network in 1975 to study local seismicity and tele-seismic *P*-wave delays.

Seismicity in the Coso Range defines a conjugate strike-slip fault system (Fig. 5), with six arms of epicenters trending away from the center of the volcanic field. A northwest trend of right-lateral strike-slip faulting is defined by three zones: Airport Lake, Haiwee, and Cactus Flats. A complementary northeast trend of left-lateral faulting is defined by the Darwin and Red Hill zones (see Fig. 5). The right-lateral motion inferred from fault plane solutions and epicenter lineations along the Haiwee and Cactus Flats zones is consistent with both mapped offsets in Owens Lake, which are associated with the great 1872 Owens Valley earthquake (CARVER, 1970), and with recent deformation studies, which show a right-lateral strain accumulating at a rate of approximately 4 mm/yr across Owens Valley (SAVAGE et al., 1975).

The strike-slip zones are interrupted in the central Coso Range. Here seismicity defines at least four north-northeast trending zones of epicenters that parallel chains of young rhyolite domes (see Fig. 6). Normal fault plane focal mechanisms are found in three of these north-striking zones. Within the central volcanic zone, six earthquake swarms were located during the first two years of recording. Along the west edge of the Coso Basin, left-lateral faulting is revealed by the focal mechanisms and the alinement of epicenters associated with an earthquake swarm in December 1975 (see swarm S1, Fig. 6). Swarm S2 at the northeast edge of the dome field and swarm S3 at the west edge of Sugarloaf Mountain both strike-slip and normal faulting focal mechanisms, whereas swarms in the central part of the rhyolite field (S4, S5, and S6) showed only normal faulting. All of the Coso swarms have been of short duration, lasting from less than one day to no more than three days.

The orientation of the principal stresses in the Coso region, as inferred from the mapped faulting pattern and fault plane solutions, is indicated in both Figs. 5 and 6. The least compressive stress, σ_3, is oriented in a general east–west direction through the region. The maximum compressive stress, σ_1, however, is oriented in a north–south direction along the strike-slip zones, and in a vertical direction in the rhyolite dome field (that is, σ_1 and σ_2 undergo a 90° rotation about the stable σ_3 direction).

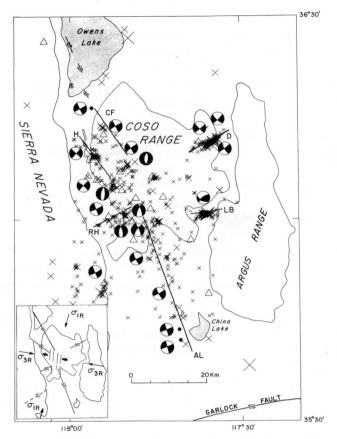

Figure 5

Summary of earthquake epicenters in the Coso Range and vicinity. Large crosses are magnitudes greater than 3.0; medium crosses, magnitudes between 2.0 and 3.0; small crosses, magnitudes less than 2.0. Heavy lines indicate earthquake trends discussed in text: CF, Cactus Flats; D, Darwin; AL, Airport Lake; RH, Red Hill; and H, Haiwee. Focal mechanisms are lower hemisphere plots, with darkened quadrants indicating compressions, white quadrants dilatations. Triangles are seismic stations. The insert shows the inferred regional principal stress directions, σ_1, being maximum compressive stress, and σ_3, being minimum compressive stress. The double arrows indicate the direction of crustal spreading.

For the strike-slip swarm earthquakes in the rhyolite dome field, σ_1 is again oriented north–south. This abrupt rotation of σ_1 and σ_2 over short distances can easily be accommodated if the difference between σ_1 and σ_2 is small throughout the volcanic field. Regardless of either mode of faulting, a slight spreading should occur in the σ_3 direction.

Danville. Two earthquake swarms have been well documented in the Danville area since 1970, yet no young volcanic rocks are exposed at the surface. The Calaveras fault, one of the main northern branches of the San Andreas system (see Fig. 2) runs north-northwest through Danville (see Fig. 7), and Quaternary faulting has been

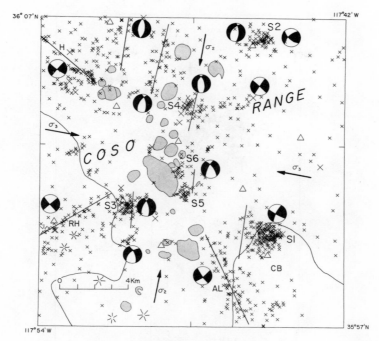

Figure 6

Detailed seismicity in the central Coso volcanic field. Shaded features are young rhyolite domes. Heavy lines show earthquake trends. Letters identify zones discussed in the text: H, Haiwee; AL, Airport Lake; RH, Red Hill; CB, Coso Basin. Earthquake swarms are identified by S1 to S6. Fault plane solutions are lower hemisphere plots, with darkened quadrants indicating compressions, and white quadrants, dilatations. Local principal stress directions indicated by σ_2 and σ_3, with σ_1 being vertical for the normal fault plane solutions.

identified along the fault just south of Danville (D. HERD, personal communication). Seismicity in the Danville region is located along the Calaveras fault south of Danville, and along the Concord fault to the northeast (see insert, Fig. 7).

In a thorough study of the June 1970 swarm, LEE *et al.* (1971) found that swarm earthquakes roughly filled a spherical zone about 2 km in diameter centered at a depth of 4 km and about 3 km northeast of the Calaveras fault (see Figs. 6 and 7, LEE *et al.*, 1971). Focal mechanisms were strike-slip, with vertical fault planes oriented parallel and perpendicular to the trend of the Calaveras (see Fig. 10, LEE *et al.*, 1971). On the basis of this first swarm, LEE *et al.* (1971) suggested that the roughly spherical hypocenter pattern could be related to either a piercement structure, such as a cold plug of mafic or ultramafic rock lubricated with serpentine, or an unidentified fault zone, not observed at the surface.

All epicenters in the central California catalog for the Danville region through January 1977 have been plotted in Fig. 7. The June 1970 swarm results are included in this plot, as are the locations for a second earthquake swarm in August 1976.

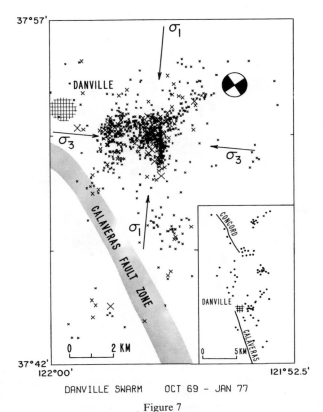

Figure 7

Seismicity in the Danville area. The dashed line outlines the August 1976 earthquake swarm. Insert shows regional seismicity. The orientation of the maximum and minimum principal stresses is indicated by σ_1 and σ_3, respectively. A representative fault-plane solution is shown.

Fault plane solutions for the larger events in the August 1976 swarm resemble those reported by LEE *et al.* (1971). Like the earlier swarm, the August 1976 epicenters cluster in a spherical zone about 2 km in diameter (see Fig. 7).

4. Discussion

The kinematics of rigid block (plate) motion requires that some form of opening or crustal spreading occur in the region of right-stepping offsets between strands of right-lateral, strike-slip faults (or left-stepping offsets between left-lateral, strike-slip faults). Such spreading centers are clearly recognized in the large-scale, step-like pattern formed by segments of oceanic ridges offset by transform faults as, for example, along the mid-Atlantic ridge.

We suggest that, by analogy, localized continental spreading centers can be recognized in dextral offsets between strands of the San Andreas fault system. These localized continental spreading centers have dimensions comparable to or less than the local crustal thickness, in contrast to the large-scale spreading centers between offsets in mid-oceanic ridges, which are many times the local crustal or lithospheric thickness.

Figure 8 shows what we feel is a likely geometry for crustal extension in local spreading centers with $(l/h) \simeq 1$. In particular, we suggest that the volume of brittle

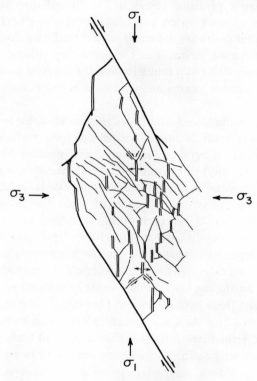

Figure 8

Schematic representation of fault pattern and opening mode for local crustal spreading centers. Single lines represent conjugate strike-slip faults. Double lines represent either dikes or normal faults (see text). Small arrows show horizontal displacement directions. Directions and regional principal stresses, σ_1 and σ_3 are shown by large arrows.

crust between offset strike-slip fault segments is broken into a series of blocks by a conjugate set of fracture planes oriented generally parallel and perpendicular to the major offset strike-slip fault planes and by an additional set of fracture planes striking in the general direction of the regional maximum principal stress, σ_1. The particular fault pattern used to illustrate this geometry in Fig. 8 is adapted from a set of mapped faults in south-central Oregon described by DONATH (1962). This geometry is basically the same as that in the model proposed by HILL (1977) relating magma injection in a system of dikes to earthquake swarms, and it is supported by lineations in the distribution of swarm earthquakes and their focal mechanisms in the Imperial Valley (JOHNSON and HADLEY, 1976; JOHNSON, 1977) and Coso Hot Springs as well as the directions of linear trends of the rhyolite domes in the Coso area (DUFFIELD and BACON, 1977).

Opening of these local spreading centers occurs in the direction of the least principal stress, σ_3, by injection of magma into dikes with their long axis in the direction of the greatest principal stress, σ_1, or by collapse along normal faults striking in the plane of the greatest (σ_1) and intermediate (σ_2) principal stresses. (Both dikes and normal faults are indicated schematically by double lines in Fig. 8.) The opening is distributed throughout the spreading volume by adjustment on the system of conjugate strike-slip faults joining the dikes or normal faults, and the adjustments tend to occur in swarm-like sequences in the manner described by HILL (1977).

Factors such as the volume and viscosity of magma available to enter the spreading volume from the lower crust or upper mantle and local variations in the greatest principal stress orientation between horizontal and vertical directions (feasible if σ_1 and σ_2 differ only slightly) will determine whether the opening will be dominated by dike expansion or normal faulting. If the opening rate is rapid enough, such local spreading centers may evolve into a center of silicic volcanism as suggested by LACHENBRUCH and SASS (1978). In their view, rapid local crustal spreading should draw sufficient low-viscosity basaltic magma from depth into the crust to initiate melting the upper silicic crust. They show that such a process can explain both the high heat flow values and extensive silicic volcanic fields found today in silicic volcanic centers that formed during the last several hundred thousand years.

The seismic results from both Coso and Danville can be readily related to our model for crustal spreading. At Coso, the three northwest-trending zones, Airport Lake, Haiwee, and Cactus Flats, form the offset transform fault system, along which focal mechanisms are strike-slip. The volcanic area, and in particular the rhyolite dome field, represents a zone of crustal extension with east–west spreading as revealed by the focal mechanisms (Fig. 6). Note that the rhyolite domes have been injected basically north–south, perpendicular to the direction of the minimum regional compressive stress in accordance with the model (Fig. 8). This relation between the orientation of volcanic lineations and regional stresses has also been emphasized by Nakamura (1977).

The Danville swarms occur in the dextral offset between the seismically active part of the Calaveras fault and the Concord fault zone (see insert, Fig. 7). We believe that the swarm activity represents the current volume of localized spreading between these offset faults in the manner described above as indicated in Fig. 8. This localized spreading may be accommodated by the intrusion of serpentine dikes following the suggestion of LEE et al. (1971) or perhaps a very recent invasion of magma from the lower crust or upper mantle.

These three examples do not, of course, make the case that earthquake swarms and volcanic and geothermal activity are exclusively associated with dextral offsets in a right-lateral strike-slip system (or left offsets for a left-lateral system). Earthquake swarms occasionally occur elsewhere in California besides the three areas discussed here. The Santa Barbara Channel swarm in 1968 is a noteworthy example (SYLVESTER *et al.*, 1970; LEE and VEDDER, 1973), and a persistent cluster of small earthquakes is closely associated with The Geysers geothermal production area (BUFE *et al.*, 1976). Although some aspects of the local spreading geometry discussed here may apply in these areas, the structural relations are either too poorly known or complicated by adjacent fault systems for a convincing case to be made that they represent an isolated spreading center. In any case, these basic relations are probably common in tectonic regimes dominated by strike-slip faulting and should be pursued in an effort to understand local tectonic processes.

5. Conclusions

We suggest a relationship between earthquake swarms and crustal spreading based on the tendency for earthquake swarms to be localized between dextral offsets of parallel, active strands of right-lateral strike-slip faults. The geometry of local crustal spreading centers is analogous to that observed in ocean floor spreading, with the offset strike-slip fault strands corresponding to transform faults and the intervening regions of local crustal spreading to an isolated spreading center. Volcanism, surface geothermal activity, and earthquake swarms are all primarily associated with the crustal spreading centers, rather than with strike-slip fault segments.

Our model for the earthquake swarm process is based on simple geometrical relations between faulting trends (either mapped or inferred from focal mechanisms and seismicity) and mapped volcanic units with the inferred directions of the principal stresses. The direction of the minimum principal stress is nearly invariant along the strike-slip fault segments and in the crustal spreading volume. Within the spreading center, the orientation of the maximum and intermediate principal stresses determine whether normal (maximum principal stress vertical) or strike-slip (maximum principal stress horizontal) faulting will dominate earthquake swarms. When combined with an appropriate thermal model, our geometrical model is capable of explaining the volcanism commonly observed in dextral offsets between parallel, active strands of right-lateral strike-slip faults.

REFERENCES

BUFE, C. G., PFLUKE, J. H., LESTER, F. W. and MARKS, S. M. (1976), *Map showing preliminary hypo-centers of earthquakes in the Healdsburg (1:100,000) Quadrangle, Lake Berryessa to Clear Lake, California, January 1969–June 1976.* U.S. Geol. Survey Open-File Rept. *76-802.*

CARVER, G. A. (1970), *Quaternary tectonism and surface faulting in the Owens Lake Basin, California*, MS. Thesis, University of Nevada, Reno.

COMBS, J. and ROTSTEIN, Y. (1976), *Microearthquake studies at the Coso geothermal area*, in Proceedings of the Second United Nations' Symposium on the Development and Use of Geothermal Resources *2*, 909–916.

DONATH, F. A. (1962), *Analysis of basin-range structure, south-central Oregon*, Geol. Soc. Am. Bull, *73*, 1–16.

DUFFIELD, W. A. and BACON, C. R. (1977), *Preliminary Geologic map of the Coso volcanic field and adjacent areas, Inyo county, California*, U.S. Geol. Survey Open-File Map *77-311*.

ELDERS, W. A., REX, R. W., MEIDOV, T., ROBINSON, P. T. and BIEHLE, S. (1972), *Crustal spreading in southern California*, Science, *178*, 15–24.

HILL, D. P., MOWINCKEL, P. and PEAKE, L. G. (1975), *Earthquakes, active faults, and geothermal areas in the Imperial Valley, California*, Science, *188*, 1306–1308.

HILL, D. P. (1977), *A model for earthquake swarms*, J. Geophys. Res., *82*, 1347–1352.

JOHNSON, C. (1977), *Swarm tectonics of the Imperial and Brawley faults of southern California* (abs), Trans. Am. Geophys. Union, *58*, 1188.

JOHNSON, C. E. and HADLEY, D. M. (1976), *Tectonic implications of the Brawley earthquake swarm, Imperial Valley, California, January 1975*, Bull. Seism. Soc. Am., *66*, 1133–1144.

KLEIN, F. W. (1976), *Earthquake swarms and the semidiurnal solid earth tide*, Geophys. R. J. Astr. Soc., *45*, 245–295.

LACHENBRUCH, A. H. (1976), *Dynamics of a passive spreading center*, J. Geophys. Res., *81*, 1883–1902.

LACHENBRUCH, A. H. and SASS, J. H. (1978), *Models of an extending lithosphere and heat flow in the basin and range province*, Geol. Soc. Am. Special Volume (in press).

LEE, W. H. K., EATON, M. S. and BRABB, E. E. (1971), *The earthquake sequence near Danville, California, 1970*, Bull. Seism. Soc. Am., *61*, 1771–1794.

LEE, W. H. K. and VEDDER, J. G. (1973), *Recent earthquake activity in the Santa Barbara Channel region*, Bull. Seism. Soc. Am., *63*, 1757–1773.

LOMNITZ, C., MOOSER, F., ALLEN, C. and THATCHER, W. (1970), *Seismicity and tectonics of the Northern Gulf of California region, Mexico, preliminary results*, Geofis. Int., *10*, 37–48.

NAKAMURA, K. (1977), *Volcanoes as possible indicators of tectonic stress orientation – principal and proposal*, Jour. Volcanology and Geothermal Res., *2*, 1–16.

OLDENBURG, D. W. and BRUNE, J. N. (1975), *An explanation for the orthogonality of ocean ridges and transform faults*, J. Geophys. Res., *80*, 2575–1585.

SAVAGE, J. C., CHURCH, J. P. and PRESCOTT, W. H. (1975), *Geodetic measurement of deformation in Owens Valley, California*, Bull. Seism. Soc. Am., *65*, 865–874.

SHARP, R. V. (1976), *Surface rupturing in Imperial Valley during the earthquake swarm of January–February, 1975*, Bull. Seism. Soc. Am., *66*, 1145–1154.

SYKES, L. R. (1967), *Mechanism of earthquakes and nature of faulting on the mid-ocean ridges*, J. Geophys. Res., *72*, 2131–2153.

SYLVESTER, A. G., SMITH, S. W. and SCHULZ, C. H. (1970), *Earthquake swarm in the Santa Barbara Channel, California, 1968*, Bull. Seism. Soc. Am., *60*, 1047–1060.

WARD, P. L. (1972), *Microearthquakes: prospecting tool and possible hazard in the development of geothermal resources*, Geothermics, *1*, 3–12.

(Received 19th January 1978, revised 24th April 1978)

Pageoph, Vol. 117 (1978/79), Birkhäuser Verlag, Basel

Variation of Continental Mantle Heat Flow with Age: Possibility of Discriminating Between Thermal Models of the Lithosphere

By V. M. Hamza[1])

Abstract – Continental mantle heat flow values are obtained by subtracting the radiogenic heat produced in the lower crust and lithosphere beneath the crust from 'reduced heat flow values' reported for various heat flow provinces. The significance of continental mantle heat flow values thus obtained is that they can be considered essentially as representing the residual heat of cooling of the continental lithosphere. A plot of these mantle heat flow values against $1/\sqrt{t}$ where 't' is the geologic age of the last thermal event suggests a linear trend. It is also found that the recently proposed relation $Q = 500 \, (1/\sqrt{t})$ for the variation of oceanic heat flow Q (in mW/M^2) with age t (in million years) provides a reasonably good fit to the mantle heat flow data. The constant thickness plate model however, is found to be unsatisfactory in explaining the variation of continental mantle heat flow with age.

Key words: Continental mantle heat flow; Radiogenic heat; Lithosphere cooling; Heat flow provinces; Thermal model of lithosphere; Heat flow-age variation.

Introduction

The inverse correlations between near surface heat flow and geologic age for continental (Polyak and Smirnov, 1968; Hamza and Verma, 1969; Verma *et al.*, 1970) and oceanic (for example Sclater and Francheteau, 1970) regions are well known. However the time scales associated with the heat flow age relations for these two regions are different by more than an order of magnitude. The decrease of heat flow with age in oceanic regions is principally governed by the residual heat of cooling of the oceanic lithosphere, while in continental regions it is determined by residual heat of cooling as well as by changes in near surface radiogenic heat. Thus to compare heat flow age relations from oceanic and continental regions it is necessary to remove the contribution by radiogenic heat in continental regions. The linear relation between heat flow and near surface radiogenic heat in continental plutonic rocks allows an accurate estimate of 'reduced heat flow' coming from below the variable radioactivity layer. Hamza (1976a, b) showed that near surface oceanic and reduced continental heat flow values, when plotted against their respective ages, indicate a continuous and

[1]) Instituto Astronômico e Geofísico, U.S.P., Av. Miguel Stéfano, 4200, São Paulo, Brasil.

overlapping trend of decreasing heat flow with age, thereby suggesting the existence of a unified heat flow age relation. In this note I present arguments suggesting that this unified heat flow age relation can possibly be used as a criterion for distinguishing between thermal models of the lithosphere.

Variation of mantle heat flow with age

The reduced heat flow values for continental regions are normally interpreted as the flux from beneath the variable radioactivity layer. The thickness of this variable radioactivity layer is uncertain and several models of vertical distribution of radiogenic heat have been suggested. Of these the most widely discussed are the step, the linear and the exponential models (LACHENBRUCH, 1970). The data and arguments presented by HEIER and ADAMS (1965), LAMBERT and HEIER (1967), SWANBERG (1972), SWANBERG and BLACKWELL (1973) and SMITHSON and DECKER (1974) support a variable distribution of radiogenic heat sources in the continental crust. HAMZA (1973, 1977) argued that the crustal and upper mantle seismic data as well as the positive correlation between U/K ratio and radiogenic heat productivity can be interpreted as evidence against the step model. Table 1 compares the thickness of the upper crust as determined from seismic data with the thickness of variable radioactivity layer as calculated from the linear model; the relation is in fact surprisingly good. The constancy of potassium and decrease of uranium and thorium in crustal layers reported by HEIER and ADAMS (1965), LAMBERT and HEIER (1967) and SWANBERG and BLACKWELL (1973) also indicate that the actual vertical distribution of radiogenic heat may lie between step and exponential models. Hence I assume here for the present purpose that the linear model is representative of the general vertical distribution of radiogenic heat in continental

Table 1

Comparison of heat producing layer with the silicic crust

(1)	(2)	(3)	(4)	(5)	
Geological regions	Thickness of Silicic crust (km)*)	Heat flow province†)	D (km)†) Step model	$2 \times D$ (km) Linear model	Difference (2)–(5)
Appalachian highland	15	New England	7.2 (0.4)‡)	14.4	0.6
Interior plains and highlands	20	Central stable region	9.4 (0.6)	18.8	1.2
Basin and range	20	Basin and range	9.4 (1.3)	18.8	1.2
Sierra Nevada	25	Sierra Nevada	10.1 (0.1)	20.2	4.8
Superior	?	Kapuskasing	13.5 (5)	27.0	?

*) From seismic data (PAKISER and ROBINSON, 1966).
†) From LACHENBRUCH (1968); ROY *et al.* (1968) and CERMAK and JESSOP (1971).
‡) The numbers in brackets are standard errors.

regions. This defines the reduced heat flow as the flux from below the upper crust. The lower crust is considered as gabbroic or as high grade metamorphic rocks. The radiogenic heat productivity of lower crust is estimated to be in the range $0.2 \ \mu W/m^3$ to $0.6 \ \mu W/m^3$. If we assume as a first approximation that the radiogenic heat productivity of the lower crust is uniform and constant at about $0.4 \ \mu W/m^3$ and that thickness of lower crust is about 20 km, the radiogenic heat produced in the lower crust is about $8.4 \ mW/m^2$. This value must then be subtracted from the reduced heat flow values to obtain the mantle heat flow in continental regions. Published values of reduced heat flow for various continental heat flow provinces are given in Table 2 along with estimated values of mantle heat flow in these regions. POLLACK and CHAPMAN (1977) have also recently calculated mantle heat flow values for various heat flow provinces. Their values are slightly higher than those given in Table 2 because of their use of the exponential model for subtracting the radiogenic heat production in continental crust. The main significance of mantle heat flow values given in Table 2 is that they can be considered as representing the residual heat of cooling of the continental lithosphere. Since radiogenic heat production in oceanic regions is extremely small in relation to surface heat flow (the basaltic oceanic crust of 5 km thick produces a heat flow of about $2.1 \ mW/m^2$) the near surface oceanic heat flow values can be considered as essentially representing the residual heat of cooling of the oceanic lithosphere. Thus a physical basis exists for comparing oceanic near surface and continental mantle heat flow values.

In Fig. 1 is shown a plot of continental mantle heat flow values (CMHF) and 'reliable averages' of oceanic heat flow values taken from SCLATER et al. (1976) and corrected for radioactivity of the basaltic crust, against $1/\sqrt{t}$. It appears from Fig. 1 that the continental and oceanic mantle heat flow values form a continuous and near linear trend. A least square fit to the combined data gives the slope and intercept values as 475 ± 34 and -2 ± 7 respectively. The slope value is in reasonably good agreement with the value of 500 reported by LISTER (1975) for near surface oceanic heat flow data unaffected by hydrothermal circulation in the ocean crust.

If CMHF values are considered as a separate data set, the least square fit gives the slope value as 164 ± 80 which is significantly different from the slope value of 517 ± 36 for the oceanic data set. However, there are several difficulties in considering the CMHF values given in Table 2 as a separate data set, the major problem being lack of sufficient data. Also care must be taken in giving equal weight to all the CMHF values in Table 2. For example, the Sierra Nevada region is believed to overly a subduction zone that was active until quite recently (BLACKWELL, 1971) and hence the calculated mantle heat flow may not be the equilibrium heat flux. Also the mantle heat flow value for the Basin and Range province may not be considered as highly reliable. The mantle heat flow for this region is calculated using the reduced heat flow given by ROY et al. (1968). However, additional new measurements have indicated considerable scatter in the heat flow–heat productivity data. This scatter is attributed to hydrothermal circulation taking place at shallow crustal depths in this region (LACHENBRUCH, 1977).

Table 2

Reduced and mantle heat flow values for various heat flow provinces

Heat flow province	N	Q_R (mW/m²)	Q_M (mW/m²)	Q_M^* (mW/m²)	Age of last thermal event	Reference
New England	8	35.1 (2.1)	26.8	24.2	Appalachian orogeny	Roy et al. (1968)
Central Stable-Grenville region	10	30.9 (0.8)	22.6	17.3	Grenvillian orogeny	Roy et al. (1968) Hamza (1976a)
SW Australia	3	26.3 (0.4)	18.0	9.6	1328–3030 m.y.	Jaeger (1970)
Kapuskasing area	3	26.3 (6.7)	18.0	8.8	Kenoran orogeny	Cermak and Jessop (1971)
Baltic shield	3	22.2 (2.5)	13.8	4.6	2000–3000 m.y.	Rao and Jessop (1975)
Sierra Nevada	6	16.7 (1.3)	8.4	7.9	Tertiary	Lachenbruch (1968)
Basin and range	12	58.5 (3.8)	50.2	49.7	Tertiary	Roy et al. (1968)
SW England	2	84.0	75.7	75.2	Tertiary	Tammemagi and Wheildon (1974)
Norwegian Caledonian orogenic belt	5	36.8 (1.3)	28.4	24.2	Caledonian orogeny	Swanberg et al. (1973) Hamza (1976a)
South Indian shield	2	33.0	24.7	14.2	2820 ± 100 m.y.	Rao et al. (1976)
North Indian region	4	38.4 (4.6)	30.1	24.7	530–893 m.y.	Rao et al. (1976)

N = Number of pairs of data for Q_R.

Q_R = Reduced heat flow. The numbers in brackets are standard errors.

Q_M = Mantle heat flow.

Q_M^* = Mantle heat flow corrected for radioactivity of lithosphere.

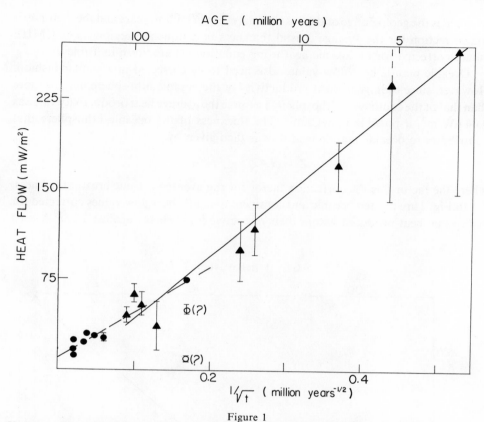

Figure 1

Variation of mantle heat flow (Q_M) with $1/\sqrt{t}$. ● continental mantle heat flow values, ▲ oceanic mantle heat flow values, —— least square fit to oceanic data, ——— least square fit to continental data excluding values for Sierra Nevada and Basin and Range provinces.

Excepting the values for Sierra Nevada and Basin and Range regions gives the slope value as 374 $\pm$ 33 which is closer but still in disagreement with the slope value for the oceanic data set.

However, a further correction is necessary for these low CMHF values before they can be identified as representing the residual heat of cooling of the continental lithosphere. It is the correction for the radiogenic heat produced within the lithosphere. Widely varying estimates of radiogenic heat productivity (0.004 to 0.04 μW/m³) has been quoted for the mantle beneath continental crust, depending on the geochemical model assumed. I assume as a first approximation a value of 0·02 μW/m³ for the present purpose, which is close to the average value for radiogenic heat generation in dunites and peridotites. The thickness of the lithospheric layer that contributes radiogenic heat to mantle heat flow is, according to the boundary layer model

$$Z = (9.4\sqrt{t} - 40) \, km \qquad (1)$$

where 't' is the geologic age of the last thermal event in million years and the factor 40 is the correction for the average crustal thickness in continental regions. The CMHF values corrected for radiogenic heat using equation (1) are given in Table 2.

Oceanic mantle heat flow values also need to be corrected in a similar fashion. However, since radiogenic heat productivity of the oceanic lithosphere may be more than that of the continental lithosphere, I assume the average heat productivity value as $0.04\ \mu W/m^3$, a possible upper limit. The thickness of the oceanic lithosphere that contributes to oceanic mantle heat flow is then given by

$$Z = (9.4\sqrt{t} - 5)\ \text{km} \qquad (2)$$

where the factor 5 is the correction factor for the average oceanic crustal thickness.

In Fig. 2 are plotted oceanic and continental mantle heat flow values corrected for radiogenic heat produced within their respective lithospheres against $1/\sqrt{t}$. A least

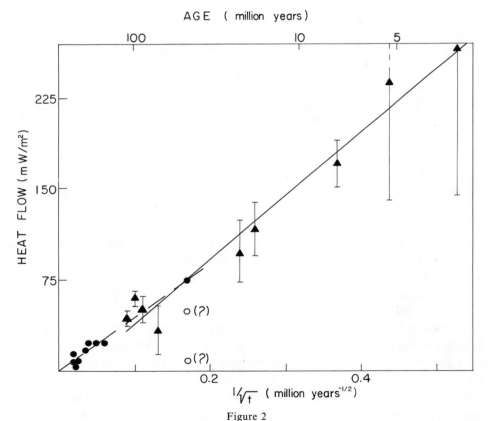

Figure 2

Variation of mantle heat flow corrected for radioactive heat production in the lithosphere beneath the crust (Q_M^*) with $1/\sqrt{t}$. ● continental data, ▲ oceanic data, —— least square fit to oceanic data, ---- least square fit to continental data excluding values for Sierra Nevada and Basin and Range provinces.

square fit to the combined data gives the slope and intercept values as 489 ± 32 and −8 ± 7 respectively. The slope value seems to be in good agreement with the value of 500 predicted by the boundary layer model. If continental data is considered as a separate set, the least square fit gives the slope as 433 ± 31, while for the oceanic data set the slope value is 520 ± 35. The correction for radiogenic heat production within the lithosphere has reduced the difference in slope values between continental and oceanic data sets. But the lack of sufficient data and uncertainties regarding the radiogenic heat production values for continental and oceanic lithospheres make comparison of slope values a difficult task.

Possibility of distinguishing between the thermal models

The variation of oceanic heat flow with distance from the ridge axis has been explained on the basis of constant thickness plate model (SCLATER and FRANCHETEAU, 1970) and boundary layer models (TURCOTTE and OXBURGH, 1972; PARKER and OLDENBURG, 1973). SCLATER and PARSONS (1976) give a brief review of the development of these two models. In the constant thickness plate model oceanic heat flow decreases from its high values near the ridge axis to a near constant value in a time interval of about 100 millions years. The boundary layer model however predicts a continuous decrease of heat flow with time. LISTER (1975) pointed out that reliable oceanic heat flow values not affected by hydrothermal circulation in ocean crust did show a relation of the type

$$Q = 500\left(\frac{1}{\sqrt{t}}\right) \tag{3}$$

where Q is in mW/m^2 and t is in million years. Both constant thickness plate model and boundary layer model explain the variation of oceanic heat flow age rather successfully. The main difference between these thermal models of lithosphere appears only at age values much in excess of 100 m.y. Thus oceanic heat flow data by itself has failed to distinguish clearly between the two thermal models. Continental heat flow values on the other hand cover a much larger age range, but no attempt has so far been made to extend these thermal models into continental regions. CROUGH and THOMPSON (1976) showed recently that decreasing heat flow in continental regions indicate a general increase in the depth to the low velocity zone. Their model assumes that there is a heat flux from asthenosphere to the lithosphere and that the low velocity zone (LVZ) is the lower limit of the lithosphere in continental regions. Thus the increase in thickness of the lithosphere with age in their model is much lower than that in the boundary layer model.

In Fig. 3 is shown a plot of continental mantle heat flow values corrected for radiogenic heat produced within the lithosphere against $1/\sqrt{t}$ along with the theoretical values of heat flow for the constant thickness plate model and boundary

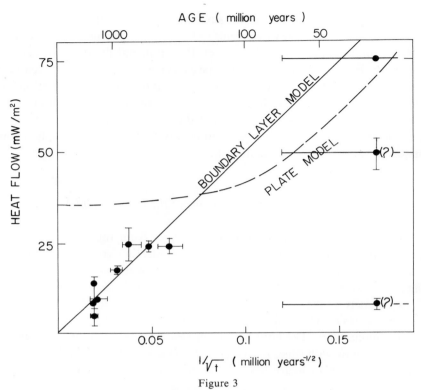

Figure 3

Comparison of theoretical values of heat flow for thermal models with values of continental mantle heat flow corrected for radioactive heat production in the lithosphere beneath the crust. —— theoretical heat flow for boundary layer model, – – – theoretical heat flow for the plate model, ● continental mantle heat flow corrected for radiogenic heat production in the lithosphere beneath the crust (Q_M^*)

layer models. The constant thickness plate model assumes a constant heat flow of about 40 mW/m² for lithospheres with ages in excess of about 100 m.y. and thus appears to be incapable of explaining the variation of continental mantle heat flow with age. The theoretical line for the boundary layer model on the other hand appears to provide a better fit to the continental mantle heat flow values.

Conclusions

On a heat flow versus $1/\sqrt{t}$ plot the continental mantle heat flow values appear to lie closer to the theoretical line for the boundary layer model rather than the constant thickness plate model. Since continental mantle heat flow values cover the range of $1/\sqrt{t}$ values where there is a clear distinction between these models, a possibility exists that they can be used as criterion for distinguishing between thermal models of the lithosphere.

Acknowledgements

I am thankful to H. N. Pollack, C. R. B. Lister and A. E. Beck for their helpful suggestions and comments. I am also thankful to I. I. G. Pacca and A. J. Melfi for their continuing support for the heat flow project at this institute, and to Conselho Nacional de Pesquisas for an operating research grant (12787/74).

REFERENCES

BLACKWELL, D. D. (1971), *The thermal structure of the continental crust*, in *The Structure and Physical Properties of the Earth's Crust* (ed. J. G. Heacock), Am. Geophys. Un. Mono. *14*, 169–184.

CERMAK, V. and JESSOP, A. M. (1971), *Heat flow, heat generation and crustal temperature in the Kapuskasing area of the Canadian shield*, Tectonophysics *11*, 287–303.

CROUGH, S. T. and THOMPSON, G. A. (1976), *Thermal model of continental lithosphere*, J. Geophys. Res. *81*, 4857–4862.

HAMZA, V. M. (1973), 'Vertical distribution of radioactive heat production in the Grenville geological province and the sedimentary sections overlying it', Unpublished Ph.D. Thesis, University of Western Ontario, London, Canada.

HAMZA, V. M. (1976a), *Possible extension of oceanic heat flow age relation to continental regions and the thermal structure of continental margins*, Anais da Acad. Brasileira de Ciências, *48* (Suplemento), 121–131.

HAMZA, V. M. (1976b), *Heat flow through lower continental and oceanic crust: A unified interpretation of the dependence on tectonic age* (Abstract) Bull. Am. Phys. Soc., Series II, *21*, 832.

HAMZA, V. M. (1977), *Possible limits on the temperature distributions in the continental lithosphere*, Anals da Academia Brasileira de Ciências, *49*, 287–295.

HAMZA, V. M. and VERMA, R. K. (1969), *The relationship of heat flow with age of basement rocks*, Bull. Volcanologique *33*, 123–152.

HEIER, K. S. and ADAMS, J. A. S. (1965), *Concentration of radioactive elements in deep crustal material*, Geochim. Cosmochim. Acta *29*, 53–61.

JAEGER, J. C. (1970), *Heat flow and Radioactivity in Australia*, Earth Planet. Sci. Lett. *81*, 285–292.

LACHENBRUCH, A. A. (1968), *A preliminary geothermal model of the Sierra Nevada*, J. Geophys. Res. *73*, 6977–6988.

LACHENBRUCH, A. H. (1970), *Crustal temperature and heat production: Implications of the linear heat flow relation*, J. Geophys. Res. *75*, 3291–3300.

LACHENBRUCH, A. H. (1977), 'High heat flow in the western U.S. and thermal effects of an extending lithosphere', Paper presented at the IASPEI/IAVCEI Joint Assembly, Durham (England).

LAMBERT, I. B. and HEIER, K. S. (1967), *Vertical distribution of Uranium, Thorium and Potassium in the Continental Crust*, Geochim. Cosmochim. Acta *31*, 377–390.

LISTER, C. R. B. (1975), 'The heat flow consequences of the Square-root law of ridge topography', Paper presented at General Assembly of Int. Union of Geod. and Geophys., Grenoble (France).

PAKISER, L. C. and ROBINSON, R. (1966), *Composition of the continental crust as estimated from seismic observations*, in *The Earth Beneath the Continents* (eds J. S. Steinhart and T. J. Smith), Am. Geophys. Union Mono. *10*, 620–626.

PARKER, R. L. and OLDENBURG, D. W. (1973), Thermal Model of Ocean Ridges, Nature Phys. Sci. **242**, 137–139.

POLLACK, H. N. and CHAPMAN, D. S. (1977), *On the regional variation of heat flow, geotherms and lithospheric thickness*, Tectonophys. *38*, 279–296.

POLYAK, B. G. and SMIRNOV, Ya. B. (1968), *Relationship between terrestrial heat flow and the tectonics of continents*, Geotectonics *4*, 205–213.

RAO, R. U. M. and JESSOP, A. M. (1975), *A comparison of the thermal characters of shields*, Can. J. Earth Sci. *12*, 347–360.

RAO, R. U. M., RAO, G. V. and HARI NARAIN (1976), *Radioactive heat generation and heat flow in the Indian Shield*, Earth Planet. Sci. Lett. *30*, 57–64.

ROY, R. F., BLACKWELL, D. D. and BIRCH, F. (1968), *Heat generation of plutonic rocks and continental heat flow provinces*, Earth Planet. Sci. Lett. *5*, 1–12.

SCLATER, J. G., CROWE, J. and ANDERSON, R. N. (1976), *On the reliability of oceanic heat flow averages*, J. Geophys. Res. *81*, 2997–3006.

SCLATER, J. G. and FRANCHETEAU, J. (1970), *The implications of terrestrial heat flow observations on current tectonic and geochemical models of the crust and upper mantle of the earth*, Geophys. J. Roy. Astron. Soc. *20*, 509–542.

SCLATER, J. G. and PARSONS, B. (1976), *Reply*, J. Geophys. Res. *81*, 4960–4964.

SMITHSON, S. B. and DECKER, E. R. (1974), *A continental crustal model and its geothermal implications*, Earth Planet. Sci. Lett. *22*, 215–225.

SWANBERG, C. A. (1972), *Vertical distribution of heat generation in the Idaho Batholith*, J. Geophys. Res. *77*, 2508–2514.

SWANBERG, C. A. and BLACKWELL, D. D. (1973), *Areal distribution and geophysical significance of heat generation in the Idaho batholith and adjacent intrusions in eastern Oregon and Western Montana*, Geol. Soc. Am. Bull. *86*, 1261–1282.

SWANBERG, C. A., CHESSMAN, M. D., SIMMONS, G., GRONLIE, G. and HEIER, K. S. (1973), *Heat flow – heat generation studies in Norway*, Tectonophysics *23*, 31–48.

TAMMEMAGI, H. Y. and WHEILDON, J. (1974), *Terrestrial heat flow and heat generation in south-west England*, Geophys. J. Roy. Astron. Soc. *38*, 83–94.

TURCOTTE, D. L. and OXBURGH, E. R. (1972), *Mantle convection and the new global tectonics*, Ann. Rev. Fluid. Mech. *4*, 33–68.

VERMA, R. K., HAMZA, V. M. and PANDA, P. K. (1970), *Further study of the correlation of heat flow with the age of basement rocks*, Tectonophysics *10*, 301–320.

(Received 25th October 1977, revised 13th April 1978)

Pageoph, Vol. 117 (1978/79), Birkhäuser Verlag, Basel

The Relationship Between Seismic Velocity and Radioactive Heat Production in Crustal Rocks: An Exponential Law[1])

By Ladislaus Rybach[2])

Abstract – In crystalline rocks seismic velocity V_p and density ρ increase, whereas radioactive heat production A decreases from acidic to basic compositions. From the velocity–density systematics for crustal rocks at different pressures an empirical $A(V_p)$ relationship has been derived for the range 5.0–8.0 km/sec which follows the exponential law: $A(V_p) = a \exp(-bV_p)$, where the numerical factors a and b depend on *in situ* pressure. A graph is given by means of which the heat production distribution $A(z)$ can be obtained for any given $V_p(z)$ structure.

Key words: Igneous and metamorphic rocks; Crustal low velocity layers; Cation packing index; Bulk density.

Introduction

Radioactive heat production, A, depends upon the amounts of uranium, thorium and potassium present in a given rock. It is a scalar and isotropic petrophysical property independent of *in situ* temperature and pressure and plays a dominant role in interpreting terrestrial heat flow patterns and in calculating thermal models for the earth's crust. Experimental determination of A is usually carried out by gamma-ray spectrometry which enables simultaneous determination of U, Th and K. Crushed or solid rock samples can be analysed (see Adams and Gasparini, 1970 or Rybach, 1971).

Only the uppermost few kilometers of the earth's crust are accessible for direct sampling (surface samples, drill cores, etc.). In order to shed light on thermal conditions and processes at depth, the distribution of radioactive heat sources in deeper parts of the crust must be inferred indirectly.

The velocity of compressional waves, V_p, and its variation with depth can be determined by seismic measurements carried out at the earth's surface. Should a relation exist between A and V_p, heat production values could be assigned to any depth z from the measured $V_p(z)$ distribution using the $A(V_p)$ relationship.

[1]) Contribution No. 207, Institute of Geophysics, ETH Zurich.
[2]) Institute of Geophysics, ETH Zurich, CH-8093 Zurich/Switzerland.

A similar procedure emerges from the interdependence of the bulk density ρ and the velocity V_p: Density values obtained from measured seismic velocities are widely used to calculate gravity models in interpreting Bouguer anomaly profiles. Several laws have been proposed for the density–velocity systematics to explain the empirical relationship found by NAFE and DRAKE (1959). Further empirical $\rho\,(V_p)$ relationships have been presented by BATEMAN and EATON (1967), GANGI and LAMPING (1971) and WOOLLARD (1975). BIRCH (1961) has shown that in crystalline rocks the $\rho(V_p)$ relationship is linear in the first approximation and is governed by the mean atomic weight of the material in question. The velocity–density relationship can be generalized by introducing the hydrodynamic velocity $V_\phi = (V_p^2 - 4V_s^2/3)^{1/2}$, see, e.g., SHANKLAND (1970) and MAO (1974).

The velocity–heat production relationship

A similar empirical relationship between A and V_p (or ρ) in crystalline rocks has been reported by RYBACH (1973a, b) giving a curve which displays the tendency of A to decrease with increasing V_p. This trend simply reflects the general petrophysical rule that with increasing percentage of mafic minerals the seismic velocity increases and the radioactivity decreases; a trend which also applies for metamorphic rocks (RYBACH, 1976a, b). Subsequently BUNTEBARTH (1975) has generalized the $A(V_p)$ relationship by introducing a quantity k, the packing index of cations and by referring to standard p/T conditions: $p_{\text{ref.}} = 4$ kb, $T_{\text{ref.}} = 20°C$. The index k is characteristic of the mineral constituents of the rock in question:

$$k = \frac{Z}{N \cdot v_M}, \tag{1}$$

where Z is the number of cations per mole, v_M the molar volumes of the corresponding minerals and N Avogadro's number.

There is a systematic variation of k in igneous and metamorphic rocks which reflects changes in chemical composition: given a rock assemblage with bulk chemistry varying from acidic (granitic) to basic compositions, k changes correspondingly from $4.7 \cdot 10^{-2}$ mol/cm^3 (granites) over intermediate values ($5.3 \cdot 10^{-2}$ mol/cm^3; gabbro-diorite) to about $6.0 \cdot 10^{-2}$ mol/cm^3 (eclogite). At the same time A decreases from about 7 HGU to 0.1 HGU (1 HGU $= 10^{-13}$ cal/cm^3 sec $= 0.417$ μW/m^3). In these rock types k and V_p are also interdependent. Since k is correlated with V_p as well as with A and since k is in causal relation with ρ, the correlation between A and V_p on one hand and ρ and A on the other (see also RYBACH, 1973, and BUNTEBARTH, 1975) are evident. A further correlation within these interdependences is the well-known $V_p(\rho)$ relationship (BIRCH, 1961; NAFE and DRAKE, 1959; WOOLLARD, 1975).

Whereas k and A are independent of *in situ* temperature and pressure, V_p shows marked pressure and temperature effects. The increase of V_p with pressure in the

range of 0–1.5 kb (corresponding to a crustal depth of 0–6 km) is especially remarkable (see, e.g., PRESS, 1966; LEBEDEV *et al.*, 1974). Thus the generally observed
increase of V_p with depth reflects changing rock type as well as changing pressure and
temperature conditions in the earth's crust. The temperature effect is, however, much
less pronounced.

In the following, the $A(V_p)$ relationship will be given for different pressures,
corresponding to different depth levels in continental crust. The $A(V_p)$ correlation
is based on $\rho(V_p)$ relationships, determined experimentally at different pressures
(VOLAROVICH *et al.*, 1966, 1967) on one hand, and on the systematic variation of A
and ρ from acidic to basic rock types on the other (Table 1). $\rho(V_p)$ relations at
different pressures are given in Fig. 1. It must be kept in mind that these Birch-type
lines (linear $\rho(V_p)$ relationship, see BIRCH, 1961) as well as the Nafe and Drake-curve
represent best fits through the sets (bands) of scattered data points. Thus the $A(V_p)$
relation suggested here will be of the same nature.

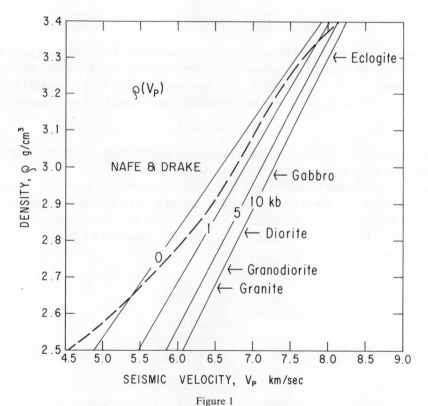

Figure 1

The seismic velocity–density relationship for crystalline rocks (arrows indicate the average density of
main rock types). Solid lines: the (V_p) relationship at different pressures (after VOLAROVICH *et al.*, 1966).
Dashed line: the Nafe and Drake relationship (see text).

Table 1

Average densities (ρ), heat production values (A) and cation packing indices (k) of major crystalline rock types*). Data from RYBACH (*1973b*) *and* BUNTEBARTH (*1975*).

Rock type	ρ (g/cm^3)	A (HGU)	A (μW/m^3)	k ($\times$ 10^{-2} mol/cm^3)
Granite	2.67	7.0	2.92	4.70
Granodiorite	2.72	3.6	1.50	4.85
Diorite	2.82	2.7	1.13	5.20
Gabbro	2.98	1.0	0.417	5.40
Eclogite	3.30	0.15	0.063	6.10

*) Ultrabasic rocks like peridotites are not included since their heat production and contribution to steady-state surface heat flow are very low.

The exponential law

Figure 2 shows the $A(V_p)$ relation at different pressures. The individual lines have been obtained by plotting V_p values (as determined from Fig. 1 for the rock types shown) against the corresponding A values (as listed in Table 1). The straight lines in the semi-logarithmic plot indicate the exponential dependence of A on V_p according to

$$A(V_p) = a \exp(-b \cdot V_p), \tag{2}$$

where a and b are numerical factors with the dimension (HGU or μW/m^3) and (km/sec), respectively. They depend on pressure as listed in Table 2.

The $A(V_p)$ relation as shown in Fig. 2 has been rearranged in Fig. 3, where the depth equivalent (in km) of the lithostatic pressure (in kb) is also given. The curves in Fig. 3 are drawn to depths for which experimental data (at corresponding pressures) are available (Fig. 2); the lines could be, however, extended to greater depths by extrapolation.

Table 2

The pressure dependence of the numerical factors a and b in equation (2) as determined from Fig. 2.

Pressure (kb)	a	b
0	7.90 · 10^4	1.735
1	6.02 · 10^5	1.951
5	5.29 · 10^6	2.193
10	3.96 · 10^7	2.483

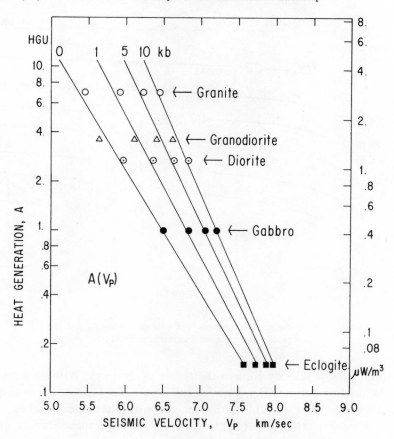

Figure 2

The seismic velocity–heat production relationship for different pressures. Heat production data from Table 1 are plotted against average V_p values as determined from Fig. 1 for the rock types marked by arrows. The lines in the semi-logarithmic plot define an exponential law: $A(V_p) = a \exp(-bV_p)$.

Discussion

Several arguments can be put forward to show that in zones of lowered seismic velocities in the continental crust ('crustal low velocity layer', first reported by MUELLER and LANDISMAN, 1966, and LANDISMAN and MUELLER, 1966) the $A(V_p)$ relationship is not valid. The lower seismic velocities call for higher heat production values in the low velocity layer. This would be, however, in contradiction with the constraints represented by the surface heat flow on one hand and by the mantle heat flow on the other (see RYBACH et al., 1977). Furthermore, a distribution of the heat producing radioisotopes with the maximum at intermediate depths (corresponding to the low velocity layer) is not likely to persist in the pressure–temperature field of the crust (RYBACH, 1976c). According to MAKRIS (1971) the $\rho(V_p)$ relationship is not

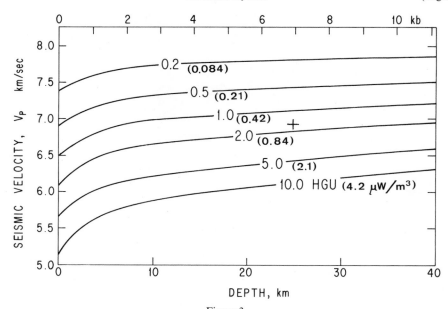

Figure 3

Diagram to determine heat production A from measured seismic velocity V_p at depth z (km). The $+$ sign
indicates a numerical example (see text).

valid within the low velocity zone in the region of the Alps. With heat production
figures obtained by interpolation between the A values at the top and at the bottom
of the low velocity layer, the above-mentioned constraints can be satisfied.

It can be expected that the $A(V_p)$ relationship holds for metamorphic rocks too.
The effect of water content of rocks on seismic velocity and particularly heat production may play an important role (e.g., higher heat production in the 'wet' amphibolite
facies rocks, low heat production in the 'dry' rocks of the granulite facies; see discussion in SMITHSON and DECKER, 1974).

The distribution of radioactive heat sources with depth, $A(z)$, can be obtained for
a given crustal section (with a known $V_p(z)$ structure), using Fig. 3. Since the $\rho(V_p)$-
lines in Fig. 1 have been determined at room temperature, the effect of temperature
on V_p (which amounts to only a few percent) can be accounted for before entering
Fig. 3, by reducing the *in situ* $V_p(z)$ values to the reference value V_p (20°C). In Table 3
the ratio $V_p(T)/V_p(20°C) = f$ is given for a few rock types (for intermediate compositions not listed in Table 3 the values sought can be found by interpolation).

The following example should demonstrate this procedure. The heat production
value is to be determined for a crustal section at a depth of $z = 25$ km, where $V_p =$
6.8 km/sec has been measured (indicating a basic rock, $\approx$ gabbro) in a continental
area with normal geothermal conditions. Thus an *in situ* temperature of T (25 km) $\approx$
400°C can be assumed to obtain a corrected value (V_p at *in situ* temperature $\rightarrow V_p$
at room temperature). With the data of Table 3 V_p (20°C) = V_p (400°C)/f =

Table 3

The effect of temperature on V_p for different rocks. $V_p(T)/V_p$ (20°C) values from FIELITZ *(1971) and* BUNTEBARTH *(1975).*

$T(°C)$	Granite	Gabbro	Peridotite
200	0.995	0.995	0.996
400	0.979	0.983	0.985
600	0.948	0.965	0.973
800	—	(0.932)	(0.957)

6.8/0.983 = 6.92 km/sec. With this figure the heat production value can be read from Fig. 3 at z = 25 km as 1.6 HGU (0.67 μW/m^3).

Conclusions

Heat production in crystalline rocks, A, varies systematically with seismic velocity, V_p, in the range of 5.0–8.0 km/sec (covering the range of most crustal rock types): A decreases with increasing V_p. The empirical relationship has the form $A(V_p) = a \exp(-bV_p)$ and fits well to average A and V_p values determined experimentally. Starting from a given velocity distribution in the crust, $V_p(z)$, which can be determined by seismic measurements, the corresponding distribution of radioactive heat sources, $A(z)$, can be found. The latter enables the calculation of thermal models as well as the determination of the mantle heat flow if the surface heat flow is known.

Experimental investigation of the $A(V_p)$ relationship, involving V_p and A determinations on the same samples, is now in progress.

Acknowledgement

This study has been partly supported by the Swiss National Science Foundation (Project No. 2.616–0.76). Thanks are due to Prof. W. Lowrie and Dr. J. Channell for critical reading of the manuscript.

REFERENCES

ADAMS, J. A. S. and GASPARINI, P. *Gamma-ray Spectrometry of Rocks* (Elsevier, Amsterdam, 1970).
BATEMAN, P. C. and EATON, G. P. (1967), *Sierra Nevada*, Science *158*, 1407–1417.
BIRCH, F. (1961), *The velocity of compressional waves in rocks to 10 kilobars*, J. Geophys. Res. *66*, 2199–2224.
BUNTEBARTH, G. (1975), 'Geophysikalische Untersuchungen über die Verteilung von Uran, Thorium und Kalium in der Erdkruste sowie deren Anwendung auf Temperaturberechnungen für verschiedene Krustentypen.' Diss. Univ. Clausthal.

FIELITZ, K. (1971), *Elastische Wellengeschwindigkeiten in verschiedenen Gesteinen unter hohem Druck und bei Temperaturen bis zu 750°C*, Z. Geophys. *37*, 943–956.

GANGI, A. F. and LAMPING, N. E. *An* in situ *method of determining the pressure dependence of phase-transition temperatures in the crust*, in *The Structure and Physical Properties of the Earth's Crust* (ed. John G. Heacock), American Geophysical Union, Geophysical Monograph 14 (Washington, 1971), pp. 185–190.

LANDISMAN, M. and MUELLER, S. (1966), *Seismic studies of the Earth's crust in continents, 2, Analysis of wave propagation in continents and adjacent shelf areas*, Geophys. J.R. astr. Soc. *10*, 539–548.

LEBEDEV, T. S., SAPOVAL, V. I. and KORCIN, V. A., *Untersuchungen der Geschwindigkeit von elastischen Wellen in Gesteinen unter den thermodynamischen Bedingungen der tiefern Erdkruste*. Veröff. Zentralinst. Phys. der Erde Nr. 22, 17–28 (Potsdam, 1974).

MAKRIS, J., *Aufbau der Erdkruste in den Ostalpen aus Schweremessungen und die Ergebnisse der Refraktions-seismik*, Hamburger Geophys. Einzelschriften *15* (De Gruyter and Co., Berlin, 1971).

MAO, N. H. (1974), *Velocity–density systematics and its implications for the iron content of the mantle*, J. Geophys. Res. *79*, 5477–5454.

MUELLER, S. and LANDISMAN, M. (1966), *Seismic studies of the earth's crust in continents, 1, Evidence for a low-velocity zone in the upper part of the lithosphere*, Geophys. J.R. astr. Soc. *10*, 525–538.

NAFE, J. E. and DRAKE, C. L. (1959), in Talwani, M., Sutton, G. A. and Worzel, J. L., *A crustal section across the Puerto Rico Trench*, J. Geophys. Res. *64*, 1545–1500.

PRESS, F., *Seismic velocities*, in *Handbook of Physical Constants* (ed. Sydney P. Clark), Geol. Soc. America Memoir 97 (Washington, 1966), pp. 197–218.

RYBACH, L., *Radiometric techniques*, in *Modern Methods of Geochemical Analysis* (ed. Richard E. Wainerdi and Ernst A. Uken), (Plenum Press, New York, 1971), pp. 271–318.

RYBACH, L. (1973a), 'Radioactive heat production of rocks from the Swiss Alps; geophysical implications.' 1st European Geophysical Society Meeting, Zurich (Abstract).

RYBACH, L. (1973b), *Wärmeproduktionsbestimmungen an Gesteinen der Schweizer Alpen*, Beitr. Geol. Schweiz, Geotechn. Ser. Liefg. 51, (Kümmerly and Frey, Bern, 1973).

RYBACH, L. (1976a), *Radioactive heat production; A physical property determined by the chemistry of rocks*, in *The Physics and Chemistry of Minerals and Rocks* (ed. R. G. J. Strens), (Wiley and Sons, London, 1976), pp. 309–318.

RYBACH, L. (1976b), *Radioactive heat production in rocks and its relation to other petrophysical parameters*, Pure and Appl. Geoph. *114*, 309–318.

RYBACH, L. (1976c), *Die Gesteinsradioaktivität und ihr Einfluss auf das Temperaturfeld in der kontinentalen Kruste*, J. Geophys. *42*, 93–101.

RYBACH, L., WERNER, D., MUELLER, S. and BERSET, G. (1977), *Heat flow, heat production and crustal dynamics in the Central Alps, Switzerland*. Tectonophysics *41*, 113–126.

SHANKLAND, T. J. (1972), *Velocity–density systematics: Derivation from Debye theory and the effect of ionic size*, J. Geophys. Res. *77*, 3750–3758.

SMITHSON, S. B. and DECKER, E. R. (1974), *A continental crustal model and its geothermal implications*, Earth Plan. Sci. Letters *22*, 215–225.

VOLAROVICH, M. P., GALDIN, J. E. and LEVIKIN, A. N. (1966), *Study of the velocity of longitudinal waves in samples of extrusive and metamorphic rocks at pressures up to 20 kb*, Izvest., Earth Phys. Ser., no. 3, engl. transl., 15–23.

VOLAROVICH, M. P., KURSKEYEV, A. K., TOMASHEVSKAYA, I. C., TUZOVA, I. L. and URAZAYEV, B. M. (1967), *Relation between the longitudinal wave propagation velocity and the rock density at high hydro-static pressures*, Izvest., Earth's Phys. Ser., no. 5, engl. transl. 276–279.

WOOLLARD, G. P. (1975), *Regional changes in gravity and their relation to crustal parameters*. Bureau Gravimétrique Intern., Bull. d'Inform. *36*, 106–110.

(Received 3rd November 1977, revised 23rd March 1978)

Pageoph, Vol. 117 (1978/79), Birkhäuser Verlag, Basel

The Degree of Metamorphism of Organic Matter in Sedimentary Rocks as a Paleo-Geothermometer, Applied to the Upper Rhine Graben

By Günter Buntebarth[1])

Abstract – The subsidence of sedimentary layers implies increasing temperature downwards within the sedimentary column, so that the degree of coalification of organic matter increases continually. Apart from temperature, the slowly reacting chemical compounds of the organic matter strongly depend on time, too.

It is shown that the coal rank is proportional to the integral of temperature and time of burial (t) for the Tertiary sedimentary rocks of the Upper Rhine Graben. This relationship is used to calculate paleo-geothermal gradients (grad T) for some boreholes in the Upper Rhine Graben, from which the rate of burial during geological history ($z(t)$) is known. The degree of coalification is measured by its mean optical reflectivity (R_m), so that the relationship between coalification and geothermal history is $R_m^2 \propto \text{grad } T \int z(t) \, dt$.

The results show high heat flow during Lower Tertiary and a decrease during Upper Tertiary at some locations of the Upper Rhine Graben. The recent high heat flow is not detectable in coalification. The young thermal anomaly is perhaps caused by ascending pore fluid and/or by heat conduction from a heat source in the lower crust.

Key words: Coal reflectivity; Metamorphism of organic matter; Burial history; Thermal history; Heat flow; Thermal conductivity.

1. Introduction

Organic inclusions are found occasionally in clays and sandstones. The size of them is of the order of a few mikrons and they can be derived from leaves, barks or other particles of vegetation. During the subsidence of sedimentary layers, pressure and temperature increase within the sedimentary rocks, and both parameters affect the organic inclusions. The organic matter begins to react. At first, the increasing pressure causes the diagenesis of the organic inclusions, removing most of the water content. Besides the effect of pressure which is assumed to be negligible during the further coalification (R. and M. Teichmüller, 1949; Huck and Patteiski, 1964), temperature gradually affects the organic matter and initiates the coal catagenesis starting in the state of brown coal and reaching finally the state of graphite. In this

---

[1]) Institut für Geophysik der TU Clausthal, Postfach 230, D-3392 Clausthal-Zellerfeld, Federal Republic of Germany.

range of metamorphism of organic matter the chemical composition is changed considerably. The higher the enrichment of carbon the higher the degree of meta-morphism. In order to analyse the chemical composition, a large amount of organic matter is necessary. Such an amount is not available generally from inclusions in sedimentary rocks. Therefore, the degree of metamorphism is determined in another way. A most useful method is its determination by the optical reflectivity coefficient which is applicable to both the coal from seams and the coaly particles dispersed in sedimentary rocks. The reflectivity coefficient gives a continuous scale for the coalification of huminite/vitrinite with values ranging from about 0.2 up to 5 and more percent (M. TEICHMÜLLER, 1970). Huminite and vitrinite are mazeral groups of humous components.

2. Empirical method for geothermal interpretation of the coalification of organic matter

The reactions of the chemical compounds depend not only upon temperature attained but also upon the time at the given temperature. Starting from this fact

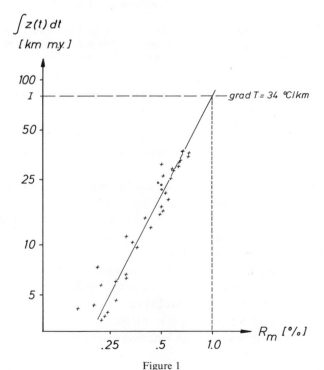

Figure 1
Integral values of subsidence and time versus the optically measured reflectivity with data from the borehole Sandhausen 1.

various methods have been developed to interpret the state of coalification for geothermal purposes (e.g. KARWEIL, 1955; TISSOT, 1969; LOPATIN, 1971; CORNELIUS, 1975). The authors presume the validity of the Arrhenius equation:

$$k = A \exp(-E/RT), \tag{1}$$

where k is the reaction-rate constant, A the frequency factor, E the activation energy, R the molar gas constant, and T the temperature.

By using constant values A and E the Arrhenius equation cannot give sufficient results over a wide range of coalification (HOOD et al., 1975). Therefore, an empirical method is used to show the relationship between the degree of coalification and temperature as well as time (BUNTEBARTH, 1977).

The actual reaction temperature of a coaly particle cannot generally be measured, because of the temperature variation $T(t)$ during the geological history caused by the subsidence of the sedimentary rocks and, very likely, by a change of the heat flow.

The temperature history during the subsidence can be calculated by summing up the temperature differences at all layers i ($i = 1, 2, \ldots, n$) of the sedimentary rock with thicknesses Δz_i, so that the temperature in the layer $i = n$ is:

$$T_n = \sum_{i=1}^{n} (\text{grad } T)_i \, \Delta z_i + T_0. \tag{2}$$

T_0 is the surface temperature and $(\text{grad } T)_i$ is the temperature gradient in the layer i. As a rough approximation, a constant temperature gradient is assumed within a column, so that the temperature history can be calculated from the burial history by

$$T_n(t) = \text{grad } T \sum_{i=1}^{n} \Delta z(t)_i + T_0. \tag{3}$$

The history of subsidence can be reconstructed in most cases to get the function $z = z(t)$.

The degree of coalification determined by the optical reflectivity coefficient (R_m) of vitrinite increases with an increase of both temperature and time. However, the effect upon the organic matter of the two parameters cannot be separated. So, the degree of coalification is assumed to be proportional to the integral of temperature and time. This proportionality is shown in a log-log graph like Fig. 1, in which the reflectivity shows the dependence on the integral of the burial history according to equation (3). The close relationship is demonstrated by a constant slope for each borehole with a value of 2.0 in the log-log graph. Hence it follows, that the relationship must be linear between the square of reflectivity and the integral of burial history as shown in Fig. 2. Therefore, an equation is obtained in terms of easily measurable parameters by

$$R_m^2 = f(\text{grad } T) \int_0^{t_1} z(t) \, dt \tag{4}$$

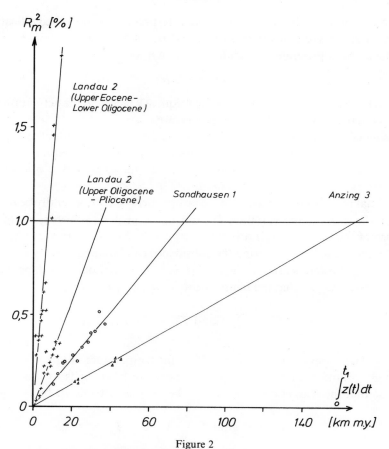

Figure 2

The square of median optical reflectivity of vitrinite (R_m^2) vs. the integral of the burial history for some boreholes.

in which $f(\text{grad } T)$ is a determinable function of the temperature gradient. The equation (4) is a function of two variables. In order to interpret it easily, the problem is considered at the plane $R_m = 1.0\%$, and the integral of the burial history and the time is called (I) at this plane. The values (I) of the integrals are taken from graphs with the square of reflectivity versus the integral like Fig. 2. The values (I) give, together with the known temperature gradients of six boreholes, the calibration function (BUNTEBARTH, 1977):

$$\text{grad } T \, [°\text{C/km}] = 98.7 - 14.6 \ln I \, [\text{km m.y.}] \tag{5}$$

This form is valid for constant geothermal conditions. If there is a change in the heat flow, the temperature gradient changes too. In general, a change in heat flow is detectable in coalification only for a decreasing heat flow, because the coalification indicates the maximum value only.

3. Application

3.1. Geothermal history of the Upper Rhine Graben

A few boreholes of the Upper Rhine Graben are considered for calculating the temperature gradients. In order to use equation (5), the value (I) must be determined for each borehole in the way reported above. Examples are shown in Figs. 1 as well as 2 where the integral values show the said dependence on the optical reflectivity. For reaching these values the data used are those reported by M. TEICHMÜLLER (1970), DOEBL et al. (1974), and M. TEICHMÜLLER (1977).

Table 1 summarizes the results of calculations for some boreholes. Assuming an average thermal conductivity for Tertiary rocks, i.e., $K = 1.9$ W/m °K the heat flow can be estimated for both geological time ranges, that is the Lower Tertiary (Upper Eocene to Lower Oligocene) and Upper Tertiary (Upper Oligocene to Pliocene). The results are given in Table 1 and in Fig. 3. The used boreholes reflect a high heat flow from Upper Eocene, there is no sample older, to the boundary Lattorf/Rupel. If more data can be used, perhaps the boundary varies within a given time range.

The highest geothermal gradient was about 75°C/km during the Lower Tertiary and decreased to 50°C/km during the Upper Tertiary. However, at present the temperature gradient shows an increased value again. Yet the high value which is measured is not detectable in coal reflectivity. So the temperature rise might exist, perhaps, since a few ten thousands of years, but couldn't have existed before Pleistocene by considering the effect of the temperature rise on the organic matter (BUNTEBARTH and TEICHMÜLLER, 1977).

When the temperature gradient during the geological history, which is calculated from coal reflectivity, is compared with that from geological estimation, the agreement is rather good (M. and R. TEICHMÜLLER, 1977). The change in the geothermal

Table 1

Geothermal history of the Upper Rhine Graben, calculated with data of some boreholes reported by
M. TEICHMÜLLER (*1970, 1977*).

Bohrung	Lower Tertiary Temperature gradient [°C/km]	Heat flow*) [mW/m²]	Upper Tertiary Temperature gradient [°C/km]	Heat flow*) [mW/m²]	Recent Temperature gradient [°C/km]	Heat flow HAENEL (1976) [mW/m²]
Hähnlein West 2	—	—	37	70	49	69
Frankenthal 10	65	123	50	95	53	—
Sandhausen 1	—	—	34	64	42	—
Dudenhofen 2	48	90	—	—	77	—
Harthausen 1	48	91	48	91	67	—
Landau 2	69	131	45	85	77	125†)
Scheibenhardt 2	78	148	49	94	77	—

*) Calculated with the thermal conductivity $K = 1.9$ W/m °K.

†) Mean value for the Landau area.

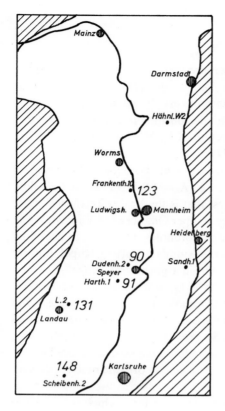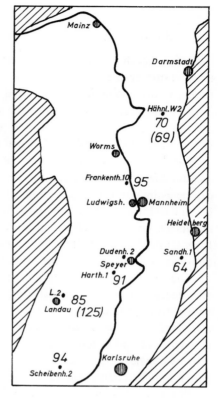

Figure 3
(Left) Heat flow values given in mW/m²-units in the Upper Rhine Graben during the Lower Tertiary.
(Right) Heat flow values given in mW/m²-units in the Upper Rhine Graben during the Upper Tertiary.
The recent heat flow is given in brackets.

conditions can be correlated with the volcanic activity in the Upper Rhine Graben.
That activity decreased about 40 m.y. ago during Uppermost Eocene (LIPPOLT et al.,
1974). It coincides well with the state of coalification of organic matter in the Upper
Rhine Graben as shown in Fig. 4.

BARTZ (1974) and ILLIES (1974) reported recent tectonic activity having started
during Pleistocene with local Quaternary sedimentation near Freiburg, Strasbourg
and Heidelberg. The recent subsidential activity is, perhaps, connected with the change
in the geothermal condition during the latest geological history.

The recent high heat flow could have existed since Pleistocene, if the stress
condition in the Upper Rhine Graben (ILLIES and GREINER, 1977) caused the pore
fluid to migrate in the direction of the surface since that time. The total heat anomaly
in the Upper Rhine Graben can be explained by water migration along faults (WER-
NER, 1975; BUNTEBARTH and SCHOPPER, 1976).

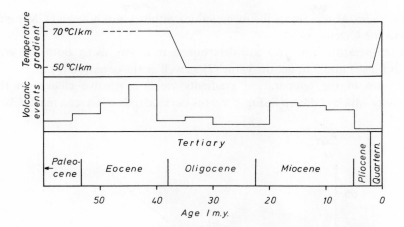

Figure 4

The change in the geothermal gradients for some boreholes in the Upper Rhine Graben and the volcanic activity after LIPPOLT *et al.* (1974) in the Rhine Graben region during Tertiary and Quaternary.

If the thermal anomaly is considered to be caused by intrusions in the lower crust or by a temperature rise at the crust/mantle boundary, the surface heat flow increases by heat conduction. Heat conduction throughout the crust needs time of the order of some million years. In this case, the recent high heat flow is very young. A thermal anomaly in the lower crust or in the mantle on one hand, or a thermal anomaly by water migration on the other hand, cannot be distinguished in the Upper Rhine Graben by means of the state of coalification.

3.2. *Influence of the burial depth on the thermal conductivity*

The results summarized in Table 1 are calculated for a constant heat flow. Such a generalization causes a variable scattering of the values within a borehole as shown in Fig. 1.

The deviations are due to the assumption of a constant temperature gradient, not considering changes in thermal conductivity, i.e., the compaction of sediments and changes in chemical composition. To evaluate the effect of compaction on thermal conductivity, a single stratigraphic layer with constant chemical composition (Cyrena marl/Chattian) is considered.

A constant heat flow Q is proportional to the thermal conductivity K and the temperature gradient:

$$Q = K \operatorname{grad} T. \tag{6}$$

Hence:

$$\frac{K_{CM}}{K_M} = \frac{(\operatorname{grad} T)_M}{(\operatorname{grad} T)_{CM}} \tag{7}$$

where the index M represents the mean values within a borehole and CM represents the values for Cyrena marl.

The temperature gradients are determined in a way using both the graph as shown in Figs. 1 and 2 to get the value (I) as well as the equation (5).

The ratio of the temperature gradients yields a relative change in thermal conductivity which is shown in Fig. 5 versus burial depth of Cyrena marl (Chattian).

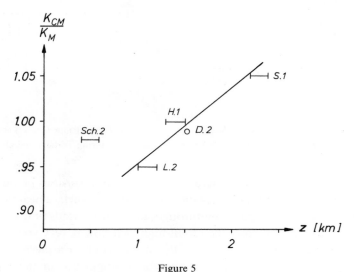

Figure 5

The ratio of the heat conductivity of Cyrena marl (K_{CM}) and the mean conductivity of each borehole (K_M) versus the burial depth of the Cyrena marl (Chattian) for the boreholes: Dudenhofen 2, Harthausen 1, Landau 2, Sandhausen 1, and Scheibenhardt 2.

The relationship reflects a linear law between subsidence and thermal conductivity in a depth range from 1 to 2.5 km. The present burial depth of Cyrena marl in the borehole Scheibenhardt 2 is not identical with the maximum subsidence, because of an unknown amount of the vertical crustal movement during the Lower Miocene.

Acknowledgment

I am grateful to Drs. M. and R. Teichmüller, Geolog. Landesamt in Krefeld, for very useful discussions.

REFERENCES

BARTZ, J. *Die Mächtigkeit des Quartärs im Oberrheingraben*, in *Approaches to Taphrogenesis* (eds. J. H. Illies and K. Fuchs), (Schweizerbart'sche Verl.-Buchh., Stuttgart, 1974), pp. 78–87.

BUNTEBARTH, G. (1977), *Eine empirische Methode zur Berechnung von paläo-geothermischen Gradienten aus dem Inkohlungsgrad organischer Einlagerungen in Sedimentgesteinen mit Anwedung auf den mittleren Oberrheingraben*, Fortschr. Geol. Rheinl. u. Westf. *27* (in print).

BUNTEBARTH, G. and SCHOPPER, J. R., *Heat flow caused by water migration along faults in dependence on petrophysical parameters*, in *Proceed. Intern. Congr. on Thermal Waters, Geothermal Energy and Volcanism of the Mediterranean Area, Oct. 5–10, Athens*, Vol. II (Athens, 1976), pp. 41–49.

BUNTEBARTH, G. and TEICHMÜLLER, R. (1977), *Zur Ermittlung der Paläotemperaturen im Dach des Bramscher Intrusivs aufgrund von Inkohlungsdaten*, Fortschr. Geol. Rheinl. u. Westf. *27* (in print).

CORNELIUS, C.-D. (1975), *Die Phasen der organischen Metamorphose im Zeit-Temperatur-Diagramm*, Compendium 74/75, Ergänzungsband zur Zeitschr. *Erdöl und Kohle, Erdgas und Petrochemie*, pp. 70–83.

DOEBL, F., HELING, D., HOMANN, W. KARWEIL, J., TEICHMÜLLER, M. and WELTE, D., *Diagenesis of Tertiary clayey sediments and included dispersed organic matter in relationship to geothermics in the Upper Rhine Graben*, in *Approaches to Taphrogenesis* (eds. J. H. Illies and K. Fuchs), (Schweizerbart'sche Verl.-Buchh., Stuttgart, 1974), pp. 192–207.

HAENEL, R. (1976), *Die Bedeutung der terrestrischen Wärmestromdichte für die Geodynamik*, Geol. Rundschau *65*, pp. 797–809.

HOOD, A., GUTJAHR, C. C. M. and HEACOCK, R. L. (1975), *Organic metamorphism and the generation of petroleum*, Bull. Am. Assoc. Petrol. Geol. *59*, pp. 986–996.

HUCK, G. and PATTEISKY, K. (1964), *Inkohlungsreaktionen unter Druck*, Fortschr. Geol. Rheinl. u. Westf. *12*, pp. 551–558.

ILLIES, J. H. *Taphrogenesis and plate tectonics*, in *Approaches to Taphrogenesis* (eds. J. H. Illies and K. Fuchs), (Schweizerbart'sche Verl.-Buchh., Stuttgart, 1974), pp. 433–460.

ILLIES, J. H. and GREINER, G., *Active movements and state of stress in the Rhinegraben*, in *Abstracts IASPEI/IAVCEI Joint General Assemblies, Aug. 8–19* (Durham, 1977), p. 75.

KARWEIL, J. (1955), *Die Metamorphose der Kohlen vom Standpunkt der physikalischen Chemie*, Z. deutsch. geol. Ges. *107*, pp. 132–139.

LIPPOLT, H. J., TODT, W. and HORN, P., *Apparent potassium–argon ages of Lower Tertiary Rhine Graben volcanics*, in *Approaches to Taphrogenesis* (eds. J. H. Illies and K. Fuchs), Schweizerbart'sche Verl.-Buchh. (Stuttgart, 1974), pp. 213–221.

LOPATIN, N. W. (1971), *Temperatura i geologicheskoe vremya kak faktory uglefikatsii (Temperature and geologic time as factors in coalification)*, Akad. Nauk SSSR, Izv., Ser. Geol. *3*, pp. 95–106.

TEICHMÜLLER, M., *Bestimmung des Inkohlungsgrades von kohligen Einschlüssen in Sedimenten des Oberrheingrabens – ein Hilfsmittel bei der Klärung geothermischer Fragen*, in *Graben Problems* (eds. J. H. Illies and St. Müller), Schweizerbart'sche Verl.-Buchh. (Stuttgart, 1970), pp. 124–142.

TEICHMÜLLER, M. (1977), *Die Diagenese der kohligen Substanzen in Gesteinen des mittleren Oberrhein-Grabens*, Fortschr. Geol. Rheinl. u. Westf. *27* (in print).

TEICHMÜLLER, M. and TEICHMÜLLER, R. (1977), *Beziehung zwischen Inkohlung und der geothermischen Geschichte des mittleren Oberrhein-Grabens*, Fortschr. Geol. Rheinl. u. Westf. *27* (in print).

TEICHMÜLLER, R. and TEICHMÜLLER, M. (1949), *Inkohlungsfragen im Ruhrkarbon*, Z. deutsch. geol. Ges. *99*, pp. 40–77.

TISSOT, B. (1969), *Premières données sur le mécanismes et la cinétique de la formation du pétrole dans les sédiments; simulation d'un schéma réactionnel sur ordinateur*, Rev. Inst. Franc. Pétrole *24*, pp. 470–501.

WERNER, D. (1975), *Probleme der Geothermik am Beispiel des Rheingrabens*, Diss., Univ. Karlsruhe.

(Received 29th October 1977, revised 17th March 1978)

Pageoph, Vol. 117 (1978/79), Birkhäuser Verlag, Basel

The Preliminary Heat Flow Map of Europe and Some of its Tectonic and Geophysical Implications

By Vladimír Čermák[1]) and Eckart Hurtig[2])

Abstract – The heat flow map of Europe was derived from 2605 existing observations, which for this purpose were supplemented by numerous results of deep borehole temperatures, gradients and local heat flow patterns. In areas without data the heat flow field was extrapolated on the basis of the regional tectonic structure and the observed correlation of heat flow and the age of the last tectono-thermal event. The heat flow pattern as obtained in the map may be described by two components: (i) regional part and (ii) local part of the measured surface geothermal activity. The regional part of the heat flow field in Europe is dominated on the whole by a general 'north-east to south-west' increase of the geothermal activity, which is an obvious consequence of the tectonic evolution, the major heat flow provinces corresponding thus to the principal tectonic units. The geothermal fine structure (local part) superimposing the former is mainly controlled by local tectonics, especially by the distribution of the deep reaching fracture zones and by the hydrogeological parameters. The correlation between the heat flow pattern and the crustal structure allows some preliminary geophysical implications: (a) areas of the increased seismicity may be connected with the zones of high horizontal temperature gradient, (b) increased surface heat flow may be generally observed in the zones of weakened crustal thickness, (c) there are considerable regional variations in the calculated temperature on the Moho-discontinuity, as well as in the upper mantle heat flow contribution.

Key words: Heat flow – age relationship; Mantle heat flow; Crustal temperature; Convective heat transfer.

1. Introduction

During recent years remarkable progress in heat flow studies has been made in Europe. The new data have been accumulating at such a rate that it is becoming quite difficult for an individual to keep up with all the results. To overcome this problem and to be able to use the measured geothermal data for further study and/or to correlate the observed surface heat flow distribution with the other geophysical and geological phenomena, a map showing the regional heat flow pattern is necessary.

The idea to prepare a detailed heat flow map of Europe was put forth at the 16th General Assembly of the International Union of Geodesy and Geophysics in Grenoble in 1975. This paper is a brief report on the work that has been done in this field and the attached map (see Fig. 4) shows the simplified version of the obtained Heat

---

[1]) Geofyzikální ústav ČSAV, 141–31 Praha-4, Boční II/1a, CSSR.
[2]) Zentralinstitut für Physik der Erde, AdW DDR, Potsdam, Telegrafenberg, German Democratic Republic.

Flow Map of Europe (ČERMÁK and HURTIG, 1977). The map itself was presented during the meeting of the International Association of Seismology and Physics of the Earth Interior in Durham in 1977 and distributed thereafter together with a more detailed Explanatory Text.[3])

2. Distribution and histogram of heat flow data

The construction of the heat flow map of Europe is based on the total of 2605 data, which include 2161 land data and 444 marine observations. In addition to the published heat flow measurements, other suitable geothermal observations such as deep temperature measurements, geothermal gradients and subsurface temperature maps were used. When local heat flow maps were available (BOLDIZSÁR, 1975; ČERMÁK, 1978; FYTIKAS and KOLIOS, 1977; GABLE, 1977; HURTIG and SCHLOSSER, 1976; KUTAS et al., 1976; LUKIĆ, 1971; TEZCAN, 1977; VELICIU et al., 1977), they were incorporated into the large final map.

The geographical distribution of the data is shown in Fig. 1. In spite of the great number of recent new heat flow measurements, the density of information is still far from being uniform and from large territories data are totally missing; these areas were left blank on the map (Ireland, Spain and Portugal, Yugoslavia, the Baltic Sea and the central part of the North Atlantic). Preliminary heat flow pattern was proposed for areas where existing deep temperature observations were completed with the estimated thermal conductivity of the surface rocks to evaluate local heat flow values (Sweden, north-central part of the USSR, Bulgaria and Turkey). The highest concentration of heat flow measurements exist in a broad longitudinal strip stretching across western, central and south-eastern Europe with additional information from Italy, Balkan area, Norway and Finland and the Ural Mts. Fairly good marine data on the geothermal activity were reported for the vicinity of Iceland, the North Sea, the Black Sea and the Eastern Mediterranean Sea.

Figure 2 shows the histogram of all the heat flow data. The mean value for land measurements 62.1 mW m^{-2} (s.d. 28.6 mW m^{-2}) equals the continental mean 62.3 mW m^{-2} as given by JESSOP et al. (1975) in the most recent world catalogue. The distribution of data and the modal value are practically identical, too. The marine data used for the present purpose only supplemented the map in the adjacent sea regions, and their frequency distribution should not be considered generally valid. These marine data show usual wide distribution with common mode of 50–60 mW m^{-2}; the second mode (30–40 mW m^{-2}) corresponds to numerous low heat flow data obtained in the Eastern Mediterranean and in the Black Sea, and also to the low values measured in the ridge areas. High and very high values from the Western

---

[3]) The sheet of the 1:5000 000 Heat Flow Map of Europe and the Explanatory Text may be obtained on request from any of the authors of this report.

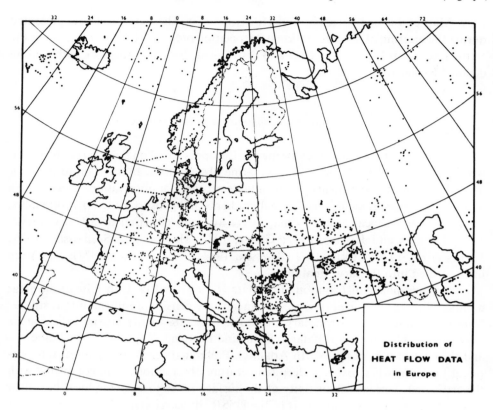

Figure 1
Distribution of heat flow data in Europe.

Mediterranean Sea and from near the active spreading centres in the Atlantic Ocean form the long 'tail' of the above histogram.

3. Preparation of the map

For the evaluation of the regional distribution of heat flow all the above information was used as check points and then extrapolated on the basis of the tectonic pattern. Generally heat flow decreases from the younger to older tectonic units and the units of the same tectonic history display similar heat flow values. Thus, knowing the age of the latest tectono-thermal event, the characteristic surface geothermal activity can be predicted (CHAPMAN and POLLACK, 1975). For the map construction the results of the correlation between heat flow and the tectonic setting by POLYAK and SMIRNOV (1968) were used, completed by similar data from several selected

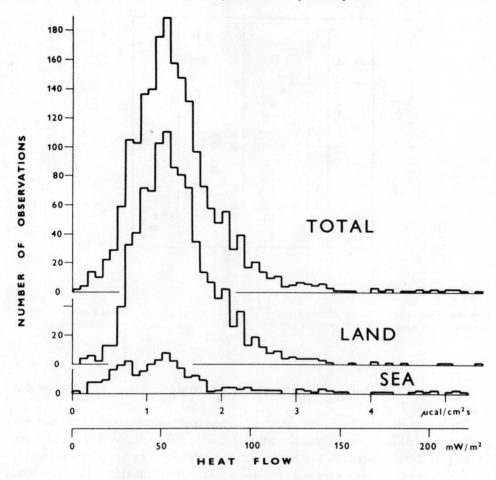

Figure 2
Histograms of the European heat flow data.

regions in Europe (Fig. 3). The course of isolines in areas with no direct observations was correspondingly adapted to follow the boundaries of principal tectonic units. Nevertheless the quality of data available is uneven and in many cases the interpretation of the observed geothermal activity is neither definite nor unambiguous. To minimize the subjective standpoint of the compilers, several working versions of the map were prepared and distributed to many specialists all over Europe for their comments and criticism. Each consecutive version was thoroughly reworked with consideration of the remarks suggested by the individual co-workers.

According to the proposal of the IHFC new heat flow units (i.e. mW m^{-2}) were used throughout the preparation of the final version of the map, and the isolines were

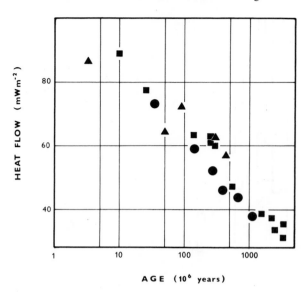

Figure 3

Heat flow versus age of the last tectono-thermal event. Dots – data after POLYAK and SMIRNOV (1968); squares – data from the USSR after KUTAS *et al.* (1976); triangles – data from Czechoslovakia after ČERMÁK (1976).

drawn at 10 mW m^{-2} intervals (intervals of 20 mW m^{-2} are used in the simplified map shown in Fig. 4).

There may be more or less pronounced influence of such geological phenomena on the observed heat flow as sedimentation, erosion, uplift, etc., as well as the effect of the past climatic changes. For further studies and for the comparison of the geo-thermal activity of various tectonic provinces and in the estimates of the total heat losses or in deep temperature calculations, attention should be paid to these per-turbations and reasonable regional corrections should be evaluated. However, such a procedure has to be applied on large territories and would require detailed knowledge of the local geological histories. For the present stage of investigation preference was given to 'uncorrected' heat flow values and the map thus shows the measured surface heat flow, i.e., its value free of any geological correction.

To make a reasonable selection of all data is also very difficult and this problem is discussed in greater detail in the explanatory text (ČERMÁK and HURTIG, 1977). No data, however, were excluded in the present stage, except for a few cases, which were agreed on by the respective authors. The course of isolines was in some cases slightly adapted in a few localities in order to smooth down the local extremes having little relation to the deep structure.

The description of oceanic heat flow measurements was motivated only to com-plete the heat flow pattern of Europe. For this purpose only the eastern part of the

Simplified version of the heat flow map of Europe
(Figures 4a and 4b)

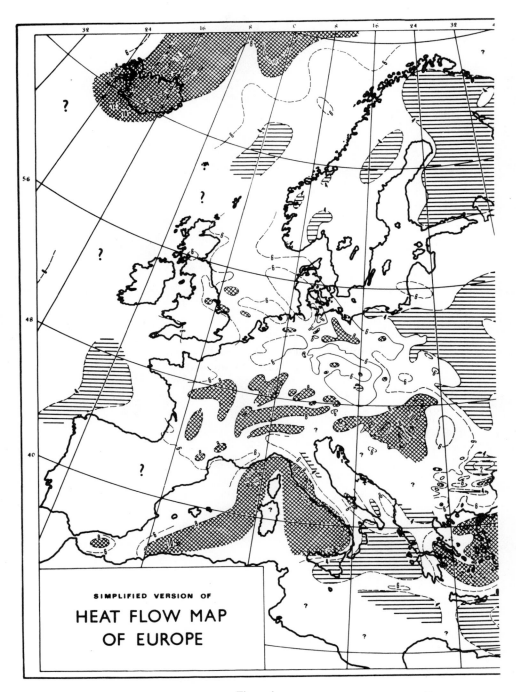

Figure 4a

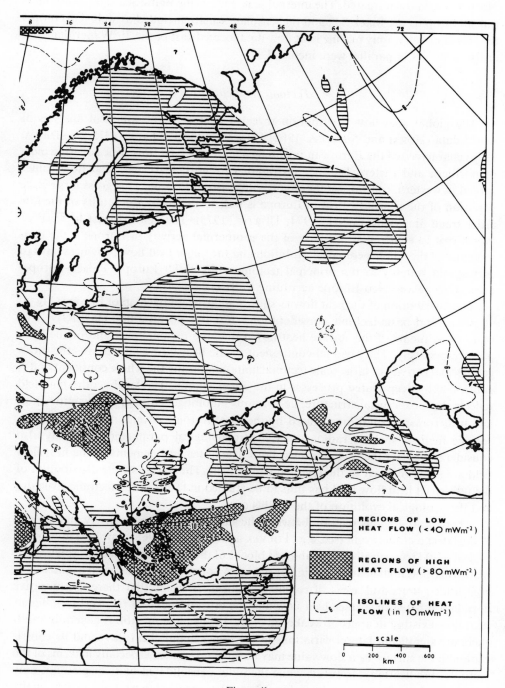

Figure 4b

Northern Atlantic is treated. The internal seas, such as the Baltic Sea and the Mediterranean Sea plus the Black Sea and the Caspian Sea are included as well. This means that for simplicity, only marine heat flow data located east of the 30°W meridian and north of the 30°N parallel were included.

4. Tectonic implications

The global heat flow distribution revealed by the recent spherical analysis of world data (HORAI and SIMMONS, 1962; CHAPMAN and POLLACK, 1975; SUETNOVA, 1976) characterized the European region by a large low heat flow zone covering most of northern and eastern Europe, surrounded by normal to high heat flow values spread in western, central, southern and southeastern Europe. This gross large-scale variation of the heat flow field in Europe is well sustained by the results of the low order trend analysis (HAENEL, 1974; HURTIG, 1975). The general 'east to west' or 'north-east to south-west' increase of the geothermal activity in Europe is the consequence of the whole geological evolution and the major heat flow provinces have to correspond roughly to the principal tectonic units of Ur-Europe, Palaeo-Europe, Meso-Europe and Neo-Europe according to STILLE (1924).

The distribution of the heat flow as shown in the heat flow map of Europe (Fig. 4) allows several basic tectonic considerations. For this purpose one may derive two main constituents of the surface heat flow value, the regional part and/or the local part, respectively. The regional component of the heat flow characterizes the mean geothermal activity of large scale tectonic units composing the whole continent and is controlled by deep-seated processes in the upper mantle. The local geothermal field reflects the local crustal structure, i.e., various inhomogeneities in rock composition and heat transfer conditions caused by the local tectonic development. Other local specific factors, such as radioactivity, hydrogeological parameters, etc. may also considerably affect the observed heat flow. The latter fine elements of the heat flow field are superimposed to the regional component. The nature of the distribution of the terrestrial heat flow field has thus to be studied in terms of the tectonic structure and the geological evolution of the earth's crust.

There is an obvious relation between the surface geothermal activity and the age of the latest tectonic event (see, e.g., POLYAK and SMIRNOV, 1968); low heat flow being typical of old tectonically well stabilized blocks of the earth's crust such as shields and platforms, and high heat flow being observed in younger orogenes, such as the Cenozoic folded units and areas of recent volcanism. It seems that there is a negative correlation between surface heat flow and the crustal thickness (see further), higher heat flow existing in areas of weakened crust. If this phenomenon is generally valid, there are significant lateral variations in the upper mantle heat flow, and its contribution to the surface heat flow value may be quite prominent especially in terms of the tectonic stabilization. However, while the older consolidated units display not only low, but relatively very uniform heat flow, the observed heat flow values in the younger orogenes vary in a broad interval. Tectonic activity, which is connected with the deep-seated processes and with the increased inflow of energy from the deeper parts, is accompanied by large crustal movements and by the re-arrangements of the individual blocks, which may disturb the heat transfer conditions. Owing to these disturbances, local (quite often near-by) anomalies of high as well as low heat flow can be observed. Usually intermontane depressions are characterized by increased geothermal activity, while frontal foredeeps display normal or sub-normal heat flow.

The heat flow pattern particularly in Central Europe is dominated by several elongated high heat flow zones, which cannot be explained by lateral variations of the crustal thickness and composition. The most probable explanation of the existence of linear, long-stretched anomalies must be assumed in additional heat transfer by convection along the deep crustal disturbances. The distribution and the depth-range of deep faults, linear fractures and graben structures may thus play an important role in the interpretation of the general heat flow pattern (MEIER et al., 1978). In most cases the maximum excess heat flow in the centre of the anomaly is of about $10\text{--}30 \text{ mW m}^{-2}$ compared with the adjacent non-anomaly region. A simple calculation may help evaluate the corresponding velocity of migrating water, if convection of water takes place and the convective heat transport q_m along the deep fracture system may be considered the source of enhanced surface geothermal activity: $q_m = v\rho cT$, where ρ is the density, c is the specific heat; for water ρ is 10^3 kg m^{-3} and c is $4.2 \text{ kJ kg}^{-1} \text{ K}^{-1}$. Only a very slow migration velocity v in the order of 10^{-8}–$10^{-10} \text{ m s}^{-1}$ (depending on the temperature difference T between the migrating water and the adjacent rock material) is necessary to cause the above mentioned heat flow excess.

The gross distribution of the heat flow field in the Mediterranean Sea revealed that the assumed plate boundaries as proposed by, e.g., MCKENZIE (1972) are in general characterized by a strong horizontal gradient of the heat flow field. The individual plates thus display different geothermal conditions. Quite outstanding is the position of the Appenninian chain, which separates the area of low to normal heat flow in eastern Italy and in the Adriatic Sea from very high heat flow values in the Tyrrhenian Sea, being the borderline between the Apulian plate and/or the African plate, respectively, on the one hand, and the European plate on the other hand. The existence of such a strong horizontal gradient is of important geodynamic meaning.

The boundary between the African and European plates following the Azores–Gibraltar ridge, touching northern Africa and crossing Sicily, Greece and Turkey is another line dividing two provinces of a different regional heat flow component. While the mean geothermal activity of about 30 mW m^{-2} characterizes the northern rim of the African plate, more than double mean heat flow ($65\text{--}70 \text{ mW m}^{-2}$) is typical of the Western Mediterranean and southern Europe. Here the local heat flow patterns correspond well to the typical general west–east strip arrangement of the Alpine–Carpathian–Caucasus folding, evidencing the north–south compression of the whole area.

5. Geophysical implications

Zones of increased subsurface temperatures and especially areas of a high horizontal temperature gradient, i.e., places where the heat flow field changes rapidly with the distance belong generally to areas with increased seismicity (KÁRNÍK, 1971; ELLENBERG, 1976). Deep earthquakes were observed along the borderline between the Apulian and the European plates and between the European and African plates. There is an assumed Benioff plane below the south-eastern Tyrrhenian Sea (CAPUTO et al., 1970). Strong earthquakes have occured in the Eastern Carpathians, as well as under the Aegean Sea or in Turkey – all these are regions characterized by quite prominent changes in their local heat flow patterns. The uneven underground temperature distribution (corresponding to the surface variations in the heat flow) may be the main reason for growing thermo-elastic stresses within the crust and in the upper mantle, which are then released in earthquakes.

Even though Hungary does not belong to the seismically active areas, systematic correlation studies were reported here (STEGENA, 1976), which show that the seismic energy was released predominantly in areas where the horizontal geothermal gradient at 1 km depth exceeds $1.3°C\ 10\ km^{-1}$. Similar phenomena can be deduced for deep temperature calculations along the deep seismic sounding profile across the Carpathians in eastern Czechoslovakia (ČERMÁK, 1975c).

The regional heat flow pattern combined with the knowledge of the crustal structure, the distribution of heat producing radioactive elements and the vertical change of the thermal conductivity of crustal rocks make the calculation of temperature–depth profile possible. Relatively low temperatures (350–500°C) at the Mohorovičić discontinuity at the depth of 45–50 km are to be expected beneath the shield and/or the East-European platform (ČERMÁK, 1977; KUTAS and GORDIENKO, 1971; BALLING, 1976). Higher Moho temperatures of 500–600°C were calculated for the Palaeozoic folded units such as the Bohemian Massif or the Ural Mountains at the depth of 35–40 km. Local temperature extremes of 600–700°C seem to exist in zones of a weakened earth's crust and/or in areas of deep faults, which manifest themselves by a relatively higher heat flow (up to 70–80 mW m^{-2}) (ČERMÁK, 1975a). Very high crustal temperatures (800–1000°C) are probable in hyperthermal regions, such as the Pannonian Basin, characterized by a relatively thin crust (25 km) and very high surface heat flow (BUNTEBARTH, 1976; ČERMÁK, 1975b). Similar high temperatures may exist in the Upper Rhine Graben valley (HAENEL, 1970). Horizontal crustal temperature changes on contact between diverse tectonic units were calculated by BALLING (1976) for the Baltic shield–Danish Embayment contact; by BUNTEBARTH (1973a) for the Alps and their foreland; by ČERMÁK (1975c) for the profile across the Carpathians and by HURTIG and OELSNER (1977) for several profiles across Europe. The regional deep temperature distribution at the depth of 30 km in north-central Europe is discussed in terms of crustal structure by HURTIG et al. (1975).

The calculated crustal temperatures are very closely connected with the extrapolation of the Moho heat flow. While beneath the shields, even if the crustal rocks are considerably depleted in radioactivity, the heat flow from the upper mantle cannot exceed 15–20 mW m^{-2}, areas such as the Pannonian Basin or the Upper Rhine Graben, even if the crustal are very rich in radioactivity (which is not likely), must have high Moho heat flow of about 50–55 mW m^{-2}. There must be substantial regional variations in the energy inflow from the upper mantle attaining up to 30–40 mW m^{-2}, which may represent the driving force for the geological evolution and the critical parameter for the geophysical understanding of deep-seated processes. If this phenomenon is of general validity, the older idea of relative constancy of the upper mantle heat flow contribution (CLARK and RINGWOOD, 1964) must be greatly re-evaluated.

The results of the deep seismic sounding together with the surface heat flow pattern suggest that there is a certain relation between the heat flow and the crustal thickness. The data from Czechoslovakia (ČERMÁK, 1976) revealed a negative character of this correlation, i.e., higher heat flows are found over local Moho elevations, while heat flow is lower over local Moho depressions. This picture is quite distinct for the young Neogene basins as well as for the relatively old Bohemian Massif (Pre-Variscan); the data from the contact zones are rather scattered. A similar relation seems to be valid for most of Central Europe, and in the Western Scandinavia, while the validity for the whole European continent requires more detailed studies in the future. It is, however, interesting to mention that a similar picture was found in Roumania (DEMETRESCU, personal communication) and within the Ukrainian shield (KUTAS, personal communication) and seems to be documented by deep temperature calculations in the Siberian platform and in Transbaikalia (LYSAK, 1976) and in the Basin and Range province in the U.S. (LACHENBRUCH, 1968).

Horizontal temperature gradients may be of importance for originating and interpreting gravity anomalies. The fields of heat flow and gravity are coupled with each other by the coefficients of thermal expansion $\alpha(T, p)$. According to the temperature–depth calculations (BUNTEBARTH, 1973b; HURTIG and WIRTH, 1974) lateral temperature variations in the lower crust of the order of 100–200°C are to be expected and will produce gravity anomalies up to 10 mgal. In the Alpine foldbelt, with strong negative Bouguer anomalies, this effect can be neglected, but it may be of importance in other areas.

Acknowledgements

This paper and the heat flow map of Europe could not have been prepared without the assistance of numerous specialists from many countries who have co-operated in the data compilation and in the map construction. Special thanks belong to B. Kunz (Austria); N. P. Balling and L. Madsen (Denmark): P. Järvimäki and M. Puranen (Finland); R. Gable and J. Goguel (France); K. Bram and R. Haenel (Federal

Republic of Germany); M. Fytikas and N. Kolios (Greece); T. Boldizsár, F. Horváth and L. Stegena (Hungary); M. Loddo and F. Mongelli (Italy); G. Grønlie and K. S. Heier (Norway); J. Majorowicz, S. Plewa and M. Wesierska (Poland); C. Demetrescu, S. Veliciu and M. Visarion (Roumania); D. S. Parasnis and S. Werner (Sweden); L. Rybach and P. G. Finckh (Switzerland); A. K. Tezcan (Turkey); P. Morgan and S. B. Smithson (USA); G. I. Buachidze, R. I. Kutas and Ya. B. Smirnov (USSR); K. Filjak and P. Lukić (Yugoslavia).

REFERENCES

BALLING, N. P. (1976), *Geothermal models of the crust and uppermost mantle of the Fennoscandian shield in south Norway and the Danisch Embayment*, J. Geophys. *42*, 237–256.

BOLDIZSÁR, T. (1975), *Research and development of geothermal energy production in Hungary*, Geothermics *4*, 44–56.

BUNTEBARTH, G. (1973a), *Model calculations of temperature–depth distribution in the area of the Alps and the Foreland*, Z. Geophys. *39*, 97–107.

BUNTEBARTH, G. (1973b), *Über die Grösse der thermisch bedingten Bouguer-Anomalie in den Alpen*, Z. Geophys. *39*, 109–114.

BUNTEBARTH, G., *Temperature calculations on the Hungarian seismic profile-section NP-2*, in *Geoelectric and Geothermal Studies* (*East-Central Europe, Soviet Asia*), (ed. A. Ádám), KAPG Geophys. Monograph (Ákadémiai Kiadó, Budapest, 1976), pp. 561–566.

CAPUTO, M., PANZA, G. F. and POSTPISCHL, D. (1970), *Deep structure of the Mediterranean Basin*, J. Geophys. Res. *75*, 4919–4923.

ČERMÁK, V. (1975a), *Combined heat flow and heat generation measurements in the Bohemian Massif*, Geothermics *4*, 19–26.

ČERMÁK, V. (1975b), *Temperature–depth profiles in Czechoslovakia and some adjacent areas derived from heat flow measurements, deep seismic sounding and other geophysical data*, Tectonophysics *26*, 103–119.

ČERMÁK, V. (1975c), *Deep temperature distribution along the deep seismic sounding profile across the Carpathians (model calculation)*, Acta Geol. Sci. Hung. *18*, 295–303.

ČERMÁK, V. *Heat flow investigation in Czechoslovakia*, in *Geoelectric and Geothermal Studies* (*East-Central Europe, Soviet Asia*), (ed. A. Ádám), KAPG Geophys. Monograph (Ákadémiai Kiadó, Budapest, 1976), pp. 414–424.

ČERMÁK, V. *Teplovoy potok i raspredeleniye glubinnykh temperatur v centralnoy i vostochnoy Evrope i ikh geodinamicheskoye prilozheniye*, in *Problemi Riftogeneza* (*Materialy k Simposiumu po riftovym zonam Zemli*). (Irkutsk, 1977).

ČERMÁK, V. *First Heat Flow Map of Czechoslovakia*, Trav. Inst. Géophys. Acad. Tchécosl. Sci., No. 461 (Geofysikální Sborník, Academia, Praha, 1978), (in press).

ČERMÁK, V. and HURTIG, E. *Preliminary Heat Flow Map of Europe, 1:5000 000* (*Map and Explanatory Text*) (Geophys. Inst. Czechoslovak Acad. Sci. and Central Earth Physics Inst., Praha – Potsdam, 1977), 58 pp.

CLARK, S. P. Jr. and RINGWOOD, A. E. (1964), *Density distribution and constitution of the mantle*, Rev. Geophys. *2*, 35–88.

CHAPMAN, D. S. and POLLACK, H. N. (1975), *Global heat flow: A new look*, Earth Planet. Sci. Letts. *28*, 23–32.

ELLENBERG, J. *Seismische Intensitäten, 1:6000 000*, in *Materialen zum tektonischen Bau von Europa* (ed. K.-B. Jubitz), Veröff. Zentralinst. Phys. Erde, No. 47 (Potsdam, 1976).

FYTIKAS, M. and KOLIOS, N. (1977), '*Preliminary Heat Flow Map of Greece (1:1,000,000)*', IGMR, Athens (unpublished).

GABLE, R. *Ébauche d'une carte du flux géothermique de la France* (Bur. rech. géol. et Minières, Orléans, 1977), 8 pp.

HAENEL, R. (1970), *Interpretation of the terrestrial heat flow in the Rhinegraben*, in *Graben Problems. Intern. Upper Mantle Proj.* (eds. J. H. Illies and St. Mueller), Scient. Rep. 27, (Schweizerbart, Stuttgart, 1970), pp. 116–120.

HAENEL, R. (1974), *Heat flow measurements in northern Italy and heat flow maps of Europe*, Z. Geophys. *40*, 367–380.

HORAI, K. and SIMMONS, G. (1969), *Spherical harmonic analysis of terrestrial heat flow*, Earth Planet. Sci. Letts. *6*, 386–394.

HURTIG, E. (1975), *Untersuchungen zur Wärmeflussverteilung in Europa*, Gerlands Beitr. Geophys. *84*, 247–260.

HURTIG, E. and OELSNER, CH. (1977), *Heat flow, temperature distribution and geothermal models in Europe: Some tectonic implications*, Tectonophysics *41*, 147–156.

HURTIG, E. and SCHLOSSER, P. Geothermal Studies in the GDR and Relations to the Geological Structure, in *Geoelectric and Geothermal Studies (East-Central Europe, Soviet Asia)*, (ed. A. Ádám), KAPG Geophys. Monograph, (Akadémiai Kiadó, Budapest, 1976), pp. 384–394.

HURTIG, E. and WIRTH, H. *Über Beziehungen zwischen geothermischen und gravimetrischen Anomalien*, Veröff. Zentralinst. Phys. Erde, No. 29, pp. 57–69 (Potsdam, 1974).

HURTIG, E., OESBERG, R.-P., RITTER, E., GRÜNTHAL, G. and JACOBS, F. (1975), *Studies of the crustal structure in the northern part of Central Europe*, Gerlands Beitr. Geophys. *84*, 317–325.

JESSOP, A. M., HOBARTH, M. A. and SCLATER, J. G. (1975), *The World Heat Flow Data Collection*, Geothermal Service of Canada, Geoth. Ser., No. 5 (Ottawa, 1975), 125 pp.

KÁRNÍK, V. *Seismicity of the European Area*, Part 2, (Academia, Praha, 1971), 218 pp.

KUTAS, R. I. and GORDIENKO, V. V. *Teplovoye polye Ukrainy*, (Naukova Dumka, Kiev, 1971), 140 pp.

KUTAS, R. I., LUBIMOVA, E. A. and SMIRNOV, YA. B. *Heat flow map of the European part of the USSR and its geological and geophysical interpretation*, in *Geoelectric and Geothermal Studies (East-Central Europe, Soviet Asia)*, (ed. A. Ádám), KAPG Geophys. Monograph (Akadémiai Kiadó, Budapest, 1976), pp. 443–449.

LACHENBRUCH, A. H. (1968), *Preliminary geothermal model of the Sierra Nevada*, J. Geophys. Res. *73*, 6977–6989.

LUKIĆ, P. (1971), 'Ein Beitrag zur Erforschung der Erdkruste im Bereich der Dinariden und in der Pannonischen Ebene,' Inaugural Diss., Univ. Köln.

LYSAK, S. V. *Heat flow, geology and geophysics in the Baikal Rift Zone and the adjacent regions*, in *Geoelectric and Geothermal Studies (East-Central Europe, Soviet Asia)*, (ed. A. Ádám), KAPG Geophys. Monograph (Akadémiai Kiadó, Budapest, 1976), pp. 455–462.

McKENZIE, D. P. (1972), *Active tectonics of the Mediterranean region*. Geophys. J. Roy. Astr. Soc., *30*, 109–185.

MEIER, R., HURTIG, E. and LUDWIG, A. (1978), *Fault tectonics and heat flow in Europe* (in preparation).

POLYAK, B. G. and SMIRNOV, YA. B. (1968), *Svyaz glubinnogo teplovogo potoka s tektonicheskim stroyeniem kontinentov*, Geotektonika *4*, 3–19.

STEGENA, L. *The Variation of temperature with depth in the Pannonian Basin*, in *Geoelectric and Geothermal Studies (East-Central Europe, Soviet Asia)*, (ed. A. Ádám), KAPG Geophys. Monograph, 425–438 (Akadémiai Kiadó, Budapest, 1976), pp. 425–438.

STILLE, H. *Grundfragen der vergleichenden Tektonik* (Bornträger, Berlin, 1924), 443 pp.

SUETNOVA, E. I. (1976), Sfericheskiy garmonicheskiy analiz dannikh o teplovikh potokakh. In *Issledovaniya teplovogo i elektromagnitnogo poley v SSSR* (eds. E. A. Lubimova and M. N. Berdichevsky), (Izd. Nauka, Moscow, 1976), pp. 37–41.

TEZCAN, A. K. (1977), 'The Geothermal Studies, Their Present Status and Contribution to the Heat Flow Contouring in Turkey,' Manuscript, MTA, Ankara.

VELICIU, S., CRISTIAN, M., PARASCHIV, D. and VISARION, M. (1977), 'Preliminary Data of Heat Flow Distribution in Roumania,' Manuscript, Inst. Geophys. Geol., Bucharest.

(Received 24th October 1977, revised 31st March 1978)

Pageoph, Vol. 117 (1978/79), Birkhäuser Verlag, Basel

Heat Flow Map of the European
Part of the U.S.S.R.

By R. I. Kutas[1], E. A. Lubimova[2] and Ya. B. Smirnov[3]

Abstract – A heat flow isoline map is presented. Low and relatively constant heat flow has been observed in the old shield areas of the East European Platform (25–40 mW/m^2). Increased heat flow (> 50 mW/m^2) has been found in the Dniepr–Donetz depression. The area south of the East European Platform is characterized by highly variable heat flow (55–100 mW/m^2). Some geophysical implications are discussed.

Key words: Heat flow map; U.S.S.R.; East European Platform; Dniepr–Donetz depression; Carpathians; Crimea peninsula; Heat generation.

The map showing the distribution of the surface heat flow on the territory of the European part of the U.S.S.R. (Fig. 1) is presented and briefly discussed. The construction of this map is based on more than 750 data, which are summarized together with coordinates, mean temperature gradients, mean thermal conductivity, etc., in several recent papers (Kutas *et al.*, 1975; Lubimova *et al.*, 1973, Subbotin and Kutas, 1974). More detailed description and some geological and geophysical interpretation can be found in, e.g., Kutas, *et al.* (1976).

Low and stable regional heat flow fields of 30–40 mW m^{-2} can be observed in the Baltic and Ukrainian shields and in most of the East-European Platform, i.e., in the areas which form the oldest part of the European continent ("Ur-Europa"). The relatively lowest values of the heat flow (25–40 mW m^{-2}) are typical for both shields and for the anticline structures within the platform; slightly elevated values (40–50 mW m^{-2}) were found in the areas of depressions.

The Dniepr–Donetz graben structure represents an interesting elongated zone of increased heat flow (more than 50 mW m^{-2}) compared with the surrounding areas. Higher surface geothermal activity corresponds to higher crustal temperatures below this structure and may be related to the consequences of the geotectonic evolution.

In the east the chain of the Ural Mountains generally closes the low heat flow area of the East-European platform. Marked decrease from values of 60–70 mW m^{-2} in the north to less than 40 mW m^{-2} in the central and southern parts of the mountains can be

---

[1]) Geophysical Institute, Acad. Sci. Ukrainian SSR, Kiev.
[2]) Institute of Physics of the Earth, USSR Acad. Sci., Moscow.
[3]) Geological Institute, USSR Acad. Sci., Moscow.

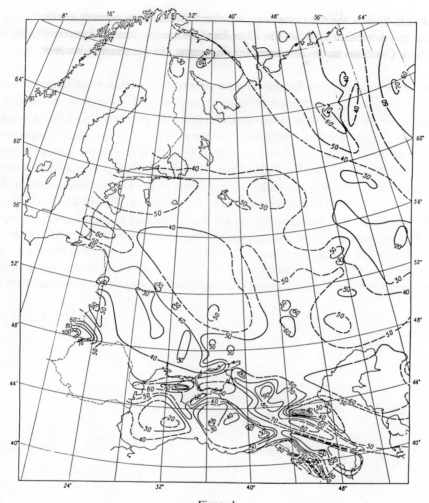

Figure 1
Map of the heat flow in the European part of the U.S.S.R. (heat flow is given in mW m^{-2}).

observed. The northern part of this area together with the western part of the West Siberian platform is relatively well known due to numerous heat flow observations, the course of isolines to the south is not quite clear and should be taken as preliminary in the map.

The area south of the East-European platform is characterized by highly variable heat flow ranging from 55 to 80 mW m^{-2} and even more in some localities. Generally, increased geothermal activity can be found within the Scythian platform, where the local, usually west–east, oriented anomalies follow the main trend of the Cenozoic folded structures. Quite prominent is the Stavropol uplift zone with heat flow over 90 mW m^{-2}. This zone stretches to the south-east and seems to be connected with the increased heat flow in the Caucasus area. Another high heat flow anomaly can be detected in the Armenian Upland. The outer foredeeps and the central depressions between both Caucasus mountain ranges are, in contrast, characterized by low heat flow.

The observed heat flow in the Crimea peninsula covers a wide range of values between 40–90 mW m^{-2}, the mean value is however relatively low. A small local positive anomaly can be seen in the central part of the peninsula with two elongated zones of increased heat flow stretching to the west and/or to the south, respectively.

Another zone of high heat flow (over 100 mW m^{-2}) is typical for the most western part of Ukraine joining the Pannonian basin. Heat flow increases rapidly in this area in the direction from the outer to inner tectonic units of the Carpathian mountain belt.

The lowest heat flow values were found in the Archaic crustal blocks composed of highly metamorphosed rocks, which are generally depleted in radioactivity, except in some rather limited zones, where the strong postorogenetic granitization was connected with increased content of radioactive elements. It seems that the main factor controlling the surface heat flow field in the Ukrainian shield is the heat generation of the near surface rocks. Figure 2 shows the experimental relation between heat flow and

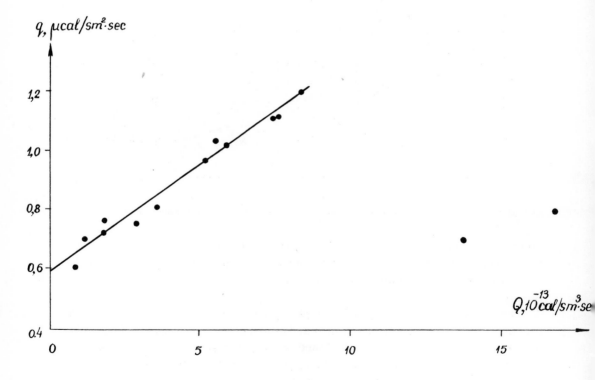

Figure 2
Heat flow versus heat generation of the surface rocks in the central part of the Ukrainian shield.

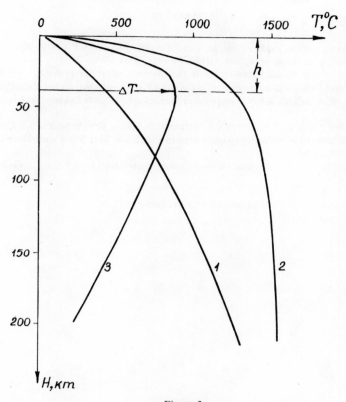

Figure 3

Temperature in the crust and upper mantle. (1) shield region; (2) geosyncline region; (3) calculated temperature difference between both models.

heat generation, however, in some cases the prevailing positive relationship is considerably distorted. This feature may be explained by the limited depth extent (1–2 km) of surface bodies of highly varying radioactive content; another solution may be sought for in the lateral variations of the upper mantle heat flow contribution, which may exceed the generally low heat production within the crust.

Using the data on crustal structure and the assumed thermal conductivity–depth variation the characteristic deep temperature profiles for the stable shield region and for the adjacent geosyncline zone were calculated (Fig. 3). While the temperature may be as low as 300–500°C at depths of 30 km below the shield, temperatures of more than 1000°C may exist at depths of 60–100 km below the tectonically active geosyncline belt. The temperature difference between both these cases reaching its maximum value of 700°C at depths of 40–50 km is decreasing at the greater depths in the upper mantle, and both curves join at depths of about 200 km.

REFERENCES

KUTAS, R. I., GORDIYENKO, V. V., BEVZYUK, M. I. and ZAVGORODNAYA, O. V. (1975). *Noviye opredelenya teplovogo potoka v Karpatskom regione*, Geofiz. Sbornik, Kiev 63, 68–71.

KUTAS, R. I., LUBIMOVA, E. A. and SMIRNOV, YA. B. *Heat flow map of the European part of the U.S.S.R. and its geological and geophysical interpretation*, in *Geoelectric and geothermal studies* (*East-Central Europe, Soviet Asia*) (ed. A. Ádám), KAPG Geophysical Monograph, 443–449 (Akadémiai Kiadó, Budapest, 1976).

LUBIMOVA, E. A., POLYAK, B. G., SMIRNOV, YA. B., KUTAS, R. I., FIRSOV, F. V., SERGIENKO, S. I. and LYUSOVA, L. N. Katalog dannykh po teplovomu potoku na territorii S.S.S.R. Mat. Mir. Centra Dannykh – B, No. 3 (Moskva, 1973), 64 pp.

SUBBOTIN, S. I. and KUTAS, R. I. *Glubinniy teplovoy potok evropeyskoy chasti S.S.S.R.* (Naukova Dumka, Kiev, 1974).

(Received 21st December 1977)

Pageoph, Vol. 117 (1978/79), Birkhäuser Verlag, Basel

Mantle Heat Flow and Geotherms for Major Tectonic Units in Central Europe

By Jacek Majorowicz[1])

Abstract – The results of seismic measurements along the deep seismic sounding profile VII and terrestrial heat flow measurements used for construction of heat generation models for the crust in the Paleozoic Platform region, the Sudetic Mountains (Variscan Internides) and the European Precambrian Platform show considerable differences in mantle heat flow and temperatures. At the base of the crust variations from 440–510°C in the models of Precambrian Platform to 700–820°C for the Paleozoic Platform and the Variscan Internides (Sudets) are found. These differences are associated with considerable mantle heat flow variations.

The calculated models show mantle heat flow of about 8.4–12.6 mW m^{-2} for the Precambrian Platform and 31 mW m^{-2} to 40.2 mW m^{-2} for Paleozoic orogenic areas. The heat flow contribution originating from crustal radioactivity is almost the same for the different tectonic units (from 33.5 mW m^{-2} to 37.6 mW m^{-2}). Considerable physical differences in the lower crust and upper mantle between the Precambrian Platform and the adjacent areas, produced by lateral temperature variations, could be expected. On the basis of 'carbon ratio' data it can be concluded that the Carboniferous paleogeothermal gradient was much lower in the Precambrian Platform area than in the Paleozoic Platform region.

Key words: Crustal geotherms; Moho temperatures; Heat flow and crustal thickness; Heat flow – age relationship; Coalification rank; Tectonic units in Poland; Heat flow transition zones.

1. Introduction

The number of heat flow data in Central Europe has increased significantly in recent years. The most numerous measurements were made in Poland, German Democratic Republic, Czechoslovakia and USSR. Heat flow data of good quality and the results of seismic measurements in Poland provide sufficient material for the calculation of the regional differences in mantle heat flow contribution and variations of crustal temperature between East European Precambrian Platform, Paleozoic Platform and Sudetic Mountains. The mantle heat flow contribution should be considered as an important information about the state of tectonophysical activity in the crust and upper mantle.

The most interesting areas for heat flow investigations are the transition zones between different geotectonical units. Until recently, few data related to heat flow transition zones were available. Transition zones between different heat flow provinces

[1]) Geological Institute, Geophysical Department, Warsaw, Rakowiecka 4, 02-519, Poland.

have been studied in USA at six locations ROY *et al.* (1971). Heat flow transition zones between 'Basin and Range' and Great Plains and between Sierra Nevada and 'Basin and Range' are well recognised.

The most interesting heat flow transition zones in Central Europe are between the old European Precambrian Platform and the adjacent younger tectonic areas. The transition zone between the Baltic Shield and the Caledonian orogenic belt was investigated by SWANBERG *et al.* (1974). In Denmark a narrow heat flow zone between the Fennoscandian Shield and the North Sea Basin has been investigated along a profile from the Precambrian of South Norway to the Danish Embayment in North Jylland by BALLING (1976). In the USSR the transition zone between the Precambrian Platform (Ukrainian Shield) and the Carpathians was investigated by KUTAS (1972). The heat flow transition zone between the East European Platform and the Paleozoic Platform which have undergone severe deformation in both the Caledonian

Figure 1

Tectonic units of Poland (according to MAREK and ZNOSKO, 1974). 1. Precambrian Platform; 2. Paleozoic blocks exposed or under a thin platform cover; 3. Paleozoic orogenic depressions; 4. Deep seated fractures of the tectonic zone of the Teisseyre–Tornquist line; 5. Presumed external boundary of Variscides; 6. Boundary of the Carpathian overthrust; 7. Deep seismic sounding profile VII.

and Variscan orogenic events was investigated in Poland by MAJOROWICZ (1976). A tectonic sketch map of this region is shown in Fig. 1.

From all these investigations it follows that the heat flow transition zones are generally narrow (less than 100 km wide). In the transition zone Old European Platform–Paleozoic Platform a thick sedimentary column exists in the 'intracratonic basin'. The mean radiogenic heat production in the sedimentary layer for the Precambrian Platform is 1.25 μW m^{-3} and 1.46 μW m^{-3} for the Paleozoic Platform on the basis of laboratory determinations.

However, the radiogenic heat in the thick sedimentary cover cannot explain the high flow anomalies in western Poland and northern Germany; on the contrary, in north-western Poland where a thick sedimentary cover exists (6–10 km), low heat flow values are observed (50 mW m^{-2}). The heat flow values for western Poland which contain numerous salt dome piercement structures were measured in wells which were not situated in zones where salt domes exist. Salt dome diapirs are excellent conductors of heat from the depth of the parent salt strata to the surface (JONES, 1970).

Narrow heat flow transition zones imply shallow depth of the partially molten upper mantle in the high heat flow provinces, with cold roots under the regions of abnormaly low heat flow (mostly Precambrian Platforms and Shields). Considering these facts it is possible to explain high heat flow provinces. Upward convection of partially molten material would result in high temperatures near the base of the crust, that are necessary to explain the heat flow distribution observed at the surface.

2. Mantle heat flow

In Central Europe there exist now about 800 heat flow determinations. In Poland heat flow investigations were made by MAJOROWICZ (1973a, b; 1976), PLEWA (1967, 1976) and WĘSIERSKA (1973).

A heat flow map of Poland and adjacent areas (Fig. 2) was constructed using published heat flow data, new geothermal measurements in deep boreholes of the Geological Institute and thermal conductivity measurements for sedimentary rocks. For Poland 110 data were used. Heat flow measurements for Czechoslovakia, USSR and German Democratic Republic were published by ČERMAK (1975), KUTAS et al. (1976), and HURTIG (1975).

To calculate mantle heat flow, the vertical distribution of radiogenic heat generation must be specified and the surface heat flow must be known. As was shown by LACHENBRUCH (1970) and ROY et al. (1971), heat generation in the crust is probably decreasing with depth. The relationship between heat flow and heat generation obtained for different tectonic areas of the world (SWANBERG et al., 1974; RYBACH, 1973; RAO and JESSOP, 1975; ROY et al., 1971; LACHENBRUCH, 1970; ČERMAK and JESSOP, 1975; MAJOROWICZ, 1976), leads to a better understanding of the vertical

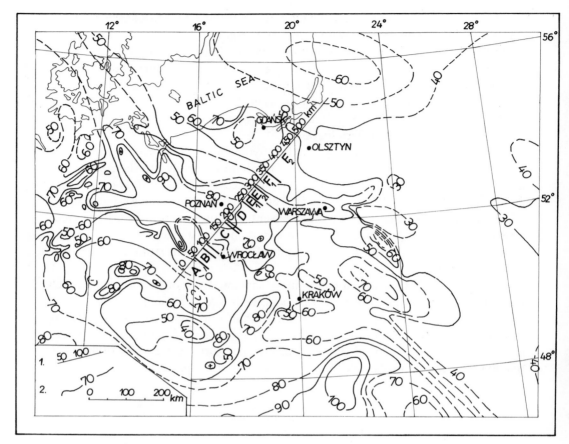

Figure 2

Heat flow map of Central Europe. 1. Position of the international deep seismic profile VII; 2. Isolines of heat flow in mW m^{-2}. Note: High heat-flow values can be observed in NW–SE direction parallel to the western border of the Precambrian Platform. The Precambrian Platform is characterised by relatively low heat flow around 45 ± 9 mW m^{-2}. Increased geothermal activity is observed in the area of Alpine folded structures in the Carpathians.

distribution of heat sources in the crust. The model indicating decrease of heat sources with depth is consistent with the observed linear relationship between the heat flow and heat generation. The hypothesis seems to hold for all continents. The relationship between heat flow Q and surface heat production of magmatic rocks A for pre-Variscan areas (including precambrian shields, Precambrian platforms and Moldanubicum of Czechoslovakia) is shown in Fig. 3. Heat flow and radiogenic heat production values for different geotectonic continental provinces form a linear relationship (reduced heat flow, $Q_0 = 23$ mW m^{-2}). These data agree very well with the result obtained by JESSOP et al. (1976) for all precambrian sites ($Q_0 = 25$ mW m^{-2}).

It is surprising that the reduced heat flow values obtained for all precambrian sites and pre-Variscan areas are lower than for precambrian shields ($Q_0 = 29.3$ mW m^{-2} after RAO and JESSOP, 1975). For precambrian shields and craton (platform) areas a Q_0 of 28.5 was found by SLACK (1974). In my opinion the differences between 23 and 29.3 mW m^{-2} are not significant in view of the error of the methods that were used to estimate heat generation and Q values.

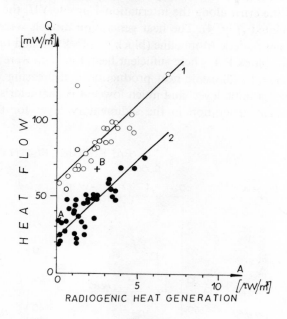

Figure 3

Heat flow–heat production relationship. 1. The upper straight line corresponds to continental Meso-Cenozoic orogenic regions (Q and A data according to ROY et al., 1971; DUCZKOW and SOKOLOVA, 1974; RYBACH, 1973; STEGENA et al., 1975). 2. The lower straight line corresponds to values for continental pre-Variscan tectonic regions (data after ARSHAVSKAYA et al., 1972; POLLACK and CHAPMAN, 1977; ČERMAK, 1975; ČERMAK and JESSOP, 1975; HYNDMAN, 1968; KUTAS, 1972; SASS, 1968; SWANBERG et al., 1974). The point marked with 'B' ($Q = 67$ mW m^{-2}, $A = 2.5$ μW m^{-3}) corresponds to mean heat flow and mean heat generation for the Variscides of south-west Poland, the point marked with 'A' ($Q = 34$ mW m^{-2}, $A = 0.1$ μW m^{-3}) represents heat flow and heat generation in an anorthosite intrusion in north-east Poland (MAJOROWICZ, 1976).

The relation between Q and A for the younger orogenic province of Basin and Range was defined by ROY et al. (1971) with $Q_0 = 58.7$ mW m^{-2}. This line is also shown in Fig. 3. The other published values from Meso-Cenozoic orogenic areas fall very close to this relation.

The model of heat generation decreasing with depth in the crust is supported by seismic data. There exists a strong correlation between the heat generation and compressional wave velocity V_p in the range from 5 to 8 km s^{-1}. Heat generation decreases with increasing V_p which reflects changes in rock type, as was shown on the basis of empirical data by RYBACH (1973, 1976). Heat generation is also related to density (HORAI and SIMMONS, 1968).

Terrestrial heat flow in crust blocks along the international profile VII is well known. (The position of the profile is shown in Fig. 1.) The velocity–depth relations are also well recognised (GUTERCH et al., 1975). From the relationship between the heat generation and compressional wave velocity (RYBACH, 1976) and from the velocity model of the crust along the international profile VII, the heat generation models were calculated (Fig. 4). The heat generation models were established for the Sudetes, the Fore-Sudetic Monocline (block C, block D), and the Old European Platform (block F_1, block F_2), where sufficient heat flow data were available. It can be seen that in general radiogenic heat production is decreasing with depth. The highest A is for the 'granitic layer' and much lower A is characteristic for the lower crust. The mean heat generation in the sedimentary layer for the Precambrian

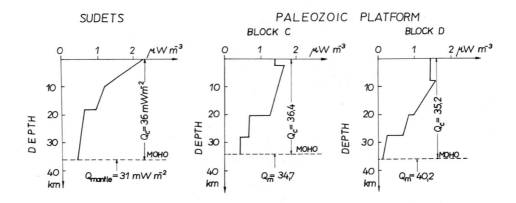

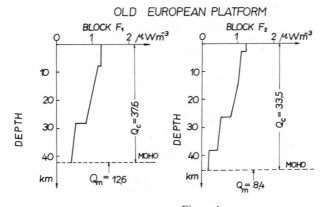

Figure 4
Heat generation models $A(z)$ for crustal blocks along the international deep seismic profile VII (Q_c is the heat flow contribution from crustal radioactivity, Q_M is the mantle heat flow).

Platform is 1.25 $\mu W\ m^{-3}$ and 1.46 $\mu W\ m^{-3}$ for Paleozoic Platform on the basis of laboratory determinations.

From the heat generation in the crust $A(z)$ and the surface heat flow Q the mantle heat flow Q_M can be calculated

$$Q_M = Q - \int_0^{z_{\text{Moho}}} A(z)\ dz. \tag{1}$$

The calculated crustal heat flow contribution and subcrustal heat flow (mantle heat flow) for Sudets, Paleozoic Platform (block C, block D) and Precambrian Platform (block F_1, block F_2) are shown in Fig. 5.

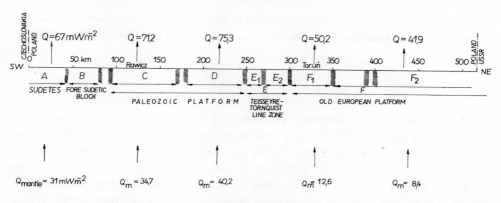

Figure 5

Mantle heat flow values Q_M for crustal blocks along the international deep seismic profile VII.

It can be seen that the crustal heat flow contributions originating from radiogenic heat generation is almost the same for the different blocks (from 33.5 mW m^{-2} to 37.6 mW m^{-2}). The observed differences in heat flow between the blocks of Precambrian Platform, Paleozoic Platform and Sudets are mainly caused by the different mantle heat flow contributions. Mantle heat flow varies from 8.4 mW m^{-2} (block F_2) to 40.2 mW m^{-2} (block D). The mantle heat flow values for Paleozoic Platform and Sudets vary from 31 mW m^{-2} to 40.2 mW m^{-2} and are much higher than Q_M for the blocks of Precambrian Platform ($Q_M = 8.4$–12.6 mW m^{-2}).

The mean reduced heat flow for pre-Variscan sites ($Q_0 = 23$ mW m^{-2}) is considerably higher than mantle heat flow values obtained for blocks F_1 and F_2 ($Q_M = 8.4$–12.6 mW m^{-2}).

The thickness of surface layer with high heat generation, interpreted from the heat flow–heat generation relationship, is 6.3 km for precambrian shields after RAO and JESSOP (1975) which is considerably lower than the thickness of 'granite layer' in the area of European Precambrian Platform (about 20 km). The upper crust exhibits high heat generation as it follows from the compressional wave velocity–heat generation relationship. The highest mantle heat flow value is obtained for block D of Paleozoic Platform, also characterised by the highest terrestrial heat flow values ($Q = 75 \text{ mW m}^{-2}$).

Considerable physical differences in the lower crust and upper mantle beneath European Precambrian Platform region, Sudets and the Paleozoic Platform region with Variscan basement, produced by lateral temperature variations, should be expected. Petrophysical properties are temperature-dependent. Especially electric conductivity is temperature dependent. Also magnetic properties of rocks change with increasing temperature. Electrical conductivity differences, of one or two orders of magnitude, could exist in the lower crust and upper mantle along the international deep seismic profile VII between the blocks of Precambrian Platform and Paleozoic Platform.

The interpretation of further geophysical and geological data confirms that different subcrustal heat flow contributions can be expected in areas of different age.

3. Mean heat flow versus the age of latest essential tectonic events

A relationship exists between heat flow and the age of consolidated basement (POLYAK and SMIRNOV, 1968). Heat flow decreases with increasing age of the latest orogenic activity of the region (Fig. 6). The differences in heat flow between geological regions of different age can be explained by, among other factors, different mantle heat flow. Higher mantle heat flow is characteristic for younger orogenic areas. The lowest mantle heat flow can be found in the shield and old platform areas. This is probably connected with different lithospheric thickness in these areas.

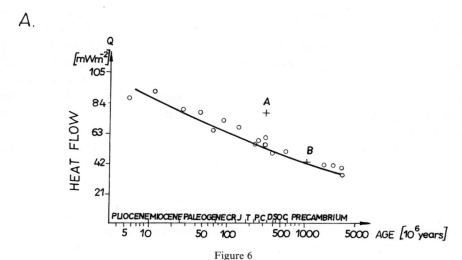

Figure 6
Mean heat flow against orogenic age for Central and Eastern Europe according to KUTAS et al. (1976). Crosses denote Polish data (A – Paleozoic Platform; B – Precambrian Platform).

The comparison of heat flow versus age for the Paleozoic Platform of Variscan age ($Q = 75$ mW m^{-3}, age $= 325 \times 10^6$ years) with that for the whole world (after POLYAK and SMIRNOV, 1968) and for Central and Eastern Europe (Point A, Fig. 6) shows that heat flow values in the European Paleozoic Platform of Variscan age area are much higher than the world's statistical average value. This fact can probably be explained by anomalous high subcrustal heat flow in the Paleozoic Platform areas like Fore-Sudetic Monocline, Paris Basin, North Sea, North German Lowland and others. The mantle heat flow in these regions is only a little lower than in the Alpine

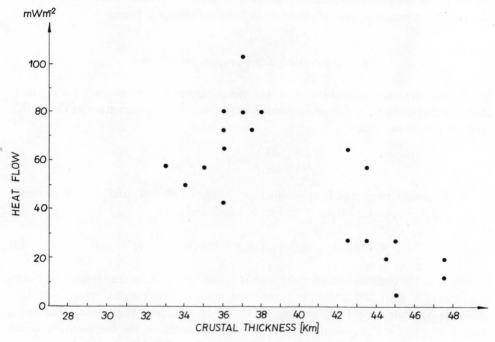

Figure 7
Correlation between the heat flow and crustal thickness for the Polish Lowland.

orogenic belt. High mantle heat flow values for the Fore-Sudetic Monocline are comparable with values obtained by BALLING (1976) for the Danish Embayment. Generally, all areas of Europe which underwent Paleozoic deformation are characterised by higher heat flow values (67–84 mW m^{-2}).

In platform areas of Europe characterised by high heat flow the Moho discontinuity is generally at an elevated position. The comparison of the heat flow with crustal thickness for the Polish Lowland of Variscan, Caledonian and Precambrian, based on 20 observations, shows that heat flow increase with decreasing crustal thickness (Fig. 7).

This relation is fairly well defined, when comparing Variscan or Caledonian orogenic zones with shields or old platforms. However, when comparing areas of Precambrian age the relation does not hold. The thermal effect of the 'elevated upper mantle', characteristic of younger orogenic zones, was dissipated a long time ago in Precambrian Platforms (CROUGH and THOMPSON, 1976). Instead, a relationship of decreasing heat flow with decreasing crustal thickness is recognised (SLACK, 1974).

High heat-flow values are connected with elevated mantle, implying shallow depth to partially molten upper mantle in the high heat-flow regions of Phanerozoic age of orogenic activity. Upward convection of partially molten material, i.e. diapiric intrusion of hot material, would result in high heat-flow contributions and elevated temperatures near the base of the crust in younger orogenic zones.

4. Temperature–depth profiles for the crust

In the one-dimensional problem, the temperature T depends only on depth z. Steady-state conditions can be assumed for the crust and the curvature of the earth's surface can be neglected

$$\frac{d}{dz}\left(K\frac{dT}{dz}\right) + A(z) = 0, \tag{2}$$

If the conductivity and heat production change to $K = K_1$ and $A = A_1$ at depth $z = z_1$, the temperature for $z > z_1$ can be expressed:

$$T = T(z_1) + \frac{1}{K_1}(Q - A_0 z_1)(z - z_1) - \frac{A_1}{2K_1}(z - z_1)^2 \tag{3}$$

where K is the thermal conductivity and A is the heat generation. Heat flow distribution ($Q = K\,dT/dz$) and heat generation models $A(z)$ for the crust in Sudetic Mountains, Paleozoic Platform (block C, block D) and Precambrian Platforms (block F_1, block F_2) presented here allow calculations of the temperature–depth profiles in these areas. The thermal conductivity K usually decreases with increasing temperature, and this dependence can be approximately by

$$K = \frac{K_0}{1 + \alpha T} \tag{4}$$

where K_0 is the thermal conductivity at 0°C and α for 'the granitic layer' is 0.32×10^{-3} °K^{-1}. The mean thermal conductivity for the granitic layer is 2.7 W m^{-1} °K^{-1} at 0°C. The conductivities of mafic rocks, forming the lower part of the crust depend only little on temperature and their conductivities may be regarded as independent upon position ($K = 1.9$ W m^{-1} °K^{-1}).

The calculated temperature–depth profiles for Sudets, Fore-Sudetic Monocline and Precambrian Platform are shown in Fig. 8. Temperature at the base of the crust in the Precambrian Platform area vary from about 440–510°C for the Precambrian

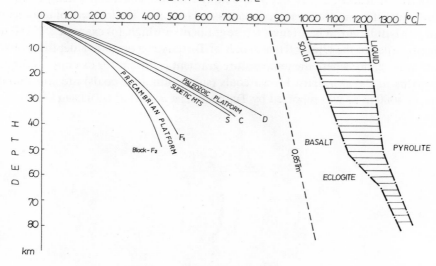

Figure 8

Temperature (°C) versus depth (km) curves for selected sites in Poland. The cross-hatched area shows the melting relations for basalt and eclogite from YODER and TILLEY (1962). 0.85 Tm line is after POLLACK and CHAPMAN (1977); Tm is mantle solidus temperature.

Platform (block F_1, block F_2) to 700–820°C for Paleozoic Platform and 700°C for Sudets.

Considerable lateral temperature variations at the crust-mantle boundary and variations in heat flow from the mantle are associated with narrow surface heat-flow transition zones. The transition zone between European Precambrian Platform and Paleozoic Platform implies considerable physical differences in the lower crust and upper mantle, i.e., differences in electrical conductivity, seismic velocities, density and magnetic properties of rocks.

Because of high subcrustal temperatures, the melting point could be reached at shallower depth in the upper mantle in the Paleozoic Platform (block C, block D) in comparison with surrounding areas. From data given by YODER and TILLEY (1962) it follows that partial melting can be reached in these areas at a depth very close to the Moho. This would be at the depth of 60–90 km as it follows from extrapolation of crustal temperature curves. Under the Precambrian Platform the geotherms do not reach the melting curve at any depth.

5. Preliminary paleogeothermal studies

The decisive influence of temperature on all chemical reactions during coalification is undeniable (TEICHMÜLLER and TEICHMÜLLER, 1968). The increase of coal

content and maturation with depth is mainly caused by increasing temperature. In
regions with different geothermal gradients and heat flow the increase of coalification
with depth is different. Where temperature gradients are high, for example 8°C 100 m^{-1}
in the rift valley of the Upper Rhine, coals of Tertiary age occur at a depth of 2300 m.
On the other hand, where the temperature gradient is low, as for example in foreland
of the Alps in Bavaria, hard brown coals (sub-bituminous coals) are still found at
depths of 4000 m as was reported by TEICHMÜLLER and TEICHMÜLLER (1968).

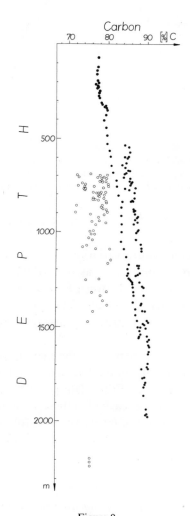

Figure 9
Carbon content (in volume %) versus depth. Open circles represent data from coal basins in the Pre-
cambrian Platform, black circles represent data from coal basins in the Paleozoic Platform area in
Central Europe.

The coalification time is also of importance. Due weight must also be given to the duration of heating. The longer that high temperature can act, the higher will be the rank of coal developed. KARSTEV *et al.* (1971) and BOSTICK (1971) developed a relationship between paleotemperature and vitrinite reflectance which is related to carbon content (increasing carbon content is related to increasing of vitrinite reflectance). For coal deposits of the same age the carbon content increases with increasing paleotemperature.

In Fig. 9 the carbon content versus depth for the Carboniferous coal basins in the Paleozoic Platform and Precambrian Platform areas are shown. The increase of carbon content with depth is much higher in the Paleozoic Platform area than in the Precambrian Platform region. The calculated paleogeothermal gradient for Carboniferous age in Paleozoic Platform is 6°C 100 m^{-1}. Present observed geothermal gradients vary from 3°C 100 m^{-1} to 4°C 100 m^{-1} in this area. It can be concluded that the Carboniferous paleogeothermal gradient was much lower in the Precambrian Platform area than in the Paleozoic Platform region.

6. Conclusions

Calculations of geothermal models show that the heat-flow transition zone between the European Precambrian Platform and the Paleozoic orogenic areas (Paleozoic Platform of Variscan basement, Sudets) is associated with considerable variations in mantle heat flow and temperature. At the base of the crust variations from 440–510°C in the models of Precambrian Platform (block F_1, block F_2) to 700–820°C in the Paleozoic Platform (Fore-Sudetic Monocline) were found.

Calculated models show heat flow from the mantle of about 8.4–12.6 mW m^{-2} for Precambrian Platform and from 31 mW m^{-2} to 37.6 mW m^{-2} for the Paleozoic orogenic areas. The heat flow contribution originating from crustal radioactivity is almost the same for the different units and varies from 33.5 mW m^{-2} to 37.6 mW m^{-2}.

Considerable physical differences in the lower crust and upper mantle beneath the Precambrian Platform and the adjacent areas, produced by lateral temperature variations, could be expected.

Acknowledgments

The author wishes to express his thanks to Drs L. Rybach (Zürich), A. H. Lachenbruch (Menlo Park, California), P. B. Slack (Denver, Colorado) and Assistant Professor Cz. Królikowski (Warszawa) for their helpful comments on the manuscript and Professor L. Stegena (Budapest) for stimulating discussions during the KAPG geothermal conferences in Liblice (Czechoslovakia) and Mamaia (Romania).

REFERENCES

ARSHAVSKAYA, N., BERZINA, I. and LUBIMOVA, E. (1972), *Geochemical and geothermal model for Pechenga and Ricolatva regions*, Geothermics, *1*, 25–30.

BALLING, N. P. (1976), *Geothermal models of the crust and upper mantle of the Fennoscandian Shield in South Norway and the Danish Embayment*, J. Geophys. *42*, 237–256.

BOSTICK, N. H. (1971), *Thermal alteration of clastick organic particles as an indicator of contrast and burial metamorphism in sedimentary rocks*, Geoscience and Man., Vol. III, Oct. 1.

CROUGH, S. T. and THOMPSON, G. A. (1976), *Thermal model of continental lithosphere*, J. Geophys. Res. *81*, 4857–4862.

ČERMAK, V. (1975), *Temperature–depth profiles in Czechoslovakia and some adjacent areas derived from heat flow measurements*, Tectonophysics *26*, 103–119.

ČERMAK, V. and JESSOP, A. M. (1975), *Heat flow, heat generation and crustal temperature in Kapuskasing area of the Canadian Shield*, Tectonophysics *26*, 103–121.

DUCZKOW, A. and SOKOLOWA, L., *Geotermiceskye issledovanya w Sibiri* (*Geothermal Investigations in Siberia*), (Nauka, Novosibirsk, 1974), pp. 275.

GUTERCH, A., MATERZOK, R., PAJCHEL, J. and PERCHUĆ, K. (1975), *Seismic structure of the earth crust along the international profile VII in the light of studies by deep seismic sounding method*, Przeglad geol. *4*, 154–163.

HORAI, K. and SIMMONS, G. (1968), *Seismic travel time anomalies due to anomalous heat flow and density*, Jour. Geophys. Res. *73*, 7577–7588.

HURTIG, E. (1975), *Untersuchungen zur wärmeflussverteilung in Europe*, Gerlands Beitrage zur Geophysik *84*, 3–4, 247–260.

HYNDMAN, R. (1968), *Heat flow and surface radioactivity measurements in the precambrian shield of western Australia*, Phys. Earth Planet. Int. *1*, 129–135.

JESSOP, A. M., HOBART, M. A. and SCLATER, J. G., *The world heat flow data collection – 1975*, Geothermal Series, Number 5 (Ottawa, Canada, 1976), 125 pp.

JONES, P. H. (1970), *Geothermal resources of the northern Gulf of Mexico basin*, Geothermics, Special Issue *2*, 14–26.

KARSTEV, A., VASSOEVICH, N., GEODEKIAN, A., NERUCHER, S. and SOKOLOV, V. , *The principal stage in the formation of petroleum*, in *Proceedings of the Eight World Petroleum Congress*, vol. 4 (Applied Science Publ. London, 1971).

KUTAS, R. I. (1972), *Investigations of heat flow anomalies in some regions of Ukraine*, Geothermics *1*, 35–39.

KUTAS, R. I., LUBIMOWA, E. A. and SMIRNOV, J. B., *Heat flow map of the European part of the USSR*, KAPG Geophysical Monograph (Akademiai Kiado, Budapest, 1976), pp. 443–449.

LACHENBRUCH, A. H. (1970), *Implication of linear heat flow relation*, J. geophys. Res. *76*, 3852–3810.

MAJOROWICZ, J. (1973a), *Heat flow data from Poland*, Nature Phys. Sc. *243*, 105.

MAJOROWICZ, J. (1973b), *Heat flow in Poland and its relation to the geological structure*, Geothermics *2*, 24–28.

MAJOROWICZ, J. (1976), *Heat flow map of Poland on the background of geothermal field of Europe and some aspects of its interpretation*, Acta geoph. pol. *24*, 147–156.

MAREK, S. and ZNOSKO, J., *Tectonic position of Kujawy and Wielkopolska*, Biuletyn Inst. Geol., v. 274, (Warszawa, 1974).

PLEWA, S. (1967), *Measurement results of the surface heat flow on the Polish Territory*, Selected Problems of upper mantle investigations in Poland, Publ. Inst. Geophys. Pol. Acad. Sc. (Materiały i Prace) *14*, 103–114.

PLEWA, S. (1976), *The new results of surface heat flow investigations of Earth's crust performed in Karpaty Mountains*, Publ. Inst. Geophys. Pol. Acad. Sc. *A-2* (101), 185–190.

POLLACK, M. N. and CHAPMAN, D. S. (1977), *On the regional variation of heat flow, geotherms, and litospheric thickness*, Tectonophysics *38*, 279–296.

POLYAK, B. G. and SMIRNOV, YA. B. (1968), *Relationship between terrestrial heat flow and the tectonics of continents*, Geotectonics *4*, 205–213.

RAO, R. V. M. and JESSOP, A. M. (1975), *A comparison of thermal characters of shields*, Can. Jour. Earth Sc. *12*, 347–360.

Roy, R. F., Blackwell, D. D. and Decker, E. R., *Continental heat flow*, in *The Nature of the Solid* (ed. Robertson), (McGraw-Hill, New York, 1971), pp. 506–543.

Rybach, L. (1973), *Wärmeproduktionsbestimmungen an Gesteinen der Schweizer Alpen*. Beitr. Geol. Schweiz, Geotech. Ser. Lfg. 51.

Rybach, L. (1976), *Radioactive heat production in rocks and its relation to other petrophysical parameters*, Pure Appl. Geophys. *114*, 309–317.

Sass, J. (1968), *Heat flow and surface radioactivity in Quirke Lake Syncline, Ontario*, Can. J. Earth Sci. *5*, 1417–1425.

Slack, P. B. (1974), *Variance of terrestrial heat flow between the North American craton and the Canadian Shield*, Geol. Soc. Amer. Bull. *85*, 519–522.

Stegena, L., Géczy, N. and Horvath, F. (1975), *Late canozoic evolution of the Panonian Basin*, Tectonophysics *26*, 71.

Swanberg, C. A., Chessman, M. D., Simmons, G., Smithson, S. B., Grönlie, G. and Heier, K. S. (1974), *Heat flow–heat generation studies in Norway*, Tectonophysics *23*, 31–48.

Teichmüller, M. and Teichmüller, A., *Geological aspect of coal metamorphism*, in *Coal and Coal-Bearing Strata* (ed. D. G. Murchison), (Oliver and Boyd, Edinburgh–London, 1968).

Węsierska, M. (1973), *A study of terrestrial heat flux density in Poland*, Publ. Inst. Geophys. Pol. Acad. Sc. (Materiały i Prace) *60*, 135–144.

Yoder, M. S. and Tilley, C. (1962), *Origin of basalt magmas, an experimental study of natural and synthetic rock systems*, J. Petrol. *3*, 342–532.

(Received 20th October 1977, revised 27th February 1978)

Pageoph, Vol. 117 (1978/79), Birkhäuser Verlag, Basel

On the Geothermal Regime of Some Tectonic Units in Romania

By Crişan Demetrescu[1])

Abstract – Heat flow values of 33–58 mW m^{-2} were found for the Transylvanian Depression, 45–57 mW m^{-2} for the crystalline nucleus of the Eastern Carpathians, and 70–120 mW m^{-2} for the Neogene volcanic area. Temperature–depth profile and some geophysical implications of the low values for the Transylvanian Depression are discussed, rendering evident clear-cut differences between this tectonic unit and other Neogene depressions. The heat flow values for the other two investigated tectonic units are usual ones for areas of their age.

A preliminary map of the heat flow distribution over the Romanian territory is presented and its relation to other geophysical fields is discussed. A positive correlation was found between gravity and heat flow, and a negative one between crustal thickness and heat flow. A general conclusion could be drawn that the heat flow distribution over the Romanian territory seems to be governed by processes taking place in the upper mantle, rather than by the radioactive decay within the crust.

Key words: Heat flow; Temperature-depth profiles; Crustal thickness; Gravity anomalies; Thermal conductivity of sediments.

1. Heat flow measurements

During the last years a growing attention has been paid to the study of the thermal field over the Romanian territory. Collecting and interpreting temperature log data and formation data resulted in useful information about the regional temperature distribution. Attempts were made to use the temperature data for deducing the heat flow; the reported values are however mere estimations based on thermal conductivities quoted from literature.

Our heat flow measurements concern the Transylvanian Depression as well as the Neogene volcanic zone and the crystalline nucleus of the Eastern Carpathians. The distribution and values of the measured heat flow are presented in Fig. 1. Details about the heat flow measurements will be given elsewhere (Demetrescu, in preparation).

The temperature measurements were taken in boreholes left to rest long enough after the boring ceased. Data about the effect of the hole boring on the natural temperature field for drilling conditions likely to be encountered in Romania were used to correct the temperature gradient in two of the cases.

---

[1]) Institute for Earth's Physics and Seismology, Bucureşti-Măgurele, Sector 6, PO Box 5202, Bucureşti, Romania.

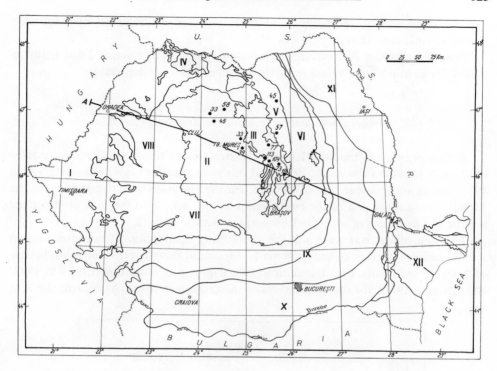

Figure 1

Distribution of the measured heat flow values (mW m^{-2}). (I) Pannonian Depression, (II) Transylvanian Depression, (III, IV) Neogene volcanic zone of the Eastern Carpathians, (V) crystalline nucleus of the Eastern Carpathians, (VI) other units of the Eastern Carpathians, (VII) Southern Carpathians, (VIII) Apuseni Mountains, (IX) Carpathian foredeep, (X) Moesian Platform, (XI) Moldavian Platform, (XII) units of the Northern and Central Dobrudja; (AA′) International deep seismic sounding profile XI. Tectonics after DUMITRESCU and SĂNDULESCU (1970).

The thermal conductivity was measured by a constant temperature difference divided bar apparatus with guard ring (BECK, 1957). Whenever possible the standard procedure with at least three water saturated discs was used. For some sedimentary rocks, disintegrating when placed in water for saturation, the oil or air saturated conductivity was measured and the water saturated conductivity was computed. In few cases measurements on fragments were done.

Examining the heat flow values, one can notice the low ones for the Transylvanian Basin (33–58 mW m^{-2}) and for the crystalline nucleus of the Eastern Carpathians (45–57 mW m^{-2}), and the high ones (70–120 mW m^{-2}) for the Neogene volcanic area.

As regards the crystalline nucleus, the complicated tectonic evolution of the area, with the last thermal event during the Hercynian orogenic movements (PAVELESCU et al., 1976) and successive Alpine subsidences and uplifts, makes it difficult to interpret the obtained heat flow values until more measurements are done.

The high values in the Neogene volcanic area are characteristic ones for such young tectonic units (LEE and UYEDA, 1965).

The values for the Transylvanian Depression need a more detailed discussion as they differ so much from values reported for other Neogene depressions.

2. *Temperature–depth profile for the Transylvanian Basin*

Recent progress in the knowledge of the structure of the crust (CONSTANTINESCU *et al.*, 1976; RĂDULESCU *et al.*, 1976) and of the sedimentary layer (VISARION *et al.*, 1973) allows some calculation of the temperature variation with depth down to the Moho discontinuity and of the contribution of the mantle to the measured surface heat flow. The results of such a calculation are shown in Fig. 2.

The step model was used and the simplifying assumption of one dimensional steady-state conduction of heat was made. Measured thermal conductivities, in case of Neogene sediments, Fig. 3, and mean values quoted in literature, (ČERMÁK, 1967; BLACKWELL, 1971), for the other layers of the crust, were used. As regards the heat

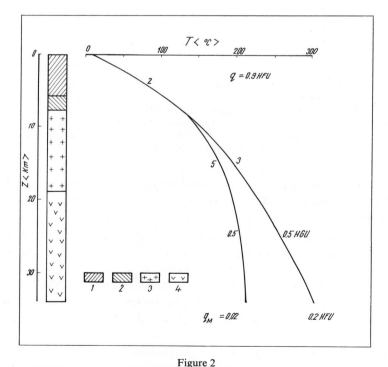

Figure 2
Temperature–depth variation for the Transylvanian Depression; (1) Tertiary sediments, (2) Mesosoic sediments, (3) granite, (4) basalt; (q) surface heat flow, (q_M) mantle heat flow, numbers on curves – heat production, 1 HFU = 41.87 mW m^{-2}, 1 HGU = 0.4187 μW m^{-3}.

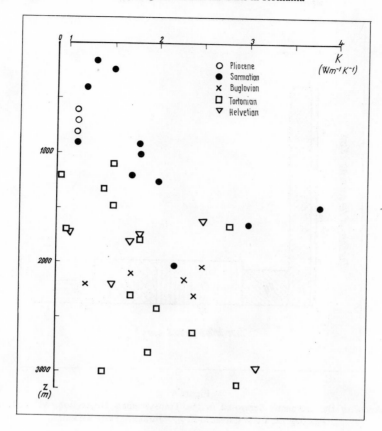

Figure 3
Thermal conductivity of Neogene sediments in Transylvanian Depression as a function of age and depth.

generation, the lack of data concerning the vertical distribution of the radioactive elements in the studied area forced us to assume different heat production values, resulting in different temperature–depth distributions.

Obvious differences are evident between the Transylvanian Depression and other Neogene depressions (Pannonian Depression, Danubian Basin, etc.), both from the point of view of the temperature–depth distribution (200–300°C compared to 1000–1200°C at the base of the crust) and of surface (38 mW m^{-2} compared to 90–100 mW m^{-2}) and mantle heat flow (8 mW m^{-2} compared to 45–60 mW m^{-2}) (STEGENA et al., 1975; ČERMÁK, 1975). Taking into account the temperature variation of the thermal conductivity and adopting a model in which the radioactivity of the granite layer decreases exponentially with depth would result in differences of the Moho temperature of only about 10–15%, not changing essentially the fact that the Moho temperature and the mantle heat flow are low ones beneath the Transylvanian Basin.

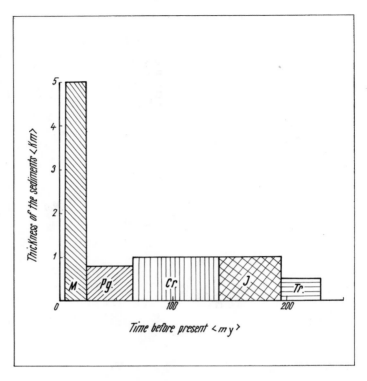

Figure 4
Mean thickness of the sediments deposited in the Transylvanian Depression. M = Miocene, Pg = Palaeogene, Cr = Cretaceous, J = Jurassic, Tr = Triassic.

Such small values could be explained by phenomena which are able to lower either the surface heat flow or the heat coming from the mantle, or both.

From the first category we shall discuss the sedimentation and the subsidence, which, for the Transylvanian Basin, were most severe during the Miocene (Fig. 4). The Miocene sedimentation lowered the pre-Miocene geothermal gradient by about 20%. The correction was calculated by Jaeger's method of the heat conduction in a moving material (JAEGER, 1965). The pre-Miocene sedimentation had no significant effects. If we also take into account the subsidence of the basin during the same interval of time, the corrected value of the heat flow will be raised to 50–60 mW m^{-2}, which is small too and does not change essentially the above statement concerning the cold Moho.

Thus a mechanism should be considered which lowers, in the area under consideration, the heat flow from the mantle. In this respect we think of a descending limb of a convection current in the mantle. The hypothesis does not contradict both the active subcrustal currents necessary to account for the strong anisostasy of the Transylvanian Depression (GAVĂT et al., 1973; CONSTANTINESCU et al., 1976) and a possible fossil subduction in the area (RĂDULESCU *et al.*, 1976). Such a convection current could also be the result of the active mantle diapirism under the Pannonian Depression (STEGENA *et al.*, 1975).

However, to decide of the possible mechanisms, more heat flow measurements are needed, having in view both the very complicated tectonics and the fact that the present data come only from areas with a very deep basement.

3. Preliminary heat flow map of Romania

A first attempt to derive the distribution of the heat flow on the Romanian territory dates back to 1970 (NEGOIȚĂ, 1970). The data used to calculate the heat flow (temperature logs taken during the drilling or immediately after the drilling ceased and thermal conductivities quoted from literature) made the heat flow values only qualitative estimations. They were used in constructing the Romanian part of the Heat Flow Map of Central and Eastern Europe, scale 1:10 000 000 (in ÁDÁM, 1974); the quality of data was taken into consideration resulting in a very simplified pattern, with isolines drawn in broken lines.

In Fig. 5 we present a new heat flow map of Romania which improves the previous image. In the construction of the map the following data were used:

– our own measurements, for the tectonic units within the Carpathian arc. The heat flow values were supplemented by geothermal gradients measured in holes with no available cores. To extrapolate the measured values into areas without any measurements, within the same tectonic unit, the relation between heat flow and tectonics was taken into account.
– the map of the distribution of the geothermal gradient at 1 km depth by PARA-SCHIV and CRISTIAN (1976), for the tectonic units on the convex side of the Carpathian arc. Though in estimating the geothermal gradient the authors made use of formation temperature data and mean air temperature, the great number of available data allows, however, a certain confidence in rendering the regional features of the phenomenon. The conductivity of the sediments was taken as $1.7–2.1$ W m^{-1} K^{-1}.

The isolines were drawn at 10 mW m^{-2} intervals, larger than the errors usually associated to heat flow determinations. Having in view the quality of available data this interval might be too small, but we felt the necessity of rendering evident the changes, sometimes very rapid, from areas with high heat flow to areas with low heat flow.

No data were available for Apuseni Mountains, Southern Carpathians, and Dobrudja.

The map, though having a preliminary character, is successful in rendering evident interesting zones. We notice the high heat flow in the Pannonian Depression, the

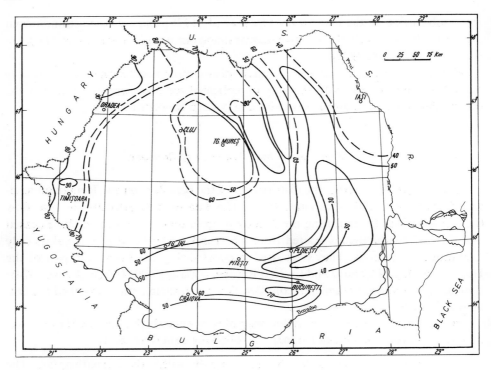

Figure 5
Preliminary heat flow map of Romania, in mW m^{-2}.

Neogene volcanic zone, the Moesian Platform; the low heat flow in the Transylvanian Basin, the crystalline nucleus of the Eastern Carpathians, the Moldavian Platform, the Carpathian foredeep; and the large variations of the heat flow between neighbouring tectonic units (Transylvanian Depression – Neogene volcanic zone – crystalline nucleus of the Eastern Carpathians; Moesian Platform – Carpathian foredeep).

This zoning of the territory is interesting from at least three points of view. Firstly, it renders evident areas in which the high heat flow is an indication of possible utilization of the geothermal energy. Secondly, the rapid variation of the heat flow indicates the existence of high horizontal thermal gradients which might be correlated with the seismicity of those areas; for instance, the contact between the high heat flow in the Moesian Platform and the low one in front of the Carpathian bend seems to coincide with an active fault on which at present earthquakes occur (CORNEA, personal communication). Thirdly, the heat flow might be used as an argument favouring some tectonic hypotheses, as discussed in the previous section; from this point of view the well developed minimum in front of the Carpathian bend is of interest, which could be the thermal consequence of an active subduction in the area.

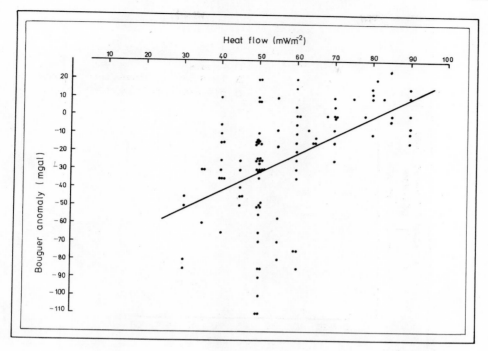

Figure 6
Correlation between heat flow and gravity.

4. Relations between heat flow and other geophysical data

The relation between gravity and heat flow was studied by comparing the maps of the regional Bouguer anomaly (SOCOLESCU and BIŞIR, 1956) and heat flow in a network of 30 minutes on longitude and 20 minutes on latitude. The result is presented in Fig. 6. The cloud of points indicates a general positive correlation between the two fields ($r = 0.5$) at this scale, a fact also reported for other territories (e.g. Czechoslovakia (ČERMÁK, 1977)).

A comparison done in the same way between the heat flow and the map of the crustal thickness based on gravity data by SOCOLESCU et al. (1964) seems to indicate a negative correlation between the two parameters. The correlation is not very clear mainly due to the crustal thicknesses in the Carpathian foredeep. Recent deep seismic sounding data (CONSTANTINESCU et al., 1976) indicate for this tectonic unit thicknesses of the crust of 40–50 km, not 33–36 km as given by gravity data. The negative correlation between heat flow and crustal thickness is a very clear one if only deep seismic sounding data are used (Fig. 7).

The fact that the heat flow is large where the crust is thin and vice versa clearly shows that the radioactive decay in the crust is not the main source of the observed

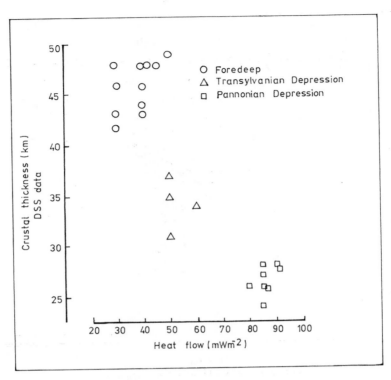

Figure 7
Correlation between crustal thickness derived from deep seismic sounding (DSS) data and heat flow.

surface heat flow, and this source is to be found in the mantle. The correlation between the gravity field, crustal thickness, and heat flow can be accounted for by having the dense masses from the mantle nearer the surface where the crust is thinner and the heat flow higher, and vice versa. These considerations and the results of the second section lead to the conclusion that the distribution of the heat flow on the Romanian territory seems to be governed mainly by processes taking place in the upper mantle.

In Fig. 8 the heat flow, the gravity field after Socolescu and Bişir (1956), and the crustal thickness are plotted, along the international deep seismic sounding profile XI for which data concerning the depth of the Mohorovičič discontinuity are rather complete (Constantinescu et al., 1976; Rădulescu et al., 1976). In the same plot the regional anomaly of the geomagnetic field, after Ciocârdel et al. (1970), and the secular variation of the horizontal component of the geomagnetic field between 1969–74 (the present author's not published data) are given.

The correlation discussed above between the heat flow, regional Bouguer anomaly and crustal thickness is evident; from this point of view a question mark can be put in the central part of the Transylvanian Depression, where the crystalline basement is in

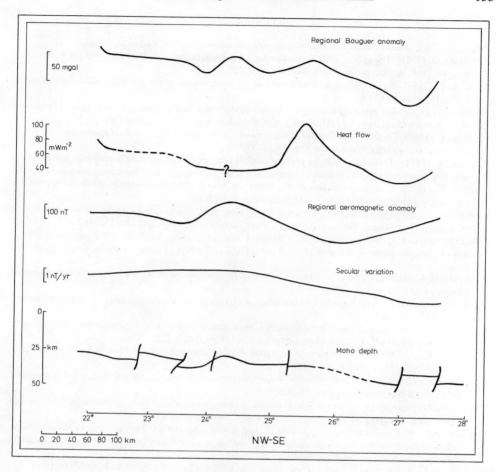

Figure 8
Heat flow, gravity, geomagnetic field and crustal thickness (with Moho faults) on the international DSS profile XI.

an elevated position and the Moho discontinuity is at a shallow depth of about 29 km. Heat flow measurements in this area are necessary to answer this question.

As regards the aeromagnetic regional anomaly, we cannot find a visible connection between the thermal phenomenon and the magnetic one. The secular variation profile seems to reflect phenomena taking place at greater depth. Models describing the influence of the thermal field on the magnetic field have been proposed (ATANASIU, 1971). The fact that the temperature at the base of the crust seems to be not close to the Curie point, makes it probable that the possible effects of high temperature on rock magnetization take place at great depth and cannot be felt at the surface by present available means.

REFERENCES

ÁDÁM, A. (Ed.), *Geoelectric and Geothermal Studies* (Akadémiai Kiadó, Budapest, 1974).

ATANASIU, G. (1971), *The geomagnetic secular variation in Romania between 1935 and 1965* (in Romanian), St. cerc. geol., geofiz., geogr., Seria geofizică, *9*, 19–26.

BECK, A. E. (1957), *A steady state method for the rapid measurement of the thermal conductivity of rocks*, J. Sci. Instr. *34*, 186–190.

BLACKWELL, D. D., *The thermal structure of the continental crust*, in *The Structure and Physical Properties of the Earth's Crust*, (ed. J. G. Heacock), Am. Geophys. Un., 1971, pp. 169–185.

ČERMÁK, V. (1967), *Coefficient of thermal conductivity of some sediments, its dependence on density and on water content of rocks*, Chemie der Erde, *26*, 271–278.

ČERMÁK, V. (1975), *Temperature–depth profiles in Czechoslovakia and some adjacent areas derived from heat flow measurements, deep seismic sounding and other geophysical data*, Tectonophysics *26*, 103–121.

ČERMÁK, V. (1977), *Geothermal models of the Bohemian Massif (Variscan) and the Western Carpathians (Alpine) and their mutual relation*, Tectonophysics *41*, 127–137.

CIOCÂRDEL, R., SOCOLESCU, M. and CRISTESCU, TR. (1970), *Sur l'origine des mouvements néotectoniques en Roumanie*, Rev. Roum. Géol., Géophys., Géogr., Série de Géophysique, *14*, 155–167.

CONSTANTINESCU, L., CONSTANTINESCU, P., CORNEA, I. and LĂZĂRESCU, V. (1976), *Recent seismic information on the lithosphere in Romania*, Rev. Roum. Géol., Géophys., Géogr., Série de Géophysique *20*, 33–41.

DUMITRESCU, I. and SĂNDULESCU, M., *Tectonic Map of Romania, Geological Atlas, Sheet no. 6* (Institute of Geology and Geophysics, 1970).

GAVĂT, I., BOTEZATU, R. and VISARION, M., *Geological Interpretation of geophysical exploration* (in Romanian), (ed. Academiei R. S. R., 1973).

JAEGER, J. C., *Application of the Theory of Heat Conduction to Geothermal Measurements*, in *Terrestrial Heat Flow*, (ed. W. H. K. Lee), (Am. Geophys. Un., 1965), pp. 7–23.

LEE, W. H. K. and UYEDA, S., *Review of heat flow data*, in *Terrestrial Heat Flow*, (ed. W. H. K. Lee), (Am. Geophys. Un., 1965), pp. 87–190.

NEGOITĂ, N. (1970), *Étude sur la distribution des temperatures en Roumanie*, Rev. Roum. Géol., Géophys., Géogr., Série de Géophysique *14*, 25–30.

PARASCHIV, D. and CRISTIAN, M. (1976), *About the geothermal regime of structural units of hydrocarbon interest* (in Romanian), St. cerc. geol., geofiz., geogr., Seria geofizică *14*, 65–75.

PAVELESCU, L., POP, G., AILENEI, G., CRISTEA, I. and SOROIU, M. (1976), *The evolution of the crystalline nucleus of the Eastern Carpathians, according to radioactive age dating* (in Romanian), St. cerc. geol., geofiz., geogr., Seria geofizică *14*, 39–49.

RĂDULESCU, D. P., CORNEA, I., SĂNDULESCU, M., CONSTANTINESCU, P., RĂDULESCU, F. and POMPILIAN, AL. (1976), *Structure de la croûte terrestre en Roumanie, Éssai d'interpretation des études sismiques profondes*, Anuarul Institutului de geologie şi geofizică *L*, 5–36.

SOCOLESCU, M. and BIŞIR, D. (1956), *The calculation of the network of pendulum stations in Romania* (in Romanian), St. cerc. fiz. *7*, 505–519.

SOCOLESCU, M., POPOVICI, D., VISARION, M. and ROŞCA, V. (1964), *Structure of the earth's crust in Romania as based on the gravimetric data*, Rev. Roum. Géol., Géophys., Géogr., Série de Géophysique *8*, 3–13.

STEGENA, L., GÉCZY, B. and HORVÁTH, F. (1975), *Late cenozoic evolution of the Pannonian Basin*, Tectonophysics *26*, 71–91.

VISARION, M., ALI-MEHMED, E. and POLONIC, P. (1973), *The integrated study of geophysical data concerning the morphology and structure of the crystalline basement in Transylvanian Depression* (in Romanian), St. cerc. geol., geofiz., geogr., Seria geofizică *11*, 193–201.

(Received 11th November 1977, revised 11th March 1978)

Pageoph, Vol. 117 (1978/79), Birkhäuser Verlag, Basel

Heat Flow in Italy

By M. Loddo[1]) and F. Mongelli[2])

Abstract – More than fifty heat flow measurements in Italy are examined. The values, corrected only for local influences (when present), are related to the main geological features with the following results: foreland areas, 55 ± 19 mW m^{-2}, foredeep areas, 45 ± 21 mW m^{-2}; folded regions and intermountain depressions, 76 ± 29 mW m^{-2}. In volcanic areas the heat flow rises to in excess of 600 mW m^{-2}. From a tectonic point of view, these values are consistent with the hypothesis that the Apennine chain is intersected by two arcuate structures: the first from Liguria to Latium is very probably a continental arc, that is an arc which occurs within a continent, and the second from Campania to Calabria is very similar from geophysical evidence to the classic island arcs.

Key words: Heat flow corrections; Heat flow and tectonic features; Mediterranean plate tectonics.

1. Introduction

The Italian region is one of the most tectonically perturbed areas in Europe. Obviously, the heat flow plays a very important role in the tectonics, and knowledge of the heat flow pattern may help to understand the developing tectonic framework. Furthermore, the volcanic activity in Italy and the existing geothermal fields of Larderello and Monte Amiata are important geothermal features which are relevant to the energy problem. Pure and applied geothermal research in Italy have different but complementary purposes (Mongelli and Loddo, 1975).

In the following sections, after some consideration of the techniques of measurement and elaboration of data, we give a summary of the heat flow data and a general discussion of the heat flow pattern in relation to the main geological, tectonic and geophysical features.

2. Review of techniques used

With the exception of the measurements in lakes, where the oceanographic technique was employed, the boreholes on land used for temperature measurements were 75 to 600 m deep, mainly in marly or sandy clays, but sometimes in limestones or

---

[1]) Istituto di Geodesia e Geofisica dell'Università di Bari, 70100 Bari, Italy.

[2]) Istituto di Geodesia e Geofisica and Osservatorio di Geofisica e Fisica Cosmica dell'Università di Bari, 70100 Bari, Italy.

Table 1
Geothermal data in Italy.

No.	Lake	Borehole	A	B	C	D	E	F	G	H	I	L	M	Refs.
1	Sirio		45-29.2	7-53.1	271	46						42.7 T, S, W	27.2 U, D	[1]
2	Viverone		45-25.0	8-01.9	230	52						41.5 T, S, W	26.4 U, D	[1]
3	Orta		45-48.8	8-23.6	290	125						69.1 T, S, W	46.9 U, D	[1]
4	Orta		45-50.6	8-23.5	290	125						68.2 T, S, W	46.5 U, D	[1]
5	Mergozzo		45-57.3	8-27.7	196	73						62.0 T, S, W	41.5 U, D	[1]
6	Maggiore		45-49.4	8-35.5	194	120						64.1 T, S, W	42.7 U, D	[1]
7	Maggiore		45-51.9	8-34.3	194	260						79.1 T, S, W	53.6 U, D	[1]
8	Maggiore		45-56.2	8-36.9	194	343						57.8 T, S, W	38.1 U, D	[1]
9	Maggiore		45-55.3	8-28.4	194	128						66.6 T, S, W	44.4 U, D	[1]
10	Maggiore		45-58.4	8-39.8	194	367						68.2 T, S, W	45.6 U, D	[1]
11	Como		45-50.4	9-05.4	198	159						71.6 T, S, W	48.6 U, D	[1]
12	Como		45-53.1	9-20.9	198	148						71.2 T, S, W	48.6 U, D	[1]
13	Como		45-57.2	9-10.6	198	398						70.8 T, S, W	47.7 U, D	[1]
14	Como		45-58.5	9-17.1	198	286						72.4 T, S, W	49.4 U, D	[1]
15	Como		46-02.0	9-16.1	198	294						61.1 T, S, W	41.5 U, D	[1]
16	Como		46-06.1	9-17.9	198	220						65.3 T, S, W	43.5 U, D	[1]
17	Como		46-08.1	9-19.5	198	179						60.7 T, S, W	40.6 U, D	[1]
18	Como		45-52.2	9-08.5	198	289						62.8 T, S, W	41.9 U, D	[1]
19	Iseo		45-42.5	10-04.8	186	245						72.0 T, S, W	49.0 U, D	[1]
20	Iseo		45-45.7	10-03.4	186	105						72.4 T, S, W	49.4 U, D	[1]
21	Garda		45-32.6	10-36.5	65	163						76.6 T, S, W	51.9 U, D	[1]
22	Garda		45-36.1	10-38.4	65	158						78.3 T, S, W	53.2 U, D	[1]
23	Garda		45-39.1	10-40.5	65	243						80.0 T, S, W	54.4 U, D	[1]
24	Garda		45-41.3	10-42.8	65	343						74.9 T, S, W	51.1 U, D	[1]
25	Garda		45-44.2	10-46.1	65	338						70.8 T, S, W	47.3 U, D	[1]
26	Garda		45-47.1	10-48.1	65	332						63.6 T, S, W	42.7 U, D	[1]
27		Imola	44-19.4	11-38.1			78	152	19 ± 2	1.38 ± 0.10	26 ± 4			[2]
28		Faenza	44-11.9	11-53.0			117	150	17.6 ± 0.4	1.30 ± 0.11				[2]
29		Forli	44-08.0	12-05.1			72	160	19.2 ± 0.7	1.30 ± 0.10	25 ± 3			[2]
30		B.S. Lorenzo	43-56.9	11-25.5			194	205	44.5 ± 6	1.42 ± 0.12	60 ± 2			[2]
31		Prato	43-42.3	11-06.7			56	567	35.6 ± 0.9	2.6 ± 0.10	93 ± 6			[2]
32		Pisa	43-42.3	10-23.1			3	100	61.5 ± 3	(1.55 ± 0.08)	95 ± 10			[2]
33		Rosignano S.	43-22.8	10-27.3			11	190	57.6	(1.66)	97	96.5 O	103 C	[3]
34		Rosignano S.	43-22.8	10-26.6			2.5	150	67.7	1.57	106	105 O	113 C	[3]
35		Rosignano S.	43-21.5	10-27.2			1	150	63.0	1.55	97	95 O	105 C	[3]

#	Name	A	B	C	D	E	F	G	H	I	L	M	References
36	Gargano	41–39.0	15–36.8		87		101	40	(2.85)	114			[4]
37	Gargano	41–36.3	15–43.9		53		79	74	(1.93)	143			[4]
38	Gargano	41–34.9	15–45.4		14		90	80	(1.51)	121			[4]
39	Fossa Bradanica	40–48.2	16–02.9		314		75	30	1.61	50			[5]
40	Fossa Bradanica	40–47.3	16–16.7		223		75	55	1.61	92			[5]
41	Fossa Bradanica	40–45.7	16–17.8		233		75	47	1.61	80			[5]
42	Fossa Bradanica	40–41.9	16–23.0		153		75	25	1.61	42			[5]
43	Fossa Bradanica	40–32.3	16–36.6		70		94	42	1.61	71			[5]
44	Fossa Bradanica	40–31.6	16–28.9		98		75	23	1.61	38			[5]
45	Fossa Bradanica	40–22.1	16–30.4		75		75	10	1.61	17			[5]
46	Fossa Bradanica	40–29.7	16–52.8		48		500	25	(1.47)	38			[6]
47	S. Demetrio	39–36.6	16–23.8		150		108	33 ± 1	(1.47)	49			[7]
48	Cutro	39–02.4	17–02.8		50		160	24 ± 1	(1.49 ± 0.01)	36 ± 2			[7]
49	S. Maria	38–52.4	16–37.3		90		120	30 ± 2	(1.44 ± 0.02)	44 ± 3			[7]
50	Etna	37–54.2	15–02.7		540		80	17	(1.61)	27	33 T		[8], [9]
51	Etna	37–48.1	15–13.3		40		80	27	(1.61)	43	46 T		[8], [9]
52	Etna	37–31.6	15–04.4		150		80	21	(1.61)	34	34 T		[8], [9]
53	Gagliano 1	37–44.2	14–31.8		700		2680	22	(1.47)	33			[6]
54	Enna 2	37–25.3	14–20.8		400		1460	20	(1.47)	29			[6]

Table legends
Table 1 – Geothermal data in Italy.

Column A – Latitude north (degree-minute)
      B – Longitude east (degree-minute)
      C – Water level (m)
      D – Depth of water (m)
      E – Collar elevation (m)
      F – Depth interval (m)
      G – Mean temperature gradient (mK m^{-1})
      H – Thermal conductivity (W m^{-1} K^{-1})
      I – Observed heat flow (mW m^{-2})
      L – Heat flow corrected (mW m^{-2}) for   T – Steady state topographic correction
                                                S – Sedimentation
                                                W – Annual temperature wave in the water
                                                O – Effect of the sea

M – Heat flow corrected (mW m^{-2}) for   U – Uplift
                                                D – Denudation
                                              C – Climatic effects

References [1] HAENEL (1974)
              [2] BOCCALETTI et al (1977)
              [3] FANELLI et al. (1974)
              [4] MONGELLI and RICCHETTI (1970a)
              [5] MONGELLI and RICCHETTI (1970b)
              [6] LODDO and MONGELLI (1975a)
              [7] LODDO et al. (1973)
              [8] MONGELLI and MORELLI (1964)
              [9] MONGELLI and LODDO (1974)

calcarenites. Thermal conductivity of the clays was mostly measured by the needle-probe method (Von Herzen and Maxwell, 1959), and occasionally by the *in situ* method (Beck *et al.*, 1956). In hard rocks the 'cut-core' method was used (Mongelli, 1968, 1969; Loddo, 1970). In some places the thermal conductivity was extrapolated from values from other similar geologic regions. Temperatures were measured using platinum resistances or thermistors connected to three or four leads compensated Wheatstone bridge circuits. The distance between measurement points ranged from 5 to 25 m. The results are quite homogenous in reliability and comparable, owing to the quality of techniques and skills of the operators.

Problems arise when we consider the local and regional influences on the heat flow and the related corrections. Discounting the geothermal fields and volcanic areas where the heat flow is so high that any correction does not significantly modify the high values, the effects of local and regional influences on the Italian heat flow data are very variable. However, it is possible to divide the conditions of these effects into a few fundamental cases:

(a) The measured temperature gradient clearly appears to be influenced. The geography and/or geology suggest the source of that influence. Generally in this case the values have been corrected.

(b) The temperature gradient measured in quite deep boreholes does not show any influence, but geography and/or geology suggest that an influence should exist. In this case, there generally was no correction applied.

(c) The temperature gradient was measured in holes too shallow to observe an influence, but geography and/or geology suggest that an influence must exist. In this case the values have generally been corrected (e.g. measurements in lakes).

(d) The temperature gradient was measured in holes too shallow to observe an influence, but geography and/or geology suggest that an influence is probable; for example, this is the case for the climatic variations, a world-wide phenomenon whose influence is sometimes absent. In this situation the values have sometimes been corrected.

In the next section we will discuss different situations in detail.

3. Heat flow data

Table 1 contains all the information on the heat flow stations in Italy, the observed and, when available, corrected values. Figure 1 is the location map for the heat flow stations. The measurements in the geothermal fields and volcanic areas are not reported in detail as they are very numerous and closely spaced. The remaining values, more than 50, may be grouped in eight distinctive areas from North to South.

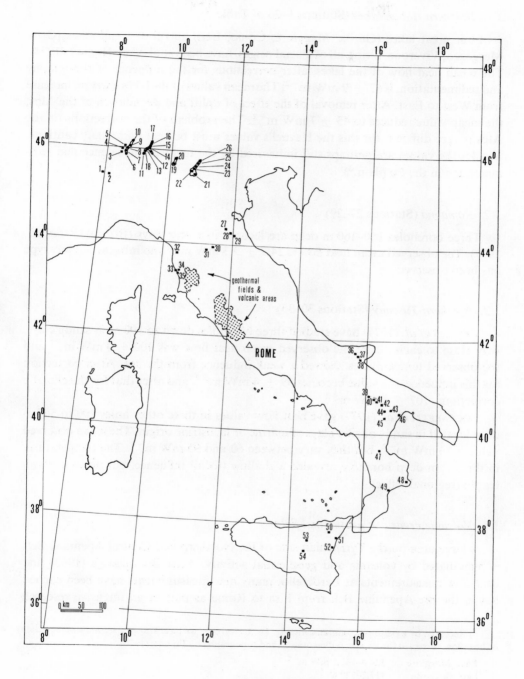

Figure 1
Location map of heat flow stations in Italy.

3.1. Northern Italian lakes (Stations 1–26 of Table 1)

Four great lakes (Garda, Maggiore, Como, Iseo) and four small lakes (Orta, Mergozzo, Viverone, Sirio), all of glacial origin, have been studied by HAENEL (1974). The mean heat flow of the lakes, after corrections for the influence of topography and sedimentation, is 67 ± 9 mW m^{-2}. The mean values of the lakes show an increase from West to East. After removal of the effect of uplift and denudation of the Alps, the mean value reduces to 45 ± 7 mW m^{-2}. The problem of the corrections on the Alps is very difficult, for this the Haenel's values must be kept with many cautions. Besides the values are perhaps still influenced by the regional subsidence and sedimentation in the Po plain.[3])

3.2. Romagna (Stations 27–29)

Three boreholes 150–160 m deep are located in marly clays (BOCCALETTI et al., 1977). The observed mean heat flow is 24.7 ± 3.3 mW m^{-2}; no influence of any type has been observed.

3.3. Northern Tuscany (Stations 30–35)

FANELLI et al. (1974) have studied three boreholes down to 190 m in sandy clays, very close to each other; the observed mean heat flow was 100 ± 4 mW m^{-2} and the observed temperatures showed a weak influence from the sea. After correction for this influence, the value becomes 99 ± 4 mW m^{-2}, and after that for the climatic corrections, 107 ± 4 mW m^{-2}.

BOCCALETTI et al. (1977) gave heat flow values in three other holes 100 to 567 m deep located in sandy or silty clays of marine or lacustrine origin. The mean observed value is 83 mW m^{-2}, but they vary between 60 and 95 mW m^{-2}. The temperatures in the 567 m deep borehole revealed a shallow (local) influence, but no long wavelength (regional) effect.

3.4. Volcanic areas and geothermal fields

A large area on the Tyrrhenian side of the Northern and Central Apennine Belt is punctuated by volcanic and geothermal activity. After BOLDIZSAR's (1963) first heat flow measurements at Larderello, many new measurements have been carried out in the pre-Apennine Belt from Pisa to Rome as part of an intensive research

---

[3]) Recently, by a personal communication by Dr. L. Rybach, we have known the following new data of heat flow corrected for ice age, sedimentation and topography by FINCKH (1976):

Lago Maggiore	124.4–131.1 mW m^{-2}
Lago di Garda	117.2–120.6
Lago d'Iseo	104.7–109.3
Lago di Como	70.8–103.4

program. The results are summarized and discussed in a recent paper by CALAMAI *et al.* (1977). The observed heat flow is very high (up to 600 mW m^{-2}), and this makes corrections unimportant, particularly in the central part of the Belt.

3.5. *Gargano Headland* (Stations 36–38)

Three closely spaced shallow holes along the Candelaro fault give a mean heat flow of 126 $\pm$ 12 mW m^{-2}, very probably influenced by local effects such as rising groundwater, subsurface variations in thermal conductivity, and perhaps magmatic intrusions (MONGELLI and RICCHETTI, 1970a).

3.6. *Bradano Trough* (Stations 39–46)

Seven shallow boreholes in quartzy clays (MONGELLI and RICCHETTI, 1970b) and another hole, 500 m deep (LODDO and MONGELLI, 1975a), gave a mean observed value of 54 $\pm$ 24 mW m^{-2}. The high s.d. is due to a regular decrease of the heat flow from NE to SW. The topographic correction was insignificant and the 500 m deep borehole revealed neither short (local) nor long wave-length (regional) effects.

3.7. *Calabria* (Stations 47–49)

LODDO *et al.* (1973) reported the observed values of three holes, 108–160 m deep, in marly clays. The mean is 43 $\pm$ 5 mW m^{-2}. These holes may be considered in the same geological and geographical conditions as those of the Bradano Trough.

3.8. *Sicily* (Stations 50–54)

The mean observed heat flow of three shallow boreholes drilled in sandy clays on Etna (avoiding lava flows) is 35 $\pm$ 7 mW m^{-2} (MONGELLI and LODDO, 1974). After steady-state topographic correction, the mean value becomes 38 $\pm$ 6 mW m^{-2}. The mean observed value of two deep oilwells in marly clays of Central Sicily is 31 $\pm$ 2 mW m^{-2} (LODDO and MONGELLI, 1975a).

4. *General discussion of the heat flow pattern*

The data presented in Table 1 are a mixture of uncorrected, partially corrected and wholly (?) corrected heat flow values, but for some values it is impossible to know what, if any corrections to apply. At this point, it is therefore very difficult to decide whether to draw a map of corrected or uncorrected values: to put together uncorrected values, some of which clearly influenced by strong local effects, or to mix corrected and uncorrected values. Both methods introduce large uncertainties.

M. Loddo and F. Mongelli   (Pageoph,

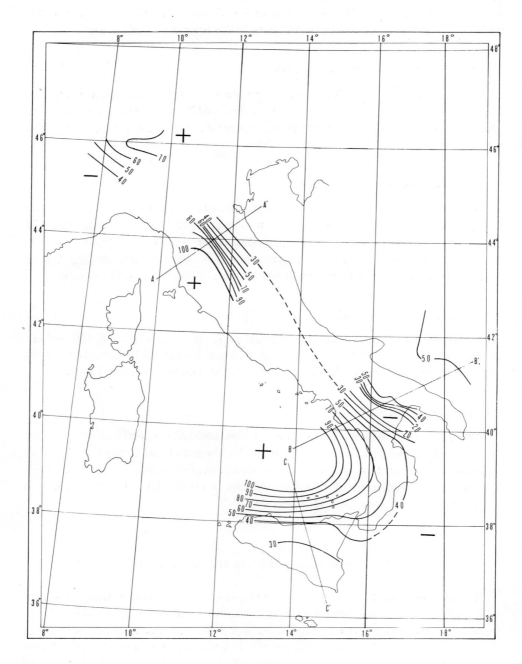

Figure 2
Contour map of heat flow in Italy (contour interval 10 mW m^{-2}).

An intermediate method has been chosen, i.e., to draw a map of the values corrected only for local effects (apparent, or geographically and/or geologically confirmed) and of values free from perturbing influences. In particular, we have considered:

- in the northern Italian lakes, the values corrected only for the local topography and sedimentation;
- in Romagna, the observed values;
- in Northern Tuscany, the values obtained near the shoreline corrected for the influence of the sea, and the other values uncorrected;
- on the Bradano Trough and Calabria, the observed values;
- in Sicily, the values obtained on the Etna corrected for the topography; other values uncorrected.

We have not considered the locally strongly influenced values of the Gargano headland.

Figure 2 is the contour map of the values described above, together with the isolines in Tyrrhenian, Ionian and Adriatic Seas as deduced from the works of ERICKSON (1970), BIRCH and HALUNEN (1966), LAVENIA (1967); it has the significance of an observed value contour map.

In spite of the data scarcity, the basic pattern of the heat flow in the Italian peninsula is quite clearly delineated. The pattern in Northern Italy is less clear where values are concentrated in the lakes, on a W–E trend, but the discussion of these values would be more appropriately considered in a general discussion on the heat flow in the Alps.

In the Italian peninsula the isolines run parallel to the Apennine chain down to Calabria and Northern Sicily. The Tyrrhenian side is intersected by high values with two maxima localized over Larderello–Monte Amiata geothermal fields and in the Tyrrhenian sea, and the Adriatic side is intersected by lower values. In the Southern Adriatic Sea, and perhaps in Apulia, the heat flow is normal. The heat flow field in the Italian peninsula is therefore definitely characterized by a strong lateral E–W gradient normal to the Apennine chain.

5. Main tectonic features of the Italian peninsula and Sicily related to heat flow

From a general point of view, the Italian peninsula and Sicily may be classified as a Cenozoic orogenic area.

In the area which extends South of the Po plain, one can recognize four fundamental structural units: they are the foreland, the foredeep and the intrapenninic basins, the Apenninic orogene and the postorogenic volcanic deposits (Fig. 3).

(a) *the foreland* consist of Jurassic-Cretaceous carbonate successions (in places covered by transgressive deposits of Paleogene, Miocene and Quaternary age);

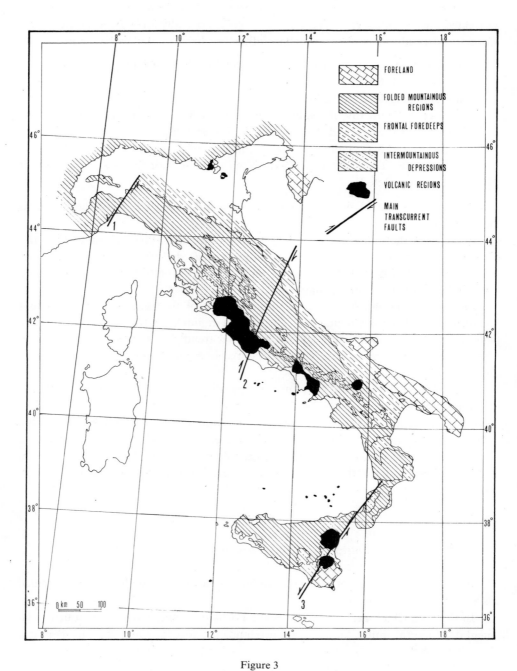

Figure 3
General tectonics of Italy. Tectonic lineaments: (1) Sestri–Voltaggio–Giudicarie, (2) Ancona–Anzio,
(3) Comiso–Messina–S.Eufemia.

the stratigraphic base of these successions consists almost exclusively of evaporitic sediments;

(b) *the foredeep* areas correspond to a large postorogenic graben, which was subsiding during the Upper Pliocene-Quaternary, and filled up by clayey-sandy sediments. The *intrapenninic troughs* represent minor physiographic bodies (which formed near the end of the orogenic phase); these troughs were filled by sediments during the Plio-Pleistocene;

(c) *the Apenninic orogene* consists of some allochthonous sedimentary successions mainly composed of clays, marls, sandstones and limestones. These successions can be related to the eugeosyncline complex (Cretaceous-Paleogene), the miogeosyncline complex (Upper Trias-Miocene) and to the miogeosyncline ridge (Jurassic-Cretaceous). The successions in the Calabro-Peloritana area also consist of crystalline rocks, which can be related to an internal massif. Finally, along the external border of the Apenninic chain, postorogenic successions outcrop: they consist of marly, clayey and arenaceous deposits, and in some places, also of gypsum deposits (Lower Miocene–Upper Miocene);

(d) *the volcanic deposits* mainly correspond to eruptive bodies, the magmatic activity of which can be related to postorogenic phases of the miogeosyncline; the age of these deposits is Pleistocene.

In Sicily, weakly differentiated basaltic magmas seem to be related with deep fractures that occur on the Iblean foreland (SW Sicily).

The tectonic structure of the Apenninic orogene is also complicated by the presence of transverse discontinuities which can be considered as transcurrent faults:

(α) the tectonic lineament Sestri–Voltaggio–Giudicarie, which sets a limit between the Alps and the Apenninic chain, shows a dextral horizontal displacement; the age of this displacement is Upper Pliocene;

(β) the tectonic lineament Ancona–Anzio indicates a sinistral horizontal displacement and the beginning of this movement can be dated as Pliocene; this lineament is still active;

(γ) the tectonic lineament Comiso–Messina–S. Eufemia shows a dextral horizontal displacement and is considered active up to now.

Table 2

Mean heat flow values and their geologic settings in Italy.

Geological feature	No. of values	Mean (s.d.) (mw m^{-2})
Foreland	11	55 $\pm$ 19
Foredeep	14	45 $\pm$ 21
Folded regions and intermountain depressions	8	76 $\pm$ 29
Volcanic areas and geothermal fields	very large	up to in excess of 600

The heat flow values are well delineated by this framework and in places they constrain the possible geological or geophysical models (Table 2).

With the exception of those values on the Gargano headland, no measurements have been made in continental foreland areas, but measurements in the Southern Adriatic Sea by LAVENIA (1967) may be indicative of this tectonic unit; they have a mean of 55 ± 19 mW m^{-2}. Measurements in foredeep areas have a mean of 45 ± 21 mW m^{-2}. The few measurements in the Apennine chain (excluding geothermal fields) and in the intra-apenninic depressions (Intermountain depressions) result in normal to high values 76 ± 29 mW m^{-2}. Measurements in volcanic areas or in geothermal fields gave locally very high values, with the exception of the Etna volcano, which has mean of 38 ± 6 mW m^{-2}.

This distribution of values is conveniently explained by modern tectonic theories. The Mediterranean area is the contact zone of the Africa and Eurasian plates, their collision producing active microplates (McKENZIE, 1972). On the basis of seismological data, LORT (1971) delineated an Apulian or Adriatic microplate between the Apennines and Dinarides. According to BOCCALETTI and GUAZZONE (1972) and many others, the Italian peninsula is composed of two arcuate structures, one from Liguria to Latium, and the other from Campania to Calabria (Fig. 4).

According to MONGELLI et al. (1975) the distinction between the arcs is well demonstrated by differences in the correlation parameters between topography and Bouguer anomalies in the two different areas. Especially the arcuate structure from Campania to Calabria is geophysically similar to the classic island arcs, such as the Japanese arc (SUGIMURA and UYEDA, 1973). In fact, (1) the existence of a Benioff zone, dipping under the Tyrrhenian Sea (PETERSCHMITT, 1956; RITSEMA, 1969), (2) the crustal structure changing from continental to an oceanic in passing from the Apulian foreland to the Tyrrhenian Sea (MORELLI et al., 1975), and (3) the high along the Calabrian coast and the low in the Bradano Trough in the isostatic anomalies (MONGELLI et al., 1975; LODDO and MONGELLI, 1975b) all favour the hypothesis of plate subduction under the Tyrrhenian marginal sea.

The contour map of the heat flow (Fig. 2) defines an arc of low heat flow, running parallel to the Calabrian arc and surrounding the high in the Tyrrhenian Sea (ERICKSON, 1970), and is also consistent with the subduction hypothesis. Specifically, the region of high heat flow coincides with those of the deep focus earthquakes and the positive isostatic anomaly; the region of low heat flow corresponds with the negative isostatic anomaly.

In contrast in the Ligurian–Tuscan arc, there is little geophysical evidence to support the subduction interpretation of BOCCALETTI and GUAZZONE (1972). The seismic activity is shallow in this area, and it is not possible to define a definite Benioff zone. From gravity data the shift between the topography and Bouguer anomalies observed along profiles crossing the Apennine chain has been interpreted as an evidence of horizontal forces acting from West to East (MONGELLI et al., 1975). Furthermore, the strong lateral heat flow gradient observed along a profile crossing

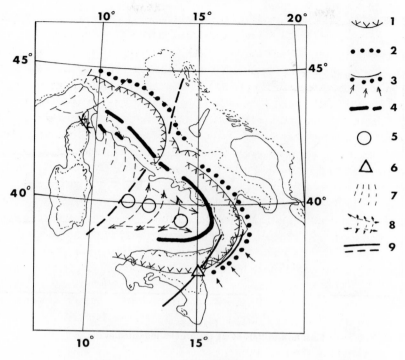

Figure 4

Geodynamic scheme illustrating a model of marginal basin–arc–trench systems for the interpretation of the Apennines arcs: (1) migration and polarity of folded arcs; (2) fore trenches almost filled and inactive; (3) oceanic type trench with active subduction; (4) post Tortonian and recent magmatic arcs; (5) more basic magmatism (plio-quarternary) in the back Calabrian arc marginal basin; (6) volcanism probably related to shear lines; (7) collapsing areas of thin sialic crust (pre-spreading); (8) mature marginal basins (multi-spread); (9) transform and transcurrent faults (according to BOCCALETTI and GUAZZONE, 1972).

the Northern Apennines (BOCCALETTI et al., 1977), is very similar to that observed in similar and more clearly defined conditions across other arcs, and clearly supports the hypothesis that this chain is a continental arc, that is an arc which occurs within a continent.

With regard to Sicily, the seismic activity in its northern part is purely superficial (GIORGETTI and IACCARINO, 1971). The fault plane solution of three earthquakes in this region indicates a thrust mechanism on E–W fault planes (McKENZIE, 1972) and deep seismic sounding profiles show that the crustal thickness increases from the Sicily channel to the northern edge of Sicily, and then sharply decreases when entering the Tyrrhenian Sea (CASSINIS et al., 1969; COLOMBI et al., 1973). BARBERI et al. (1974), on the basis of paleomagnetic and volcanological evidence, maintain that Sicily is a part of the African plate, and since the Mesozoic the African and European plates have converged with the consumption of the lithosphere. The observed low heat flow in Sicily correspond to the Sicily Trough.

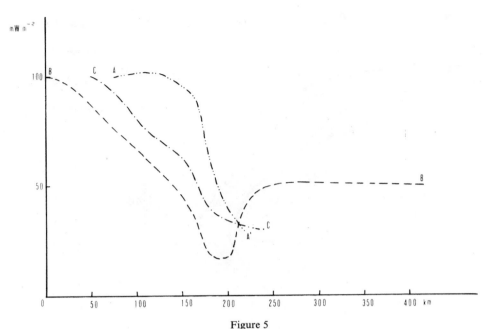

Figure 5
Heat flow in Italy along the profiles indicated in Fig. 2.

Figure 5 presents the heat flow profiles across the three main geotectonic structures discussed above. The existing geothermal fields of Italy, and the hope of increasing the geothermal power potential is closely related to the general heat flow/tectonic framework (MONGELLI and LODDO, 1975).

REFERENCES

BARBERI, F., CIVETTA, L., GASPARINI, P., INNOCENTI, F., SCANDONE, R. and VILLARI, L. (1974), *Evolution of a section of the Africa–Europe plate boundary: paleomagnetic and volcanological evidence from Sicily*, Earth Planet. Sci. Lett. 22, 123–132.

BECK, A. E., JAEGER, J. C. and NEWSTEAD, G. N. (1956), *The measurement of the thermal conductivities of rocks by observations in boreholes*, Austr. J. Phys. 9, 286–296.

BIRCH, F. S. and HALUNEN, A. J. (1966), *Heat flow measurements in the Atlantic Ocean, Indian Ocean, Mediterranean Sea, and Red Sea*, J. Geophys. Res. 71, 583–586.

BOCCALETTI, M. and GUAZZONE, G. (1972), *Gli archi appenninici, il Mar Ligure ed il Tirreno nel quadro della tettonica dei Bacini marginali retro-arco*, Mem. Soc. Geol. It. 11, 201–216.

BOCCALETTI, M., FAZZUOLI, M., LODDO, M. and MONGELLI, F. (1977), *Heat flow measurements on the Northern Apennines arc*. Tectonophysics 41, 101–112.

BOLDIZSAR, T. (1963), *Terrestrial heat flow in the natural steam field at Larderello*, Geof. Pura Appl. 56, 115–122.

CALAMAI, A., CATALDI, R., LOCARDI, E. and PRATURLON, A. (1977), *Distribuzione delle anomalie geotermiche nella fascia preappenninica Tosco-Laziale (Italia)*, ENEL, Roma, Relazione di studio e ricerca n. 321, 1–50.

CASSINIS, R., FINETTI, I., GIESE, P., MORELLI, C., STEINMETZ, L. and VECCHIA, O. (1969), *Deep seismic refraction research on Sicily*, Boll. Geof. Teor. Appl. 10, 140–160.

COLOMBI, B., GIESE, P., LUONGO, G., MORELLI, C., RIUSCETTI, M., SCARASCIA, S., SCHÜTTE, K. G., STROWALD, J. and de VISINTINI, G. (1973), *Preliminary report on the seismic refraction profile Gargano–Salerno–Palermo–Pantelleria (1971)*, Boll. Geof. Teor. Appl. *15*, 225–252.

ERICKSON, J. A. (1970), 'The measurement and interpretation of heat flow in the Mediterranean and Black Sea', Ph.D. Thesis Massachusetts Institute of Technology, Cambridge, Massachusetts, 272 pp.

FANELLI, M., LODDO, M., MONGELLI, F. and SQUARCI, P. (1974), *Terrestrial heat flow measurements near Rosignano Solvay (Tuscany), Italy*, Geothermics, *3*, 65–73.

FINCKH, P. (1976), *Waermeflussmessungen in Randalpenseen*, Diss. ETH Zurich, Nr. 5787.

GIORGETTI, F. and IACCARINO, E. (1971), *Seismicity of the Italian region*, Boll. Geof. Teor. Appl. *13*, 143–154.

HAENEL, R. (1974), *Heat flow measurements in Northern Italy and Heat flow maps of Europe*, Z. Geophys. *40*, 367–380.

LAVENIA, A. (1967), *Heat flow measurements through bottom sediments in the Southern Adriatic Sea*, Boll. Geof. Teor. Appl. *9*, 323–332.

LODDO, M. (1970), *Temperature in an insulated slab heated by a plane source. Application to thermal conductivity measurements of rocks*, Geothermics–Special Issue *2*, 437–442.

LODDO, M. and MONGELLI, F. (1975a), *Heat flow in Southern Italy and surrounding seas*, Boll. Geof. Teor. Appl. *16*, 115–122.

LODDO, M. and MONGELLI, F. (1975b), *Gravity of sinking lithosphere in Mediterranean*. Paper read at the 4th Congress of A.G.I., Rome.

LODDO, M., MONGELLI, F. and RODA, C. (1973), *Heat flow in Calabria, Italy*, Nature Phys. Sci. *244*, 91–92.

LORT, J. M. (1971), *The tectonics of the Eastern Mediterranean: a geophysical review*, Rev. Geophys. *9*, 189–216.

MCKENZIE, D. P. (1972), *Active tectonics of the Mediterranean region*, Geophys. J.R. Astron. soc. *30*, 109–185.

MONGELLI, F. (1968), *Un metodo per la determinazione in laboratorio della conducibilità termica delle rocce*, Boll. Geof. Teor. Appl. *10*, 51–58.

MONGELLI, F. (1969), *Thermal conductivity measurements of some Apulian limestones by the 'cut core' method*, Atti XVIII Convegno annuale 'Associazione Geof. Italiana', Napoli, Italy, pp. 137–154.

MONGELLI, F. and LODDO, M. (1974), *The present state of geothermal investigations in Italy*, Acta Geodaet. Geophys. et Montanist. *9*, 449–456.

MONGELLI, F. and LODDO, M., *Regional heat flow and geothermal fields in Italy*, in *Proceedings Second United Nations Symposium on the Development and Use of Geothermal Resources*, (San Francisco, 1975), vol. 1, pp. 495–498.

MONGELLI, F. and MORELLI, C. (1964), *Studio geotermico preliminare dell' Etna*, Rivista Mineraria Siciliana n. 85–87.

MONGELLI, F. and RICCHETTI, G. (1970a), *Heat flow along the Candelaro Fault–Gargano Headland (Italy)*, Geothermics–Special Issue *2*, 450–458.

MONGELLI, F. and RICCHETTI, G. (1970b), *The earth's crust and heat flow in the Fossa Bradanica, Southern Italy*, Tectonophysics, *10*, 103–125.

MONGELLI, F., LODDO, M. and CALCAGNILE, G. (1975), *Some observations on the Apennines gravity field*, Earth Planet. Sci. Lett. *24*, 385–393.

MORELLI, C., GIESE, P., CASSINIS, R., COLOMBI, B., GUERRA, I., LUONGO, G., SCARASCIA, S. and SCHÜTTE, K. G. (1975), *Crustal structure of Southern Italy. A Seismic refraction profile between Puglia–Calabria–Sicily*, Boll. Geof. Teor. Appl. *18*, 183–210.

PETERSCHMITT, E. (1956), *Quelques données nouvelles sur les seismes profonds de la mer Thyrrhenienne*, Annali di Geofisica *9*, 305–334.

RITSEMA, A. R. (1969), *Seismic data of the Western Mediterranean and the problem of oceanization*, Verhandl. K. Nederl. Geol. Mijnbouwk. Genootsch. *26*, 105–120.

SUGIMURA, A. and UYEDA, S., *Island Arcs Japan and Its Environs*, (Elsevier Scientific Pub. Co., Amsterdam, 1973), 247 pp.

VON HERZEN, R. and MAXWELL, A. E. (1959), *The measurement of thermal conductivity of deep sea sediments by a needle-probe method*, J. Geophys. Res. *64*, 1557–1563.

(Received 3rd October 1977, revised 1st March 1978)

Pageoph, Vol. 117 (1978/79), Birkhäuser Verlag, Basel

Review of Heat Flow Data from the Eastern Mediterranean Region[1])

By Yoram Eckstein[2])

Abstract – Heat flow data from the eastern Mediterranean region indicates an extensive area of low heat flow, spreading over the whole basin of the Mediterranean east of Crete (Levantine Sea), Cyprus, and northern Egypt. The average of the marine heat flow measurements in the Levantine Sea is 25.7 ± 8.4 mW/m^2, and the heat flow on Cyprus is 28.0 ± 8.0 mW/m^2. The estimated values of heat flow in northern Egypt range from 38.3 ± 7.0 to 49.9 ± 9.3 mW/m^2, apparently with no consistent trend. To the east, on the coast of Israel, the heat flow values increase, ranging from 36.6 ± 22.4 to 56.7 ± 14.2 mW/m^2 along a SSE trend. The trend apparently correlates with an increase in crustal thickness, which is about 23 km at the north-west base of the Nile-Delta-cone, and close to 40 km beneath Israel.

Key words: Levantine Sea; Cyprus; Crete; Egypt; Israel; Gulf of Suez; Crustal thickening; Oceanic and continental lithosphere.

Introduction

During the past few years there was, and still continues, a quite live discussion on the origin of the present crustal and tectonic pattern of the Middle East in general, and of such elements like the Eastern Mediterranean, Jordan – Dead Sea and Red Sea Rift System in particular. Theories proposed are varied, but all are ultimately dependent upon terrestrial heat to provide energy for uplift, subsidence, rifting and other types of plate motions. Knowledge of terrestrial heat flow and hence of the temperature distribution in the subsurface of the region provides an important constraint on the proposed models for the origin of the present crustal structure. Measurements of terrestrial heat flow in the Middle East are neither abundant nor uniformly distributed. There were no marine heat flow measurements available from the eastern Mediterranean before ERICKSON (1970). The first heat flow measurements on land from the region were made in Cyprus (MORGAN, 1973). ROSENTHAL and ECKSTEIN (1968) reported first measurements of the geothermal gradients in Israel, and ECKSTEIN and SIMMONS (1975; 1978) published detailed data and regional interpretation of 68 heat flow measurements in boreholes in Israel. MORGAN *et al.* (1976) reported preliminary heat flow estimates, based on bottom hole temperatures recorded during various oil well logging procedures, and mean surface temperatures

[1]) Contribution No. 157, Department of Geology, Kent State University, Kent, Ohio, USA.
[2]) Division of Hydrogeology, Geological Survey of Israel and Department of Geology, Kent State University, Kent, Ohio, 44242, USA.

obtained from meteorologic data, combined with an assumed mean thermal conductivity for rocks in northern Egypt. No data on heat flow are available from Turkey as well, although extensive geothermal exploration has provided a number of deep borehole temperature observations (KURTMAN and SAMILGIL, 1976; ALPAN, 1977; TEZCAN, 1976; TAN, 1976; ESDER and SIMSEK, 1976). CERMAK and HURTIG (1976) incorporated in their maps the heat-flow pattern proposed by TEZCAN (1977), based on estimates of thermal conductivity of the rock formations and measurements of the bottom hole temperature taken in 22 areas.

Heat flow data

The marine heat flow data presented by ERICKSON (1970) are probably the most evenly and systematically distributed in the region (Table 1a). The locations are spaced rather uniformly throughout the eastern Mediterranean and include all of the major topographic provinces. The thermal gradient data were obtained using an Ewing thermograd described by LANGSETH (1965). The penetrations achieved with the piston corer varied between 2.5 and 13.5 meters. Thermal conductivity was measured on the sediments obtained at the same location as the thermal gradient measurement using the needle-probe technique (Von HERZEN and MAXWELL, 1959). The observed data were corrected for an estimated rate of sedimentation of 4.3 cm/1000 years.

The data on heat flow on Cyprus (MORGAN, 1973 and 1975) are based on in-hole measurements of the temperature gradient, carried out with a thermistor probe in conjunction with Wheatstone bridge. Thermal conductivity was obtained from measurements carried out on core- and rock-fragment specimens. The observed results were corrected for steady-state influence of topography.

Table 1

Heat flow data from the Eastern Mediterranean Region.
(a) Marine data (from ERICKSON, 1970)

Lat. N	Long. E	Heat flow (mW/m²)	Lat. N	Long. E	Heat flow (mW/m²)
34° 15′	25° 01′	25.7 ± 9.8	32° 58′	32° 24′	43.8 ± 6.3
34° 17′	26° 13′	24.8 ± 12.3	33° 11′	32° 29′	15.8 ± 5.6
33° 46′	27° 54′	18.7 ± 3.3	33° 50′	32° 16′	25.0 ± 5.4
34° 19′	27° 11′	10.3 ± 2.1	35° 41′	32° 58′	25.6 ± 4.8
34° 12′	28° 58′	26.6 ± 7.5	33° 52′	33° 17′	34.5 ± 9.3
35° 31′	28° 12′	30.2 ± 6.1	34° 11′	33° 37′	10.7 ± 2.2
33° 44′	30° 07′	26.8 ± 3.3	34° 19′	34° 42′	35.7 ± 18.6
35° 20′	30° 07′	38.1 ± 8.3	34° 53′	34° 53′	27.9 ± 3.0
33° 00′	31° 06′	20.6 ± 7.7	35° 56′	34° 14′	23.0 ± 16.0
34° 49′	31° 04′	24.8 ± 7.6	35° 32′	35° 26′	20.4 ± 5.3
36° 01′	31° 46′	32.1 ± 8.2			

Table 1 (*contd.*)
(b) Data from Israel (from ECKSTEIN and SIMMONS, 1978)

Lat. N	Long. E	Heat flow (mW/m²)	Group*)	n†)	$\bar{x} \pm s$‡) (mW/m²)
29° 31'	34° 56'	93.3	a		
35'	58'	56.5	a		
38'	58'	41.4	b		
38'	59'	31.8	b		
45'	59'	45.2	b	5	53.7 ± 23.9
29° 47'	35° 01'	47.3	b		
53'	03'	45.6	b		
30° 00'	05'	62.0	a	3	51.6 ± 9.0
30° 07'	34° 52'	51.5	a		
36'	53'	37.7	c		
48'	46'	32.7	c	2	35.2 ± 3.5
30° 00'	35° 05'	62.0	a		
21'	10'	66.1	a		
21'	10'	77.5	a		
39'	14'	45.2	b		
48'	17'	67.4	a		
52'	19'	40.2	b		
57'	01'	63.2	a		
58'	12'	54.4	a		
58'	11'	34.3	b		
59'	13'	463.4§)	a	9	56.7 ± 14.2
31° 19'	34° 25'	47.3	c		
26'	30'	64.0	c		
10'	33'	60.7	c		
23'	34'	36.1	c		
23'	50'	57.3	c		
29'	32'	84.6	c		
29'	36'	44.8	c		
31'	52'	18.4	b		
33'	44'	30.6	c		
38'	34'	44.4	c		
39'	41'	83.7	c		
42'	36'	36.0	c		
42'	41'	33.5	c		
44'	46'	45.2	c		
55'	58'	38.5	c		
59'	53'	79.5	c	16	50.3 ± 19.7
31° 06'	35° 02'	37.7	c		
15'	01'	42.7	b		
16'	16'	73.2	c		
19'	12'	68.2	c		
21'	21'	92.5	a		
32'	07'	66.6	c		
44'	20'	24.3	b		
46'	09'	56.9	c		
46'	27'	73.3	a		
48'	26'	59.4	a		
49'	02'	12.6	b		

Table 1 (*contd.*)

(b) Data from Israel (from ECKSTEIN and SIMMONS, 1978)

Lat. N	Long. E	Heat flow (mW/m^2)	Group*)	n†)	$\bar{x} \pm s$‡) (mW/m^2)
31° 52'	15'	23.0	b		
52'	27'	23.9	b	13	50.3 ± 24.7
32° 02'	35° 26'	45.2	b		
06'	16'	7.1	b		
37'	13'	31.4	b		
37'	15'	46.5	b		
37'	15'	45.6	b		
41'	20'	87.1	a		
45'	04'	13.0	b		
46'	07'	23.4	b		
54'	24'	46.5	b	9	38.4 ± 23.7
32° 58'	35° 35'	28.0	b		
59'	34'	37.3	b		
33° 00'	34'	22.2	b	3	29.1 ± 7.6
33° 04'	35° 13'	70.7	a		

*) Division corresponding with the three groups of data presented by ECKSTEIN and SIMMONS (1978).

†) Number of averaged values.

‡) Average heat flow and standard deviation.

§) Values not included in averaging.

The data on heat flow in Israel (ECKSTEIN and SIMMONS, 1975, 1978) are based on an extensive study, and measurements of temperature gradient carried out in sixty-eight boreholes (Table 1b). The measurements were made with thermistor probe, and were corrected for steady-state effects of topography. Thermal conductivity of the rock formations was obtained from an extensive network of measurements, using the divided bar technique (BIRCH, 1950; MISENER and BECK, 1960; BECK, 1965) and the needle-probe technique (VON HERZEN and MAXWELL, 1959), applied to core samples and unconsolidated sediments. ECKSTEIN and SIMMONS (1978) divide their data into three groups, distinguishing heat flow values obtained from:

(a) boreholes located along deep fault structures associated directly or indirectly with the Jordan–Dead Sea–Gulf of Elat trough, and characterized by thermal influence of convecting hot ground water;

(b) boreholes located in the inner parts of the Jordan–Dead Sea–Gulf of Elat trough and other regions affected by movement of cool groundwater in the recharge areas;

(c) boreholes located in sites unaffected by hydrologic processes.

The values presented in the third group are considered by the authors as representative of the regional heat flow by conduction. The average regional heat flow based on this definition is 52.4 ± 17.2 mW/m^2. ECKSTEIN and SIMMONS (1978) point however, to an interesting distribution of their heat flow values in the third group.

Table 1 (*contd.*)

(c) Data from Egypt (based on the data presented by MORGAN *et al.*, 1976)

Lat. N	Long. E	n	Heat flow mW/m^2
31°	25°		
31°	26°	7	45.8 $\pm$ 11.3
30°	25°		
30°	26°	8	49.9 $\pm$ 9.3
29°	25°		
29°	26°	10	44.7 $\pm$ 11.4
28°	25°		
28°	26°	3	40.2 $\pm$ 8.4
31°	27°		
31°	28°	6	41.5 $\pm$ 12.3
30°	27°	9	49.2 $\pm$ 11.2
29°	27°	4	44.7 $\pm$ 11.9
30°	28°	13	45.6 $\pm$ 10.3
29°	28°	10	46.2 $\pm$ 10.7
30°	29°	9	44.3 $\pm$ 9.5
29°	29°	11	43.0 $\pm$ 10.6
28°	29°		
28°	30°	5	42.6 $\pm$ 10.1
31°	30°	12	40.6 $\pm$ 9.9
30°	30°	5	38.3 $\pm$ 8.9
29°	30°	6	47.3 $\pm$ 8.9
31°	31°	8	39.8 $\pm$ 10.7
30°	31°	2	39.3 $\pm$ 7.0

(d) Data from Cyprus (after MORGAN, 1975)

n	Heat flow (mW/m^2)	
	Range	$\bar{x} \pm s$
18	5.4 $-$ 46.0	28.0 $\pm$ 8.0

Out of the 23 values of the heat flow in this group, 13 locations range from 30.6 to 47.3 mW/m^2, with an average of 39.3 $\pm$ 5.9 mW/m^2, whereas the remaining 10 locations range from 56.9 to 84.6 mW/m^2, with an average of 69.5 $\pm$ 10.5 mW/m^2. They imply that the 10 values of higher heat flow indicate locations with deep thermal anomalies superimposed over the regional, comparatively low heat flow of about 40 mW/m^2 suggested by the 13 measurements.

The data from Egypt (Table 1c, MORGAN *et al.*, 1976) are based on geothermal gradients calculated from bottom hole temperatures from over 150 oil wells in northern Egypt and the Gulf of Suez. The temperature readings were taken with maximum recording thermometer during various logging proceedings at an interval, or at the end of drilling. The gradients were calculated by subtracting the mean surface temperature from the bottom hole temperature, and dividing the result by the depth of the borehole. Thermal conductivity values were estimated from various lithological drilling records.

Discussion

The most striking feature on the heat flow map of the Eastern Mediterranean Region is an overall low heat flow area spreading all over the sea east of Crete and apparently encroaching onto the north-east Africa and the coastline of Israel (Fig. 1).

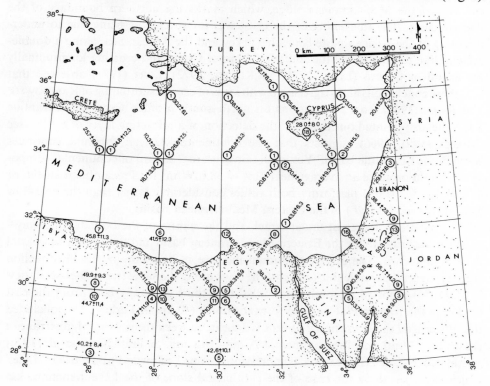

Figure 1

Heat flow map of the Eastern Mediterranean Region. Heat flow values averaged in mW/m² for 1° × 1° or 1° × 2° areas. Numbers in circles represent the number of averaged values for respective area. Sources: marine data – ERICKSON, 1970; Cyprus – MORGAN, 1975; Egypt – values calculated for data presented by MORGAN *et al.*, 1976; Israel – ECKSTEIN and SIMMONS, 1978.

The range of 10.3–43.8 mW/m² with an average of 25.7 ± 8.4 mW/m² reported from the marine heat flow measurements east of Crete by ERICKSON (1970) is surprisingly low if oceanic crust and lithosphere occur beneath the area. ERICKSON et al. (1977) suggests that the low heat flow is compatible with the assumption that the eastern Mediterranean is a relict piece of the oceanic crust of the former Tethys. Although the value of 25.7 mW/m² is lower than the commonly accepted equilibrium heat flow (46 to 50 mW/m²; SCLATER and FRANCHETAU, 1970) through the oldest ocean floors, the difference is attributed to the extensive accumulation of sediments in the Eastern Mediterranean. That argumentation, however, would be invalid on the northern coasts of Egypt, or on Cyprus and the eastern border of the Mediterranean, which emerged from the Tethys in the late Eocene. In fact the heat flow range of 38.3 to 49.9 mW/m² reported from northern Egypt (MORGAN et al., 1976) is much the range one would expect from a foreland of a Precambrian shield. WOODSIDE and BOWIN (1970) reported a gravity profile along 31°E, postulating that the depth of the Moho increases from 23 km off the Egyptian coast in the south to 34 km beneath the axis of the Mediterranean ridge, which marks the northern boundary of the African plate. The thickening of the crust was interpreted as being due to under-thrusting of the African plate beneath Turkey and the Aegean Sea, along a double-arcuate line passing along the southern shore of Crete, and crossing longitudinally the island of Cyprus (LORT, 1971; MCKENZIE, 1972). LORT (1971) observed that '. . . these conclusions are in agreement with ideas of crustal shortening of RABINOWITZ and RYAN (1969) from calculations on north–south profiles across the Levantine Sea.' She also pointed out that '. . . the detection of a refracting layer of 6.1 km/sec (MOSKALENKO, 1966) suggests that the continental African foreland continues beneath the Levantine basin.' Yet, such thin crust (ca. 25 km) of continental composition would produce an average heat flow of 38 mW/m² for Precambrian shield, or 43 mW/m² for stable platform – both values considerably larger than the heat flow range observed (op. cit.) in the Eastern Mediterranean basin.

The nature of the crust under Israel, and the nature of the boundary – if any – between the crust under the Eastern Mediterranean basin and the Levantine coastal zone is unknown. The apparent gradual increase in heat flow in the SE direction (Fig. 1) suggests possible thickening of the crust along the same trend. BEN MENAHEM et al. (1976) reported a 'continental crust' that thins gradually both from the east (50 km in the Iranian Plateau) and from the north (40 km in Turkey). He concluded that the crustal thickness in the Levant is a mean between the eastern Mediterranean crust (ca. 23 km) and the continental Asian crust (ca. 50 km). NEEV et al. (1976) postulated '. . . a NE trending compressional zone,' named 'The Pelusium Line,' which stretches along the base of the continental slope of the Levant some 60 km off the coast of Israel.' The zone is, according to NEEV et al. (1976), 'of regional significance, since it forms a sharp divide between,' among other elements, different geophysical properties, i.e., the dominance of positive Bouguer anomalies in the eastern Mediterranean as compared with the onshore, which may indicate an oceanic type of crust west of the Pelusium Line and continental type east of it." NEEV *et al.* (1976) also refers to GERGAWI and KHASHAB (1968), who suggest that the crust under the Nile Delta is 24 km thinner than the crust south of it at Helwan, SE of the Pelusium Line. A continental crust of the thickness postulated east and south-east of the Pelusium Line, by GERGAWI and KHASHAB (1968) in north-east Egypt, and by BEN MENAHEM *et al.* (1976) in Israel, would produce an average heat flow of 40 mW/m^2 for Precambrian shield, or 50 mW/m^2 for stable platform. The values are in good agreement with the heat flow estimates from northern Egypt (MORGAN *et al.*, 1976) and with the average regional heat flow of 52.4 $\pm$ 17.2 mW/m^2 reported from Israel (ECKSTEIN and SIMMONS, 1978).

MORGAN (1975) and MORGAN *et al.* (1976) suggested that the lower than the world average heat flow (61.5 mW/m^2; LEE, 1970) focusing in the Levantine Sea is resulting from a stable mantle heat flow contribution, added to a very low heat production within the upper crust. This model seems to be partly supported by the data on heat production of only 0.67 $\pm$ 0.49 μW/m^3 in the granites of the Precambrian basement of north-eastern Sudan (EVANS and TAMMEMAGI, 1974), suggesting a substantial regional depletion in the radiogenic heat producing elements within the upper crust. Yet, while that model might explain generally low heat flow in Egypt and in Israel, the marine heat flow data from the eastern Mediterranean and the heat flow on Cyprus are bafflingly equal, or even lower than the commonly accepted heat flow at the base of either oceanic or continental lithosphere (25–28 mW/m^2; SCLATER and FRANCHETAU, 1970). Therefore low rate of heat generation within the upper crust of the eastern Mediterranean alone, cannot explain the anomaly. An alternative hypothetical model was suggested by ERICKSON *et al.* (1977). Assuming that thermal and compositional differences in the upper mantle beneath oceans and continents may extend to the depth of more than 400 km (JORDAN, 1975; SIPKIN and JORDAN, 1975) they postulate that '. . . if the temperature and rate of radiogenic heat production in the asthenosphere beneath the continents are lower than beneath the oceanic lithosphere, then one of the possible effects of plate movements, and in particular of continent–continent collisions, might be the displacement of oceanic lithosphere onto or over continental asthenosphere (or vice versa).' As a consequence, cool and poorly radiogenic 'continental' asthenosphere from beneath Africa would flow northward under north Egypt and the eastern Mediterranean, displacing warmer oceanic asthenosphere, which was originally beneath the Tethys, and is presently flowing northwards and downwards beneath Turkey and the Aegean Sea, producing there high heat flow (TEZCAN, 1977; CERMAK and HURTIG, 1977) and triggering recent volcanic activity. The relationship between this model and the apparent thickening of the crust on the eastern and the south-eastern board of the Mediterranean, and thus higher heat flow in Israel and in Egypt, is yet to be studied.

References

ALPAN, A., *Geothermal energy exploration in Turkey*, in *Proc. 2nd U.N. Symp. on the Dev. and Use of Geoth. Res.* (San Francisco, California, 1976), vol. 1, pp. 25–28.

BECK, A. E. (1965), *Techniques of measuring heat flow on land*, in *Terrestrial Heat Flow*, (ed. by W. H. K. Lee), Geoph. Monogr. 8, (A.G.U., 1965), pp. 24–50.

BEN MENAHEM, A., NUR, A. and VERED, M. (1976), *Tectonics, seismicity and structure of the Afro-Eurasian junction – the breaking of an incoherent plate*, in *Physics of the Earth and Planet Interiors, 12,* 1–50.

BIRCH, F. (1950), *Flow of heat in the front range, Colorado*, Bull. Geol. Soc. Am. *61,* 567–630.

CERMAK, V. and HURTIG, E. (1977), *Preliminary Heat Flow Map of Europe 1:5000000*, (IASPEI-Heat Flow Commission, Potsdam, GDR., 1977)

ECKSTEIN, Y. and SIMMONS, G. (1975), *Terrestrial Heat Flow in Israel*, in *Geol. Soc. Am. Ann. Meeting*, Abstr. vol. 7, no. 7, p. 1064.

ECKSTEIN, Y. and SIMMONS, G. (1978). *Measurements and interpretation of terrestrial heat flow in Israel*, manuscript submitted for publication by *Geothermics*.

ERICKSON, A. J., 'The Measurements and Interpretation of Heat flow in the Mediterranean and Black Sea,' Ph.D. Thesis (M.I.T., Cambridge, Massachusetts, 1970).

ERICKSON, A. J., SIMMONS, M. G. and RYAN, W. B. F., *Review of Heat Flow from the Mediterranean and Aegean Seas*, in *Int. Symp. on the Struc. Hist. of the Medit. Basin, Split, Yugoslavia, Oct. 1976*, Proc., pp. 263–280 (ed. by B. Biju-Duval and L. Montadert), (Editions Technip, Paris, 1977)

ESDER, J. and SIMSEK, S., *Geology of Izmir-Seferihisar Area, Western Anatolia of Turkey; Determination of reservoirs by means of gradient drilling*, in *Proc. 2nd U.N. Symp. on the Dev. and Use of Geoth. Res.* (San Francisco, California, 1976), vol. 1, pp. 349–361.

EVANS, T. R. and TAMMEMAGI, H. Y. (1974), *Heat flow and heat production in Northeast Africa*, Earth and Planet. Sci. Letters *23,* 349–356.

GERGAWI, A. and EL-KHASHAB, H. M. A. (1968), *Seismicity of Egypt*, Helwan Obs. Bull. No. 76,

JORDAN, T. H. (1975), *The Continental Tectonosphere*, Rev. Geoph. and Space Physics. *13* (13), 1–12.

KURTMAN, F. and SAMILGIL, E., *Geothermal energy possibilities, their exploration and evaluation in Turkey*, in *Proc. 2nd U.N. Symp. on the Dev. and Use of Geoth. Res.* (San Francisco, California, 1976), vol. 1, pp. 447–458.

LANGSETH, M. G., *Techniques of measuring heat flow through the ocean floor*, in *Terrestrial Heat Flow*, Geoph. Monogr. 8 (ed. by W. H. K. Lee), (A.G.U., 1965), pp. 58–77.

LEE, W. H. K. (1970), *On the global variations of terrestrial heat flow*, Phys. Earth Planet. Int. *2,* 332–341.

LORT, J. M. (1971), *The tectonics of the Eastern Mediterranean: A geophysical review*, Reviews of Geoph. and Space Physics, *9* (2), 189–216.

McKENZIE, D. (1972), *Active tectonics of the Mediterranean region*, Geoph. J. Roy. Astr. Soc. *30,* 109–185.

MISENER, A. D. and BECK, A. E., *The measurements of heat flow over land*, in *Methods and Techniques in Geophysics* (S. K. Runcorn, ed.), (Interscience Publ., New York, 1960), pp. 10–61.

MORGAN, P. (1973), 'Terrestrial Heat Flow Studies in Cyprus and Kenya,' Ph.D. Thesis, University of London.

MORGAN, P., *Cyprus heat flow with comments on the thermal regime of the Eastern Mediterranean*, Abstr. Gen. Assembly IUGG, (Grenoble, France, 1975), pp. 23–24.

MORGAN, P., BLACKWELL, D. D., FARRIS, J. C., BOULOS, F. K. and SALIB, P. G. (1976), *Preliminary geothermal gradient and heat flow values for Northern Egypt and the Gulf of Suez from oil well data*, in *Proc. IAHS Congr. on Thermal Waters, Geoth. Energy and Vulcanism of the Mediterr. Area.* (Athens, Greece, 1976), vol. 1, pp. 424–438.

MOSKALENKO, V. N. (1966), *New data on the structure of the sedimentary strata and basement in the Levant Sea*, Oceanology *6,* 828–836.

NEEV, D., ALMAGOR, G., ARAD, A., GINZBURG, A. and HALL, J. K. (1976), *The geology of the Southeastern Mediterranean*, Bull. Geol. Survey of Israel, *68,* 1–51.

RABINOWITZ, P. D. and RYAN, W. B. F. (1969), *Gravity anomalies in the Eastern Mediterranean*, Eos, Trans. A.G.U. *50,* 208.

ROSENTHAL, E. and ECKSTEIN, Y. (1968), *Temperature gradients in the subsurface of the Dead Sea area, Israel*, Israel J. Earth Sci. *17,* 131–136.

SCLATER, J. G. and FRANCHETEAU, Y. (1970), *The implications of terrestrial heat flow observations on current tectonic and geochemical models of the crust and upper mantle of the earth*, Geoph. J. Roy. Astr. Soc. *20*, 509–542.

SIPKIN, S. A. and JORDAN, T. H. (1975), *Lateral homogeneity of the upper mantle determined from the travel times of ScS*, J. Geoph. Res. *80* (11), 1474–1484.

TAN, E., *Geothermal drilling and well testing in the Afyon Area, Turkey*, in *Proc. 2nd U.N. Symp. on the Dev. and Use of Geoth. Res.* (San Francisco, California, 1976), vol. 2, pp. 1523–1526.

TEZCAN, A. K., *Geophysical studies in Sarayköy-Kizyldere geothermal field, Turkey*, in *Proc. 2nd U.N. Symp. on the Dev. and Use of Geoth. Res.* (San Francisco, California, 1976), vol. 2, pp. 1231–1240.

TEZCAN, A. K. (1977), *The geothermal studies, their present status and contribution to the heat flow contouring in Turkey*, Manuscript quoted in Cermak and Hurtig, 1977.

VON HERZEN, R. P. and MAXWELL, A. E. (1959), *The measurements of thermal conductivity of deep-sea sediments by a needle-probe method*, J. Geoph. Res. *64* (10), 1557–1563.

WOODSIDE, J. and BOWIN, C. (1970), *Gravity anomalies and inferred crustal structure in the Eastern Mediterranean Sea*, Bull. Geol. Soc. Am. *81* (4), 1107–1122.

(Received 21st December 1977, revised 14th March 1978)

Pageoph, Vol. 177 (1978/79), Birkhäuser Verlag, Basel

Geothermal Resource Assessment of the United States

By L. J. Patrick Muffler and Robert L. Christiansen[1])

Abstract – Geothermal resource assessment is the broadly based appraisal of the quantities of heat that might be extracted from the earth and used economically at some reasonable future time. In the United States, the Geological Survey is responsible for preparing geothermal assessments based on the best available data and interpretations. Updates are required every few years owing to increasing knowledge, enlarging data base, improving technology, and changing economics. Because geothermal understanding is incomplete and rapidly evolving, the USGS complements its assessments with a broad program of geothermal research that includes (1) study of geothermal processes on crustal and local scales, (2) regional evaluations, (3) intensive study of type systems before and during exploitation, (4) improvement of exploration techniques, and (5) investigation of geoenvironmental constraints.

Key words: Resource base; Hot igneous systems; Hydrothermal convection systems; Geopressurized systems.

1. Introduction

Geothermal resource assessment can be defined as the broadly based estimation of supplies of geothermal heat that might become available for use, given reasonable assumptions about technology, economics, governmental policy, and environmental constraints. This assessment implies not merely the determination of how heat is distributed in the upper part of the earth's crust but also the evaluation of how much heat could be extracted for man's use. Heat in place in the earth's crust (above a reference temperature) is the *geothermal resource base*; that fraction of the resource base that could be extracted economically and legally at some reasonable future time is the geothermal *resource* (Muffler, 1973; White and Williams, 1975; Muffler and Cataldi, 1978).

2. Periodic Resource Assessments

In the United States, the U.S. Geological Survey (USGS) is the government agency responsible for assessing mineral and energy resources, including geothermal energy. The goal of the Survey's geothermal assessment is to provide a knowledge

[1]) U.S. Geological Survey, 345 Middlefield Rd., MS 18, Menlo Park, CA 94025 USA

of the Nation's geothermal resource in sufficient breadth and detail to allow optimum energy planning, to encourage systematic exploration, and to support appropriate development of geothermal resources by private industry.

The first systematic effort to estimate the geothermal resources of the entire United States was carried out by the USGS in the first 6 months of 1975, and was published as USGS Circular 726 (WHITE and WILLIAMS, 1975). This study evaluated the geothermal resources in several categories: (a) regional conductive environments (DIMENT et al., 1975), (b) igneous-related geothermal systems (SMITH and SHAW, 1975), (c) hydrothermal convection systems (RENNER et al., 1975), and (d) geopressured systems (PAPADOPULOS et al., 1975). In each category, the USGS study used a two-step procedure: (1) estimation of the geothermal resource base to a specified depth, and (2) evaluation of the part of this resource base that might be recovered under reasonable technological and economic assumptions (NATHENSON and MUFFLER, 1975; PAPADOPULOS et al., 1975; PECK, 1975).

In estimating the resource base in *regional conductive environments*, Circular 726 divided the United States into large heat-flow provinces. Distributions of temperature with depth were then calculated, using a model for the variation of heat-producing elements with depth. This allowed the calculation of the thermal energy in rocks above a 15°C reference temperature to a depth of 10 km. This huge amount of heat (Table 1) provides an upper limit for the geothermal resource base of the United States, but the technology for its use remains to be demonstrated.

Superimposed on the regional conductive environments are extensive areas, particularly in the western United States, of *young igneous systems* that are attractive exploration targets. The resource base associated with these targets was evaluated in Circular 726 by considering the size and age of the igneous systems and estimating the amount of heat still remaining in the intrusions and adjacent country rock (Table 1). Heat loss from the igneous intrusions was assumed to be entirely conductive; cooling by hydrothermal convection was assumed to be offset by magmatic pre-heating and additions of magma after the assumed time of emplacement.

Circular 726 estimated the thermal energy contained in the reservoirs of identified *hydrothermal convection systems* to a depth of 3 km. The areas and the depths to the tops and bottoms of reservoirs were estimated from available drillhole, geologic, and geophysical data; nominal values were assumed where data were lacking. Reservoir temperatures were determined from drillhole data where possible, but generally had to be estimated by applying the SiO_2 and Na—K—Ca chemical geothermometers to thermal waters from the hot springs or wells. This procedure gave the identified resource base for hydrothermal convection systems above a 15°C reference temperature; the undiscovered resource base was estimated to be 3 or 5 times the identified resource base. Finally, a recovery factor of 10 to 25 percent was used to calculate the geothermal resource recoverable from hydrothermal convection systems.

For *geopressured systems*, time and manpower constraints in 1975 allowed

Table 1

Summary of geothermal resource estimates reported in *USGS Circular 726* (WHITE and WILLIAMS, 1975).

	Resource base (10^{18} joules)	Recovery factor for heat	Resource (heat recoverable, 10^{18} joules)	Electricity[m] conversion factor	Electricity ($MW \cdot c$)	Utilization efficiency	Beneficial heat (10^{18} joules)
Conduction-dominated environments	33 000 000[a]	0	0	—	—	—	—
Hot igneous systems							
Evaluated	100 000	0	0	—	—	—	—
Unevaluated	310 000	0	0	—	—	—	—
Hydrothermal convection systems							
Vapor-dominated							
Identified	110[b]	0.1	8.0[c]	0.2	494	—	—
Undiscovered	100	0.1	10.0	0.2	625	—	—
Hot-water (>150°C)							
Identified	1550[a, d]	0.25	250[c]	0.08–0.12	7506	—	—
Undiscovered	~5150	0.25	~1290	0.08–0.12	~38 000	—	—
Hot-water (90–150°C)							
Identified	1440[e]	0.25	360	—	—	0.24	86.5
Undiscovered	4420	0.25	~1100	—	—	0.24	~260
Geopressured systems							
Evaluated – Plan 1	45 710[f]	0.021	959[g, h]	0.08	24 380[i, j]	—	—
Plan 2		0.033	1510[g]	0.08	38 140[i]	—	—
Plan 3		0.005	230[g, h]	0.08	5690[i, k]	—	—
Unevaluated	140 000[l]	>0.023	>3200[g, h]	0.08	~75 000[i]	—	—

a To 10 km.
b Includes 30 × 10^{18} joules in National Parks.
c Excludes National Parks.
d Includes 557 × 10^{18} joules in Yellowstone National Park.
e Bruneau-Granview area Idaho, 1100 × 10^{18} joules.
f 'Fluid resource base' to 7 km in Louisiana and 6 km in Texas.
g Plus thermal energy from combustion of dissolved methane.
h Plus thermal equivalent of mechanical energy.

i Plus any electricity generated from methane.
j Plus 9970 $MW \cdot c$ from mechanical energy.
k Plus 3560 $MW \cdot c$ from mechanical energy.
l Offshore Gulf Coast, deeper parts of Gulf Coastal Plain, and other geo-pressured basins.
m Ratio of electricity generated to the enthalpy of the fluid at the indicated temperature (relative to a reference enthalpy of 15°C water).

evaluation only of the geopressured resource in onshore Tertiary deposits of the Gulf Coast. Using the abundant drilling data in that area, the temperature and amount of geopressured fluid were estimated, thus allowing calculation of thermal energy in the water alone (the 'fluid resource base'). Various production modes were analyzed and the amount of recoverable energy calculated. This energy consists of the thermal energy (Table 1), an almost equal amount of energy from dissolved methane, and a small amount of kinetic energy from the high pressures in the reservoirs.

Any geothermal resource assessment should be periodically updated in response to changing conditions. Among these are (a) expanding exploration and drilling activity, (b) development of improved and new technologies for exploration, evaluation, extraction, and utilization, (c) rapid evolution of geothermal knowledge, and (d) the increasing role of geothermal energy in response to changing economic, social, political, and environmental conditions (in particular, an increasing awareness of the limits to petroleum and natural gas resources, both domestic and international).

Accordingly, the U.S. Geological Survey plans to complete an updated and expanded geothermal resource assessment of the United States by the end of 1978. Aspects to be given increased emphasis in 1978 include the following:

(a) Evaluation of available data on low-temperature (<90°C) hydrothermal convection systems.

(b) Refinement of areas, thicknesses, temperatures, and recoverabilities of high-temperature (>150°C) and intermediate-temperature hydrothermal convection systems, in part using data acquired and compiled in the course of systematic evaluation of Known Geothermal Resource Areas (ISHERWOOD and MABEY, 1978).

(c) Utilization of the recently operational USGS system of computer-based storage and retrieval of geothermal data (GEOTHERM).

(d) Assessment of geopressured resources not inventoried in 1975 (offshore Tertiary deposits and onshore Mesozoic deposits of the Gulf Coast, and geopressured resources in other sedimentary basins).

(e) Refinement of the size and age of young igneous systems and more thorough evaluation of the effects of hydrothermal convection on the cooling of plutons.

It is clear from the literature that the geothermal community is far from consensus on the methodology, assumptions, and terminology that should be used in estimating geothermal potential (MUFFLER, 1973; BARNEA, 1975; SHANZ, 1975). With this in mind, and in preparation for the 1978 geothermal resource assessment of the United States, one of us (L.J.P.M.) has cooperated this past year with scientists and engineers of the National Electric Agency of Italy (ENEL) in evaluating techniques for geothermal resource assessment, under the sponsorship of the U.S. Energy Research and Development Administration (ERDA), now the Department of Energy (DOE). Recommendations for uniform terminology and methodology were presented in

September 1977 at the ENEL-ERDA Larderello Workshop on Geothermal Resource Assessment and Reservoir Engineering (MUFFLER and CATALDI, 1978) along with a test application to central and southern Tuscany (CATALDI *et al.*, 1978).

3. *Application of Geothermal Research to Geothermal Resource Assessment*

Geothermal resource assessment is much more than a mere inventory of existing data. It must incorporate all pertinent available data, of course, but in addition it requires analysis and interpretation of many geological, geochemical, geophysical, and reservoir-engineering factors such as (a) regional tectonic setting, (b) regional heat flow, (c) late Cenozoic volcanic history, (d) lithostratigraphic and structural framework, (e) chemical composition of fluids, (f) regional and local hydrologic setting, (g) heat-extraction mechanisms, and (h) reservoir behavior and longevity.

In general, the quantitative relations of these factors to geothermal resources are poorly known. Consequently, the USGS carries out a broad program of geothermal research aimed at understanding the processes that govern the origin, transport, and storage of crustal heat. Major components of this research are (1) study of geothermal processes on crustal and local scales, (2) regional geothermal evaluations, (3) intensive study and characterization of representative geothermal systems, (4) improvement of geothermal exploration techniques, and (5) investigation of geoenvironmental constraints.

In addition to studies carried out by its own staff, the USGS supports resource-related geothermal studies by various universities and private groups. The geothermal resource assessment of the USGS also draws heavily on complementary studies supported by the DOE Division of Geothermal Energy (DGE) and carried out under contract to universities, national laboratories (e.g., Los Alamos, Lawrence Berkeley, etc.), and private industry. These resource-related studies, only part of the program of DOE/DGE, are directed towards specific aspects of geothermal systems that concern the reservoir, including the study of exploration and evaluation methods and the assessment of reservoir potential on a site-by-site basis.

Data and interpretations resulting from these geothermal investigations, mainly sponsored by the USGS and DOE, have direct application to geothermal resource assessment of the United States. In the following paragraphs we cite a few illustrative examples, in part drawn from the literature and in part taken from ongoing research that, in our opinion, will have significant impact on future geothermal resource assessment. Specific areas discussed are indicated on Fig. 1.

Areal studies

Regional geological and geophysical research and studies of the chemistry of thermal fluids have direct application to geothermal resource assessment. Examples

Figure 1
Map showing geothermal areas and regions of the western United States.

include (a) volcanological and water-chemistry studies in the High Lava Plains of Oregon (MacLeod et al., 1975; Mariner et al., 1975a), (b) hydrologic, geologic and water-chemistry studies in the Great Basin of Nevada (Hose and Taylor, 1974; Olmstead et al., 1975; Mariner et al., 1975b), and (c) volcanologic and water-chemistry investigations in Alaska (Miller and Barnes, 1976; Smith and Shaw, 1975). Ongoing regional investigations that will affect the 1978 resource assessment include systematic studies of New Mexico (University of New Mexico), of the Snake River Plain, Idaho (USGS and others), and of the Gulf Coast (University of Texas, USGS, and others).

Information from intensive studies or representative geothermal systems is important both for assessment of those systems themselves and for extrapolation of similar systems not yet drilled or studied in detail. Significant information has emerged from detailed study of several geothermal systems, including (a) Long Valley, California (a series of 13 papers in vol. 81, no. 5, of the *Journal of Geophysical Research*; also Fournier et al., in press), (b) The Geysers, California (McLaughlin and STANLEY, 1978; HEARN *et al.*, 1976; ISHERWOOD, 1976; LIPMAN *et al.*, 1978), (c) the Raft River Valley, Idaho (WILLIAMS *et al.*, 1976; MABEY *et al.*, 1978), and the Coso Mountains, California (DUFFIELD, 1975; LANPHERE *et al.*, 1975; DUFFIELD and BACON, 1977). Detailed investigations are continuing at these and at other representative geothermal systems, including Roosevelt Hot Springs in Utah (University of Utah), the east flank of Kilauea Volcano in Hawaii (University of Hawaii), Mt. Drum in the Wrangell Mountains Alaska (USGS), the San Francisco volcanic field of Arizona (USGS), Buena Vista and Grass Valleys in Nevada (Lawrence Berkeley Laboratory), Klamath Falls in Oregon (Oregon Institute of Technology, USGS), and several systems in the Imperial Valleys (University of California at Riverside and Lawrence Livermore Laboratory).

Heat flow and hydrology

Regional heat-flow studies have obvious application to geothermal resource assessment; combined with models for the distribution of heat-generating elements in the crust, they permit calculation of temperature–depth curves and thus estimation of the amount of heat contained in various depth zones of the crust (DIMENT *et al.*, 1975). These studies also provide a useful exploration tool, particularly for systems that have no associated hot springs (BLACKWELL and BAAG, 1973). Continuing heat-flow investigations by many institutions (USGS, Southern Methodist University, University of Utah, etc.) will aid in refining the geothermal resource base in regional conductive environments and may help delimit some hydrothermal systems. Studies of the areal distribution of heat-generating elements will aid in assessing low-temperature geothermal resources, particularly in the coastal plain of the eastern United States (Virginia Polytechnic Institute).

The role of water circulating in the crust and its effect on regional heat flow is a research topic just beginning to be addressed seriously (OLMSTED *et al.*, 1975; LACHENBRUCH and SASS, 1977). On a local scale, however, combined heat-flow and hydrologic studies have proved useful for evaluating the coupled movement of heat and water in geothermal systems (e.g., LACHENBRUCH *et al.*, 1976; SOREY *et al.*, 1977).

Volcanology

Studies of the size, age, and evolution of late Cenozoic volcanic systems provide a powerful means of estimating the amount of heat likely to be stored in major high-level crustal thermal anomalies (SMITH and SHAW, 1975). This approach is not yet rigorous or precise, and the need for geologic, geochronologic, and geophysical investigations on both regional and local scales is obvious. In addition to comprehensive areal investigations (e.g., BAILEY *et al.*, 1976), research topics that bear directly on this method of resource assessment include (a) the use of teleseismic *P*-delays to delimit the geometry of crustal and mantle thermal anomalies (STEEPLES and IYER, 1976), (b) the improvement of electromagnetic methods for delimiting hot or molten zones at depth (University of Utah, University of Texas, Brown University, USGS, etc.), (c) development of new techniques for dating young volcanic rocks (e.g., thermoluminescence; MAY, 1977), (d) petrologic study of the thermo-dynamic parameters of magma chambers (CARMICHAEL et al., 1974), and (e) analyses of the effect of hydrothermal circulation on the cooling history of plutons (CATHLES, 1977; NORTON, 1977; FOURNIER, 1977b).

Geochemistry of fluids

Geochemical studies of waters and gases from hot springs, fumaroles, and wells are critical to geothermal resource assessment, and they provide the only means of estimating temperatures in hydrothermal convection systems at depths not yet reached by drilling (TRUESDELL, 1976). In the 1975 resource assessment of the United States, the SiO_2 and Na—K—Ca chemical geothermometers were used extensively to estimate reservoir temperatures (RENNER et al., 1975). Recently developed models that allow evaluation of the mixing of waters and loss of steam (FOURNIER and TRUESDELL, 1974; TRUESDELL and FOURNIER, 1976; FOURNIER, 1977a) should be applied systematically in future resource assessments. Also, certain isotopic geo-thermometers (such as fractionation of ^{18}O between H_2O and dissolved SO_4) equilibrate more slowly than the conventional chemical geothermometers and thus are capable of indicating temperatures in deeper and hotter parts of hydrothermal convection systems (MCKENZIE and TRUESDELL, 1976; TRUESDELL, 1976).

Exploration technology

The development today of new and improved exploration techniques will affect future geothermal resource assessments through increased ease of location and more precise delineation of geothermal anomalies. Several effective techniques new to geothermal exploration have emerged during the past few years, including (a) rapid audio-frequency magnetotelluric surveying (HOOVER and LONG, 1976), (b) self-potential surveying (ZABLOCKI, 1976), (c) dispersion of mercury above geothermal systems (MATLICK and BUSECK, 1976), and (d) dispersion of helium above geothermal systems (ROBERTS et al., 1975). The continuing rapid improvements of equipment, processing techniques, and interpretation methods for electromagnetic and seismic investigations also enhance future geothermal resource assessments through the expected deeper penetration, more rapid coverage, increased resolution, and more meaningful and reliable interpretation in terms of geologic models.

Reservoir mechanics

Various investigations in the general field of reservoir mechanics bear directly on the estimation of how much heat can be recovered from a given reservoir, be it a hydrothermal-convection, geopressured, or hot-dry-rock system (NATHENSON, 1975; PAPADOPULOS et al., 1975). Such investigations also enable prediction of the behavior and longevity of a geothermal system under exploitation. Direct contributions of reservoir mechanics to geothermal resource assessment include:

(a) Development of mathematical models to simulate natural geothermal systems and their evolution under exploitation (e.g., MERCER et al., 1975; BROWNELL et al., 1977; TURCOTTE et al., 1977; SCHUBERT and STRAUS, 1977).
(b) Application of well-testing methods to geothermal systems (BARELLI et al., 1975; RAMEY, 1976; ATKINSON et al., 1978).
(c) Analysis of steam-production mechanisms and the temperatures of sub-surface boiling in vapor-dominated reservoirs through ^{18}O and D studies of produced steam and water (TRUESDELL and FRYE, 1977).
(d) Evaluation, by repetitive gravity and levelling surveys, of fluid recharge to a hydrothermal-convection reservoir (HUNT, 1977; ISHERWOOD, 1977).
(e) Investigation of processes and determination of the feasibility of fracturing hot impermeable rock and extracting heat from it (SMITH, 1979; GRINGARTEN et al., 1975; WEERTMAN and CHANG, 1977; LOCKNER and BYERLEE, 1977).

Environmental investigations

Only that geothermal heat which is producible in an environmentally acceptable and thus in a legal manner can be considered a resource. Accordingly, the evaluation of potential environmental problems is a prerequisite to any geothermal resource assessment.

Environmental acceptability should be demonstrated before extensive new geothermal development. Although all environmental effects must be considered, certain earth-science aspects are especially critical to resource assessment. Comprehensive reviews of these aspects were presented recently by AXTMANN (1976) for chemical problems (including H_2S pollution; also see ALLEN and McCLUER, 1976) and by SWANBERG (1976) for physical problems (including land subsidence due to fluid withdrawal and the triggering of microearthquakes by fluid reinjection).

4. Conclusion

Geothermal resource assessment is not merely an inventory of production and drilling data but involves an evaluation of a broad spectrum of geologic, geophysical, technologic, and economic factors. In the case of an emerging nontraditional energy source such as geothermal energy, there are few case histories and little experience upon which to base resource and reserve estimates. Consequently, there is an important role for research, particularly in earth sciences, directed both toward the nature and distribution of heat in the earth's crust and toward the ways in which this heat can be produced economically and in an environmentally acceptable manner.

References

ALLEN, G. W. and MCCLURE, H. K. (1976). *Abatement of Hydrogen Sulfide Emissions from The Geysers Geothermal Power Plant*, Proc. 2nd United Nations Symp. Dev. Use geotherm. Resour. 1313–1315.

ATKINSON, P., BARELLI, A., BRIGHAM, W., CELATI, R., MANETTI, G., MILLER, F., NERI, G. and RAMEY, H. (1978), *Well-testing in the Travale-Radicondoli Geothermal Field*, Geothermics (in press).

AXIMANN, R. C. (1976), *Chemical Aspects of the Environmental Impact of Geothermal Power*, Proc. 2nd United Nations Symp. Dev. Use geotherm. Resour. 1323–1327.

BAILEY, R. A., DALRYMPLE, G. B. and LANPHERE, M. A. (1976), *Volcanism, Structure, and Geochronology of Long Valley Caldera, Mono County, California*, J. geophys. Res. *81*, 725–744.

BARELLI, A., MANETTI, G., CELATI, R. and NERI, G. (1976), *Build-up and Back-pressure Tests on Italian Geothermal Wells*, Proc. 2nd United Nations Symp. Dev. Use geotherm. Resour. 1537–1546.

BARNEA, J. (1975), *Assessment of Geothermal Resources of the United States – 1975 – a Critique*, Geotherm. Energy *3*, 35–38.

BLACKWELL, D. D. and BAAG, C. (1973), *Heat Flow in a "Blind" Geothermal Area near Marysville, Montana*, Geophys. *38*, 941–956.

BROWNELL, D. H., GARG, S. K. and PRITCHETT, J. W. (1977), *Governing Equations for Geothermal Reservoirs*, Wat. Resour. Res. *13*, 926–934.

CARMICHAEL, I. S. E., TURNER, F. J. and VERHOOGEN, J., *Igneous Petrology* (McGraw-Hill, New York 1974), 739 pp.

CATALDI, R., LAZZAROTTO, A., MUFFLER, P., SQUARCI, P. and STEFANI, G. C. (1978), *Assessment of Geothermal Potential of Central and Southern Tuscany*, Geothermics (in press).

CATHLES, L. M. (1977), *An Analysis of the Cooling of Intrusives by Ground-water Convection which Includes Boiling*, Econ. Geol. *72*, 804–826.

DIMENT, W. H., URBAN, T. C., SASS, J. H., MARSHALL, B. V., MUNROE, R. J. and LACHENBRUCH, A. H. (1975), *Temperatures and Heat Contents Based on Conductive Transport of Heat*, Circ. U.S. geol. Surv. *726*, 84–103.

DUFFIELD, W. A. (1975), *Late Cenozoic Ring Faulting and Volcanism in the Coso Range Area of California*, Geol. *3*, 335–338.

DUFFIELD, W. A. and BACON, C. R. (1977), *Preliminary Geologic Map of the Coso Volcanic Field and Adjacent Areas, Inyo County, California*, Open-file Rep. U.S. geol. Surv. *77–311*, 2 pls.

FOURNIER, R. O. (1977a), *Prediction of Aquifer Temperatures, Salinities, and Underground Boiling and Mixing Processes in Geothermal Systems*, Proc. 2nd internat. Symp. water-rock Interaction, Strasbourg, France, August 1977, Sec. III (High-temp. water-rock Interaction), 117–126.

FOURNIER, R. O. (1977b), *Constraints on the Circulation of Meteoric Water in Hydrothermal Systems Imposed by the Solubility of Quartz* (abs.), Abstr. with Programs, Geol. Soc. America *9*, 979.

FOURNIER, R. O., SOREY, M. L., MARINER, R. H. and TRUESDELL, A. H. (1977), *Chemical and Isotopic Prediction of Aquifer Temperatures in the Geothermal System at Long Valley, Calif.*, J. volc. geotherm. Res. (in press).

FOURNIER, R. O. and TRUESDELL, A. H. (1974), *Geochemical Indicators of Subsurface Temperature-Part 2, Estimation of Temperature and Fraction of Hot Water Mixed with Cold Water*, J. Res. U.S. geol. Surv. *2*, 263–270.

GRINGARTEN, A. C., WITHERSPOON, P. A. and OHNISHI, Y. (1975), *Theory of Heat Extraction from Fractured Hot Dry Rock*, J. geophys. Res. *80*, 1120–1124.

HEARN, B. C., DONNELLY, J. M. and GOFF, F. E. (1976), *Geology and Geochronology of the Clear Lake Volcanics, California*, Proc. 2nd United Nations Symp. Dev. Use geotherm. Resour. 423–428.

HOOVER, D. B. and LONG, C. L. (1976), *Audio-Magnetotelluric Methods in Reconnaissance Geothermal Exploration*, Proc. 2nd United Nations Symp. Dev. Use geotherm. Resour. 1059–1064.

HOSE, R. K. and TAYLOR, B. E. (1974), *Geothermal Systems of Northern Nevada*, Open-file Rep., U.S. geol. Surv. *74–271*, 27 pp.

HUNT, T. M. (1977), *Recharge of Water in Wairakei Geothermal Field Determined from Repeat Gravity Measurements*, N.Z. J. Geol. Geophys. *20*, 303–317.

ISHERWOOD, W. F. (1976), *Gravity and Magnetic Studies of The Geysers-Clear Lake Geothermal Region, California, U.S.A.*, Proc. 2nd United Nations Symp. Dev. Use geotherm Resour. 1065–1073.

ISHERWOOD, W. F. (1977), *Reservoir Depletion at The Geysers, California, in Geothermal – State of the Art*, Trans. geotherm. resour. Coun. *1*, 149.

ISHERWOOD, W. F. and MABEY, D. R. (1978), *Evaluation of Known Geothermal Resource Areas in the Western United States*, Geothermics (in press).

LACHENBRUCH, A. H., SOREY, M. L., LEWIS, R. E. and SASS, J. H. (1976), *The Near-Surface Hydrothermal Regime of Long Valley Caldera*, J. geophys. Res. *81*, 763–768.

LACHENBRUCH, A. H. and SASS, J. H. (1977), *Regional Heat Flow, Thermal State and Crustal Processes in the United States*, *in* Heacock, J. G., ed., The Earth's Crust: its Nature and Physical Properties, Monogr. Am. geophys. Un. *20*, 626–675.

LANPHERE, M. A., DALRYMPLE, G. B. and SMITH, R. L. (1975), *K—Ar Ages of Pleistocene Rhyolitic Volcanism in the Coso Range*, California, Geol. *3*, 339–341.

LIPMAN, S. C., STROBEL, C. J. and GULATI, M. S. (1978), *Reservoir Performance of The Geysers Field*, Geothermics (in press).

LOCKNER, D. and BYERLEE, J. D. (1977), *Hydrofracture in Weber Sandstone at High Confining Pressure and Differential Stress*, J. geophys. Res. *82*, 2018–2026.

MABEY, D. R., HOOVER, D. B., O'DONNELL, J. E. and WILSON, C. W. (1978), *Reconnaissance Geophysical Studies of Geothermal Systems in the Southern Raft River Valley, Idaho*, Geophys. (in press).

MACLEOD, N. S. WALKER, G. W. and McKEE, E. H. (1976), *Geothermal Significance of Eastward Increase in Age of Upper Cenozoic Rhyolitic Domes in Southeastern Oregon*, Proc. 2nd United Nations Symp. Dev. Use geotherm. Resour. 465–474.

MARINER, R. H., RAPP, J. B., WILLEY, L. M. and PRESSER, T. S. (1974a), *The Chemical Composition and Estimated Minimum Thermal Reservoir Temperatures of Selected Hot Springs in Oregon*, Open-file Rep. U.S. geol. Surv. 27 pp.

MARINER, R. H., RAPP, J. B., WILLEY, L. M. and PRESSER, T. S. (1974b), *The Chemical Composition and Estimated Minimum Thermal Reservoir Temperatures of the Principal Hot Springs of Northern and Central Nevada*, Open-file Rep. U.S. geol. Surv. 32 pp.

MATLICK, J. S., III and BUSECK, P. R. (1976), *Exploration for Geothermal Areas Using Mercury: a New Geochemical Technique*, Proc. 2nd United Nations Symp. Dev. Use geotherm. Resour. 785–792.

MAY, R. J. (1977), *Thermoluminescence Dating of Hawaiian Alkalic Basalts*, J. geophys. Res. *82*, 2023–2029.

MCKENZIE, W. F. and TRUESDELL, A. H. (1976), *Geothermal Reservoir Temperatures Estimated from the Oxygen Isotope Compositions of Dissolved Sulfate and Water from Hot Springs and Shallow Drill Holes* Geothermics *5*, 51–61.

MCLAUGHLIN, R. J. and STANLEY, W. D. (1976), *Pre-Tertiary Geology and Structural Control of Geothermal Resources, The Geysers Steam Field, California*, Proc. 2nd United Nations Symp. Dev. Use geotherm. Resour. 475–485.

MERCER, J. W., PINDER, G. F. and DONALDSON, I. G. (1975), *A Galerkin-Finite Element Analysis of the Hydrothermal System at Wairakei, New Zealand*, J. geophys. Res. *80*, 2608–2621.

MILLER, T. P., BARNES, I. and PATTON, W. W., Jr. (1975), *Geologic Setting and Chemical Characteristics of Hot Springs in West-Central Alaska*, J. Res. U.S. geol. Surv. *3*, 149–162.

MILLER, T. P. and BARNES, I. (1976), *Potential for Geothermal-Energy Development in Alaska – Summary*, Mem. Am. Ass. petrol. Geol. *25*, 149–153.

MILLER, T. P. and SMITH, R. L. (1977), *Spectacular Mobility of Ash Flows Around Anikchak and Fisher Calderas, Alaska*, Geol. *5*, 173–176.

MUFFLER, L. J. P. (1973), *Geothermal Resources*, Prof. Pap. U.S. geol. Surv. *820*, 251–261.

MUFFLER, L. J. P. and CATALDI, R. (1978), *Methods for Regional Assessment of Geothermal Resources*, Geothermics (in press).

NATHENSON, M. (1975), *Physical Factors Determining the Fraction of Stored Energy Recoverable from Hydrothermal Convection Systems and Conduction-Dominated Areas*, Open-file Rep. U.S. geol. Surv. *75–525*, 38 pp.

NATHENSON, M. and MUFFLER, L. J. P. (1975), *Geothermal Resources in Hydrothermal Convection Systems and Conduction Dominated Areas*, Circ. U.S. geol. Surv. *726*, 104–121.

NORTON, D. (1977), *Fluid Circulation in the Earth's Crust*, *in* Heacock, J. G., ed., The Earth's Crust: its Nature and Physical Properties, Geophys. Monogr. Am. geophys. Un. *20*, 693–704.

OLMSTED, F. H., GLANCY, P. A., HARRILL, J. R., RUSH, F. E. and VAN DENBURGH, A. S. (1975), *Preliminary*

Hydrogeologic Appraisal of Selected Hydrothermal Systems in Northern and Central Nevada, Open-file Rep. U.S. geol. Surv. *75–76*, 267 pp.

PAPADOPULOS, S. S., WALLACE, R. H., Jr., WESSELMAN, J. B. and TAYLOR, R. E. (1975), *Assessment of Onshore Geopressured-Geothermal Resources in the Northern Gulf of Mexico Basin*, Circ. U.S. geol. Surv. *726*, 125–140.

PECK, D. L. (1975), *Recoverability of Geothermal Energy Directly from Molten Igneous Systems*, Circ. U.S. geol. Surv. *726*, 122–124.

RAMEY, H. J., Jr. (1976), *Pressure Transient Analysis for Geothermal Wells*, Proc. 2nd United Nations Symp. Dev. Use geotherm. Resour. 1749–1757.

RENNER, J. L., WHITE, D. E. and WILLIAMS, D. L. (1975), *Hydrothermal Convection Systems*, Circ. U.S. geol. Surv. *726*, 5–57.

ROBERTS, A. A., FRIEDMAN, I., DONOVAN, T. J. and DENTON, E. H. (1975), *Helium Survey, A Possible Technique for Locating Geothermal Reservoirs*, Geophys. res. Lett. *2*, 209–210.

SCHANZ, J. J., Jr. (1975), *Resource Terminology: An Examination of Concepts and Terms and Recommendations for Improvement*, Electric Power Research Institute Res. Project 336, August 1975, 116 pp.

SCHUBERT, G. and STRAUS, J. M. (1977), *Two-Phase Convection in a Porous Medium*, J. geophys. Res. *82*, 3411–3421.

SMITH, M. C. (1978/9), *Heat Extraction from Hot, Dry Crustal Rock*, Pure appl. Geophys. *117*, 290–296.

SMITH, R. L. and SHAW, H. R. (1975), *Igneous-Related Geothermal Systems*, Circ. U.S. geol. Surv. *726*, 58–83.

SOREY, M. L., LEWIS, R. E. and OLMSTED, F. H. (1977), *The Hydrothermal System of Long Valley Caldera, California*, Open-file Rep. 77–437, 188 pp.

STEEPLES, D. W. and IYER, H. M. (1976), *Teleseismic P-Wave Delays in Geothermal Exploration*, Proc. 2nd United Nations Symp. Dev. Use Geotherm. Resour. 1199–1206.

SWANBERG, C. A. (1976), *Physical Aspects of Pollution Related to Geothermal Energy Development*, Proc. 2nd United Nations Symp. Dev. Use geotherm. Resour. 1435–1443.

TRUESDELL, A. H. (1976), *Summary of Section III: Geochemical Techniques in Exploration*, Proc. 2nd United Nations Symp. Dev. Use geotherm. Res. liii–lxxix.

TRUESDELL, A. H. and FOURNIER, R. O. (1976), *Calculation of Deep Temperatures in Geothermal Systems from the Chemistry of Boiling Spring Waters of Mixed Origin*, Proc. 2nd United Nations Symp. Dev. Use geotherm. Res. 837–844.

TRUESDELL, A. H. and FRYE, G. A. (1977), *Isotope Geochemistry in Geothermal Reservoir Studies*, J. petrol. Technol. (in press).

TURCOTTE, D. L., RIBANDO, R. J. and TORRANCE, K. E. (1977), *Numerical Calculation of Two-Temperature Thermal Convection in a Permeable Layer with Application to the Steamboat Springs Thermal Area, Nevada*, in Heacock, J. G., ed., The Earth's Crust: its Nature and Physical Properties, Geophys. Monogr. Am. geophys. Un. *20*, 722–736.

WEERTMAN, J. and CHANG, S. P. (1977), *Fluid Flow through a Large Vertical Crack in the Earth's Crust*, J. geophys. Res. *82*, 929–931.

WHITE, D. E. and WILLIAMS, D. L. (eds.) (1975), *Assessment of Geothermal Resources of the United States – 1975*, Circ. U.S. geol. Surv. *726*, 155 pp.

WILLIAMS, P. L., MABEY, D. R., ZOHDY, A. A. R., ACKERMANN, H., HOOVER, D. B., PIERCE, K. L. and ORIEL, S. S. (1976), *Geology and Geophysics of the Southern Raft River Valley Geothermal Area, Idaho, U.S.A.*, Proc. 2nd United Nations Symp. Dev. Use geotherm. Resour. 1273–1282.

ZABLOCKI, C. J. (1976), *Mapping Thermal Anomalies on an Active Volcano by the Self-Potential Method, Kilauea, Hawaii*, Proc. 2nd United Nations Symp. Dev. Use geotherm. Resour. 1299–1309.

(Received 7th November 1977)

Pageoph, Vol. 117 (1978/79), Birkhäuser Verlag, Basel

Possible Geothermal Resources in the Coast Plutonic Complex of Southern British Columbia, Canada[1])

By T. J. Lewis[2]), A. S. Judge[2]) and J. G. Souther[3])

Abstract – In southern British Columbia the terrestrial heat flow is low (44 mW m^{-2}) to the west of the Coast Plutonic Complex (CPC), average in CPC (50–60 mW m^{-2}), and high to the east (80–90 mW m^{-2}). The average heat flow in CPC and the low heat generation (less than 1 μW m^{-3}) indicate that a relatively large amount of heat flows upwards into the crust which is generally quite cool. Until two million years ago the Explorer plate underthrust this part of the American plate, carrying crustal material into the mantle. Melted crustal rocks have produced the inland Pemberton and Garibaldi volcanic belts in the CPC.

Meager Mountain, a volcanic complex in the CPC 150 km north of Vancouver, is a possible geothermal energy resource. It is the product of intermittent activity over a period of 4 My, the most recent eruption being the Bridge River Ash 2440 y B.P. The original explosive eruption produced extensive fracturing in the granitic basement, and a basal explosion breccia from the surface of a cold brittle crust. This breccia may be a geothermal reservoir. Other volcanic complexes in the CPC have a similar potential for geothermal energy.

Key words: Coast Plutonic Complex; Quaternary volcanism; Plate tectonics; Heat flow; Heat production

Introduction

Regions within Canada having quite different thermal characteristics are being evaluated for potential to produce geothermal energy (SOUTHER, 1976). From deep sedimentary basins the production of large volumes of hot water could yield large amounts of space heat and process heat in the Prairies (JESSOP, 1976). In southern British Columbia the area west of the Rocky Mountain Trench has geothermal characteristics similar to the Basin and Range province (JUDGE, 1976; JUDGE *et al.*, in prep.) and could have a similar potential for geothermal energy production. In this paper we examine the type of geothermal resource to be expected in the Coast Plutonic Complex of southern British Columbia (see Fig. 1).

The Coast Plutonic Complex (CPC) is a belt extending along the coast of British Columbia from approximately 49°N into the Yukon, north of 60°N (RODDICK and

[1]) Earth Physics Contribution No. 704.
[2]) Earth Physics Branch, Energy, Mines and Resources, 1 Observatory Crescent, Ottawa, Canada K1A 0Y3.
[3]) Geological Survey of Canada, 100 W. Pender Street, Vancouver, B.C.

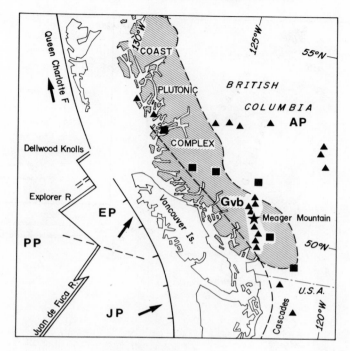

Figure 1

Tectonics near the Coast Plutonic Complex: AP – American Plate, EP – Explorer Plate, JP – Juan de Fuca Plate, and PP – Pacific Plate, GVB – Garibaldi Volcanic Belt. Arrows indicate plate movements relative to AP. Square symbols indicate Miocene volcanic centres, and the star which is Meager Mountain and triangles indicate Quaternary volcanics. The dashed line is the western limit of volcanism and the buried trench is indicated west of Vancouver Island. (After RIDDIHOUGH and HYNDMAN, 1976.)

HUTCHISON, 1974). The Insular belt to the west of CPC is represented by Vancouver Island, and the Intermontane and Hinterland belts lie to the east. The rugged CPC consists mainly of granodiorite, diorite and quartz diorite with associated gneiss. Potassium–argon ages indicate a linear zonation: the oldest rocks, up to 146 My, are on the western side, and the youngest rocks, as young as 40 My, are on the eastern side. HUTCHISON (1970) suggested that successive uplift and unroofing from west to east produced this pattern of ages. MONGER *et al.* (1972) incorporated this process in a plate tectonic synthesis for the Cordillera.

Much younger epizonal plutons and volcanics cover a much smaller part of this area. A row of Quaternary volcanic centres forming the Garibaldi volcanic belt is shown in Figs. 1 and 2 extending north through CPC. In the Meager Creek area this belt is intersected by the older, north-westerly trending belt of late Tertiary and Quaternary high-level plutons, the Pemberton belt. Both belts are thought to be related to subduction of the Juan de Fuca Plate, now much less extensive than in the late Miocene when the Pemberton belt was active. RIDDIHOUGH and HYNDMAN

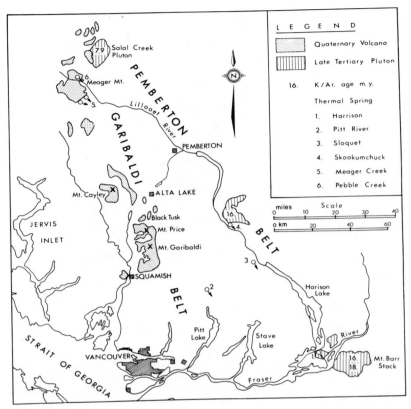

Figure 2
The Pemberton and Garibaldi volcanic belts.

(1976) studied the ocean floor magnetic lineaments and concluded that subduction of the small Explorer plate under CPC may still be happening, but at a relatively slow speed.

Heat flow and heat production

HYNDMAN (1976) measured heat flow in coastal inlets of southern British Columbia using oceanic techniques. Corrections to such measured heat flows are required to remove the effects of temperature changes of the bottom waters, sedimentation, thermal refraction by bottom sediments, topography, microclimatic effects, uplift and erosion. Unfortunately the parameters for calculating these corrections are not well known, and the corrections can be large. After making the necessary corrections, Hyndman obtained low heat flows, increasing towards the east, as shown in Fig. 3. JESSOP and JUDGE (1971) measured a high heat flow at Penticton, east of the CPC. More recent heat flow determinations (LEWIS, 1977) in and near the southern CPC

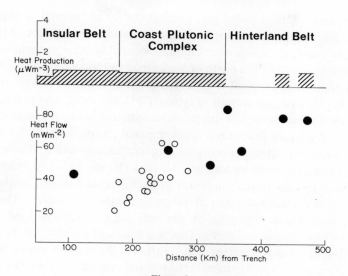

Figure 3

Profiles of heat flow and surficial heat production across the Coast Plutonic Complex, as a function of distance east of the buried trench off Vancouver Island. Open circles indicate measurements made in the inlets and solid circles indicate land measurements.

are also shown in Fig. 3, details of these measurements will be published elsewhere (LEWIS *et al.*, in prep). None of these measurements is near the younger plutons or volcanics. Therefore, the regional heat flow increases from west to east across CPC from about 40 mW m^{-2} in the Insular belt to 80 mW m^{-2} in the Intermontane and Hinterland belts.

Heat production measurements have been made on fresh looking, representative surface samples of bedrock using a gamma-ray spectrometer. Published values (LEWIS, 1976) and new results show a low heat production ranging from 0.5 to 1.0 µW m^{-3}. Average values are shown in Fig. 3. There is no apparent variation in potassium content nor heat production across CPC. The amount of heat generated in the upper crust, 10 mW m^{-2}, is small, and thus the majority of the measured heat flow is coming from beneath the upper crust. In most areas of CPC where there is no mass movement within the crust the crustal temperatures will be normal: approximately 240°C at 10 km depth. However, of the total area, the small proportion formed by young plutons and volcanics probably has much higher temperatures.

Meager Mountain

The initial assessment of the geothermal resource potential of western Canada, including the compilation of data on locations and ages of Quaternary volcanoes and young, high-level plutons and a geochemical survey of thermal springs, led to the selection of the Meager Creek area for more detailed investigation. Meager Creek, near which hot surface springs were found, is located at the upper end of the Lillooet Valley approximately 150 km north of Vancouver in moderately well foliated quartz diorite and granodiorite typical of CPC (RODDICK and WOODSWORTH, 1975). Meager Mountain lies on a rugged surface of typical CPC basement, and partially on one of several small satellitic bodies of quartz monzonite similar to the nearby Salal Creek pluton. The Salal Creek pluton is a small epizonal pluton dated at 7.9 My.

Meager Mountain itself is a complex of several closely related dacite and andesite lava domes and associated pyroclastic deposits (READ, 1977; ANDERSON, 1975). A basal breccia, exposed on the south side of the complex contains jumbled blocks of quartz diorite of dimensions up to 20 metres in a tuffaceous matrix. The initial eruption was an explosive discharge of gas-rich magma accompanied by extensive fracturing of the granitic basement. This was followed successively by dacite flows, the deposition of up to 500 metres of acid tuff, and the events forming the main mass of the complex, a porphyritic andesite (LEWIS and SOUTHER, 1978). Potassium–argon ages of 4.2 My and 2.1 My indicate that this main second phase was the product of a long episode of intermittent volcanism. The dacites of the third phase were deeply dissected prior to eruption of the Bridge River Ash, suggesting that the complex was dormant for a period prior to the most recent eruption. The Bridge River Ash is younger than 2240 y. The centres of the phases of activity moved progressively northwards with time.

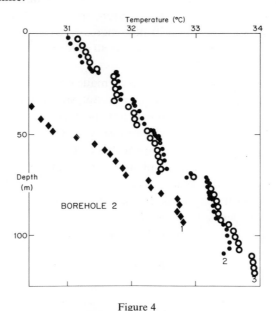

Figure 4
Temperature logs of a shallow borehole near Meager Creek. The logs were taken on successive days before drilling started, following a night in which no drilling nor circulation took place.

Temperatures were measured in six shallow boreholes drilled in the granodiorite and quartz diorite of the Meager Creek area. The holes were uncased, and water flowed through some portion of each hole. Figure 4 shows three temperature logs run to the bottom of a hole at three different pauses in the drilling. The sudden changes in temperature at many depths indicate where water is entering this hole. Using bottom hole temperatures it was possible to calculate the original conductive heat flows at four of the six sites before water started flowing through the holes. The conductive heat flow varied from 100 to 930 mW m^{-2} (LEWIS and SOUTHER, 1979), or 2 to 16 times the world average heat flow.

The results from other geophysical surveys support the claim that this area contains a geothermal reservoir. The excess silica in the Meager Creek hot springs might indicate that the water has been at a much higher temperature (HAMMERSTROM and BROWN, 1977). A deep resistivity survey indicated a large anomaly extending downwards and northwestwards from Meager Creek (NEVIN SADLIER-BROWN GOODBRAND LTD., 1975). The present observations suggest that hot water is available in the brecciated basement rocks, and these rocks may form a reservoir which can be economically exploited.

The type of geothermal resource

In this tectonic setting, it is expected that steam or hot water may be available to form an economical energy source. The cool, brittle crust overlying a slowly subducting oceanic crust has been subjected to vertical forces, causing block faulting. Lack of seismicity indicates no movement at present. Although in general crustal temperatures may be low, heat is brought to or near the surface from great depth by magma moving along major crustal faults. The magma is generated from material with a low melting point which previously was part of the oceanic crust. The volcanic belts are the surface indication of intrusions formed by the ascending magma solidifying before it reached the surface.

To remove heat economically from rock requires a hydrological system which will absorb the heat at high temperatures from a large mass of hot rock, and carry it at a useful temperature to a transfer point on the surface. If such a system operated naturally, it may have already dissipated most of the energy. Usually large amounts of water flow through the contacts of volcanic flows in young rocks (e.g. PALMASON, 1967), and these rocks are quickly cooled, and chemically altered.

At Meager Mountain the first explosive eruption has formed a large volume of coarsely brecciated rock, part of which may form a reservoir (see Fig. 5). Erosional debris and further lava flows have covered parts of this unit. Later eruptions have emplaced other hot rock in contact with parts of the breccia, increasing the probability that it contains hot water. Therefore, parts of this breccia unit may form an economic geothermal energy resource.

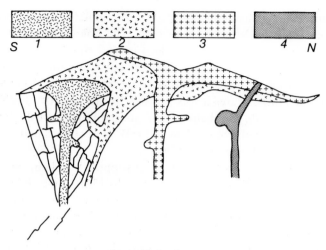

Figure 5

Cartoon showing the four successive phases of eruptions which formed Meager Mountain. The initial phase, 1, produced the large volume of breccia; the following phases erupted progressively northwards.

Meager Mountain is an example of the type of geothermal occurrence to be expected in the Coast Plutonic Complex. Other similar volcanic complexes in CPC such as Mt. Cayley, the Franklin Glacier and Silverthrone complexes are being studied since they have a similar potential for producing geothermal energy.

References

ANDERSON, R. G. (1975), The geology of the volcanics in the Meager Creek Map-area, southwestern British Columbia, U. of British Columbia B.Sc. Thesis.

HAMMERSTROM, L. T. and BROWN, T. H. (1977), *Geochemistry of thermal waters in the Mt. Meager Hot Springs Area, British Columbia* in *Rept. of Activities*, Geol. Surv. Can., Paper 77-1A, 283–285.

HUTCHISON, W. W. (1970), *Metamorphic framework and plutonic styles in the Prince Rupert Region of the central Coast Mountains, British Columbia*, Can. J. Earth Sci. 7, 376–405.

HYNDMAN, R. D. (1976), *Heat flow measurements in the inlets of southwestern British Columbia*, J. Geophys, Res. *81*, 337–349.

JESSOP, A. M., *Geothermal Energy from Sedimentary Basins*, Geothermal Series No. 8, *Energy, Mines and Resources* (Ottawa, 1976).

JESSOP, A. M. and JUDGE, A. S. (1971), *Five measurements of heat flow in southern Canada.* Can. J. Earth Sci. *8*, 711–716.

JUDGE, A. S. (1977), *Terrestrial heat flow and the thermal structure of the Eastern Canadian Cordillera* (abstr.). 1977 Annual Meetings of G.A.C., M.A.C., S.E.G. and C.G.U., Vancouver, Program with Abstracts, Vol. 2, 28.

JUDGE, A. S., LEWIS, T. J. and JESSOP, A. M. (in prep.) *The heat flow, heat production and geothermal energy potential of southeastern British Columbia.*

LEWIS, T. J. (1976), *Heat generation in the Coast Range Complex and other areas of British Columbia*, Can. J. Earth Sci. *13*, 1634–1642.

LEWIS, T. J. (1977), *Terrestrial heat flow in the Southern Coast Plutonic Complex* (abstr.). 1977 Annual Meetings of G.A.C., M.A.C., S.E.G. and C.G.U., Vancouver, Program with Abstracts, Vol. 2, 32.

LEWIS, T. J., JUDGE, A. S. and JESSOP, A. M. (in prep.) *The geothermal structure of the Coast Plutonic Complex.*

LEWIS, T. J. and SOUTHER, J. G. (1979), *Meager Mountain, B.C. – A possible geothermal energy resource*, Geothermal Series No. 9, *Energy, Mines and Resources*, Ottawa.

MONGER, J. W. H., SOUTHER, J. G. and GABRIELSE, H. (1972), *Evolution of the Canadian Cordillera: A plate-tectonic model*, Am. J. Sci. *272*, 577–602.

NEVIN SADLIER-BROWN GOODBRAND LTD., (1975), *Detailed Geothermal Investigation at Meager Creek*, Report to B.C. Hydro and Power Authority.

PALMASON, G., *On heat flow in Iceland in relation to the mid-Atlantic ridge*, in BJORNSSON, J. (ed.), *Iceland and Mid-Ocean Ridge*, pp. 111–127, Rept. of Symp. (Geoscience Soc., Reykjavik 1967).

READ, P. B. (1977), *Meager Creek volcanic complex, southwestern British Columbia*, in *Report of Activities Pt. A, Geol. Surv. Can.* Paper 77-1A.

RIDDIHOUGH, R. P. and HYNDMAN, R. D. (1976), *Canada's active western margin – the case for subduction*, Geoscience Canada *3*, 269–278.

RODDICK, J. A. and HUTCHISON, W. W. (1974), *Setting of the Coast Plutonic Complex, British Columbia*, Pacific Geology *8*, 91–108.

RODDICK, J. A. and WOODSWORTH, G. J. (1975), *Coast Mountains Project: Pemberton (92 J West Half) Map-area*, B.C. Geol. Surv. Can. Paper 75-1, part A, 37–40.

SOUTHER, J. G. (1976), *Geothermal Potential of Western Canada*, in *Proc. Second U.N. Symp. on the Development and Use of Geothermal Resources, San Francisco*, Vol. 1, pp. 259–267.

(Received 3rd January 1978)

Pageoph, Vol. 117 (1978/79), Birkhäuser Verlag, Basel

Geothermal Energy Prospects in Brazil: A Preliminary Analysis

By V. M. Hamza[1]), S. M. Eston[2]) and R. L. C. Araujo[1])

Abstract – Results of geothermal gradient measurements in 44 localities in Brazil are presented. The Precambrian shield areas are found to be characterized by relatively low temperature gradients in the range 6 to 20°C/km while younger sedimentary basins are characterized by gradients in the range 15 to 35°C/km. An inverse correlation between geothermal gradient and tectonic age has been observed. This as well as the favourable hydrological conditions suggest that the best sites for extraction of geothermal energy in Brazil are the younger sedimentary basins. The Parana Basin is found to offer at present the best site for extraction of geothermal energy in Brazil. Preliminary examination of the temperature distributions in the major aquifer (Botucatu sandstone) suggest that this aquifer contains substantial quantities of warm waters in the temperature range 40 to 90°C. The water layer in this confined aquifer is in convective motion and can be considered as a low enthalphy geothermal system. Many of the routine uses to which geothermal waters are put, such as space heating and soil warming, are not applicable in Brazil mainly because of the favourable climatic conditions. Conversion of this geothermal energy into electrical energy is also unlikely to be economical. Hence we do not consider the Parana Basin geothermal system as an independent economically exploitable energy resource. However, a few other applications are pointed out where geothermal waters can be used as a supplementary or supporting energy source in increasing the efficiency of economically viable systems utilizing hot waters.

Key words: Precambrian shield; Parana sedimentary basin; Low enthalpy geothermal system; Geothermal gradients.

Introduction

Our objective is to present a preliminary analysis of the prospects for the utilization of geothermal energy in Brazil on the basis of existing geothermal data. The geothermal investigations that have been carried out so far in Brazil are of a preliminary character, and very little is known about the vertical distribution of temperature beneath major tectonic units. The picture is further complicated because of the highly uneven distribution of geothermal measurements. An appreciable number of temperature measurements have been made in southeastern Brazil, but vast areas in the western, central and northern parts are without any measurements at all. Thus the general conclusions reached in the present work should be considered as tentative and as applying only to the eastern parts of Brazil. A more complete picture can be obtained only after a sufficiently large number of geothermal measurements have been carried out in the remaining parts of Brazil.

[1]) Instituto Astronômico e Geofísico, Universidade de São Paulo, C.P. 30.627 – São Paulo, Brasil.
[2]) Instituto de Pesquisas Tecnológicas, Cidade Universitária, São Paulo, Brasil.

Previous work

Some preliminary estimates of geothermal gradients for a few localities in Brazil were reported by UYEDA and WATANABE (1970). A more detailed study of gradients using bottom hole temperatures in oil wells was reported by MEISTER (1973). Table 1 gives a summary of these earlier measurements. The main difficulty with these earlier investigations is that they make use of bottom-hole temperatures measured in oil wells immediately after drilling has ceased. Since drilling activity causes considerable disturbances in the *in situ* rock temperatures, the gradient values derived from these measurements can be considered as indicating only approximate values of geothermal gradients. For example gradient values quoted by UYEDA and WATANABE (1970) for the Parana Basin are in the range 14 to 19°C/km while MEISTER (1973) reports gradient values of 17 to 26°C/km for a major part of the basin. In view of the larger

Table 1

Summary of earlier measurements of geothermal gradients in Brazil

Locality	Latitude	Longitude	Geothermal gradient (°C/km)	Reference
Água Grande	12°24′	38°22′	36.0	UYEDA and WATANABE, 1970
Miranga	12°21′	38°12′	30.3	,,
Buracica	12°14′	38°28′	26.0	,,
Parana Basin I	26°50′	51°50′	14.2	,,
II	23°35′	50°30′	14.7	,,
III	21°45′	54°40′	19.2	,,
Continental Shelf	11°08′	37°03′	25.3	,,
Eastern Brazil I	10°07′	36°27′	25.6	,,
II	9°42′	35°41′	27.3	,,
III	10°00′	36°10′	25.3	,,

Sedimentary basin	Number of wells	Average gradient (°C/km)	Reference
Sergipe/Alagoas	121	25	MEISTER, 1973
Barreirinhas a Ceará	54	23	,,
Recôncavo	210	22	,,
Limoeiro/Foz do Amazonas	15	22	,,
Alamada	1	22	,,
Maranhão	17	17 to > 26	,,
Camuruxatiba/Mucuri	8	17 to 26	,,
Acre e Amazonas	43	20	,,
Espirito Santo	17	17 to 26	,,
Bragança e São Luiz	6	17 to 26	,,
Jequitinhonha	6	19	,,
Campos	5	17 to 26	,,
Santos	3	17 to 22	,,
Paraná	39	17 to > 26	,,
Tucano	49	17	,,

number of measurements as well as better accuracy, MEISTER's (1973) values may be considered as more reliable.

Recent work

Since 1974 however a large number of accurate geothermal gradient values have been obtained using calibrated thermistor thermometers capable of measuring temperature to an accuracy of better than 0.01°C. Measurements have so far been carried out in more than 100 boreholes, 20 water wells, 3 oil wells and 3 underground mines. Repeat measurements were made, whenever possible, in order to correct for any possible effects due to drilling activities. On the basis of these data, regional geothermal gradients have been obtained for about 44 localities in Brazil shown in Fig. 1. Table 2 gives a summary of these regional temperature gradients, rock types

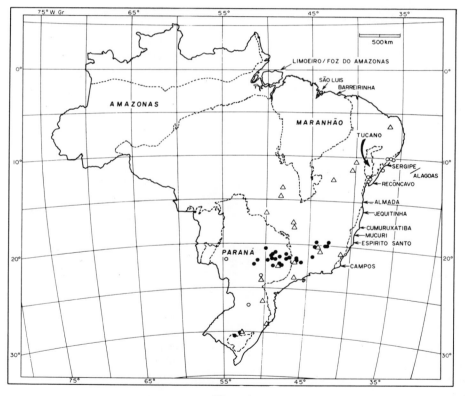

Figure 1

Locations of earlier and recent geothermal gradient measurements in Brazil. Closed cirles: recent measurements reported in the present work; open triangles: recent measurements by ARAUJO *et al.* (1976), HAMZA *et al.* (1976a, b) and VITORELLO *et al.* (1976, 1977); open circles: earlier measurements by UYEDA and WATANABE (1970). The names refer to the sedimentary basins where measurements of geothermal gradients were made by MEISTER (1973).

Table 2

Summary of recent measurements of geothermal gradients in Brazil

Number and locality	Latitude	Longitude	Tectonic Age (m. y.)	Depth Interval (m)	Rock type	Temperature gradient (°C/km)	Reference
1. Cassia de Coqueiros (S. Paulo)	21°17'	47°10'	Upper cretaceous	50– 80	Diabase Sill	15.7	Present work
				80–140	Sandstone (BT)	– –	
2. Batatais (S. Paulo)	20°54'	47°35'	,,	20– 80	Basalt (SG)	15.3	,,
				102–262.5	Sandstone (BT)	– –	
3. Brotas	22°16'	48°06'	,,	0–150	Sandstone (BT)	– –	HAMZA *et al.* (1976b)
4. Dourados (S. Paulo)	22°07'	48°19'	,,	60– 80	Basalt (SG)	13.4	,,
				80–120	,,	20.3	
5. Serra Azul (S. Paulo)	21°19'	47°34'	,,	24– 55	Diabase	– –	Present work
				55–196	Sandstone (BT)	– –	
6. Jau (S. Paulo)	22°18'	48°33'	,,	130–200	Basalt (SG)	18.4	VITORELLO *et al.* (1977)
7. Bauru (S. Paulo)	22°19'	49°04'	,,	0– 60	Sandstone (BA)	– –	HAMZA *et al.* (1976b)
				160–190	Sandstone (BT)		
				190–310	Sandstone (PR)	– –	
				310–364	Sand and Silt-stone (EN)	– –	
8. Itápolis (S. Paulo)	21°36'	48°49'	,,	90–130	Basalt (SG)	25.2	HAMZA *et al.* (1976b)
				130–140	,,	49.8	
				140–180	,,	24.5	
9. Jaboticabal (S. Paulo)	21°16'	48°19'	,,	50–200	Basalt (SG) Fractured	1.3	,,
10. Olímpia (S. Paulo)	20°45'	48°55'	,,	200–480	Basalt (SG)	23.7	Present work
				480–670	Basalt (Fractured ?)	19.3	
				700–810	Basalt (Highly fractured)	6.4	,,
				865–900	Sandstone (BT)	– –	
11. Uchoa (S. Paulo)	20°58'	49°10'	,,	100–130	Basalt (SG)	19.7	,,
				130–170	,,	33.8	
				170–180	,,	40.0	
12. Ibirá (S. Paulo)	21°05'	49°15'	,,	0– 70	Basalt (SG)	– –	,,
13. Lins (S. Paulo)	21°40'	49°44'	,,	0–130	Basalt (SG)	– –	,,
14. Novo Horizonte (S. Paulo)	21°29'	49°13'	,,	60–140	Basalt (Compact, fractured)	20.3	HAMZA *et al.* 1976b)
				140–170	Basalt (Compact)	25.3	
				180–320	Basalt (Compact and fractured)	18.3	
				320–380	Basalt (Fractured)	11.8	
				380–420	Basalt (Highly fractured)	6.7	
				420–460	Sandstone (BT)		

(BA = Bauru; BT = Botucatu; SG = Serra Geral; PR = Piramboia; EN = Estrada Nova).
– – Indicates low or near zero gradients.

Table 2—(*continued*)

Number and locality	Latitude	Longitude	Tectonic Age (m. y.)	Depth Interval (m)	Rock type	Temperature gradient (°C/km)	Reference
15. Votuporanga (S. Paulo)	20°25′	49°58′	Upper cretaceous	0– 50 220–380 390–420 420–470	Sandstone (BA) Basalt (SG) ,, Sandstone (BT)	– – 31.0 21.3 – –	Present work
16. Lucélia (S. Paulo)	21°43′	51°01′	,,	0–160		– –	,,
17. Presidente Prudente (S. Paulo)	22°08′	51°24′	,,	130–230	Basalt (SG)	37.0	,,
18. Butiá-Rio Pardo (R. G. Sul)	30°05′	52°30′	180–230		Shales and Limestones	30–42	Hamza *et al.* (1976a) Vitorello *et al.* (1977)
19. Cachoeira do Sul (R. G. Sul)	30°00′	52°55′	180–230		Shales and Limestones	25–35	Present work
20. Araranguá-I., Muller (S. Catarina)	28°40′	49°30′	230–270		Shales and Limestones	26–40	Vitorello *et al.* (1977)
21. Papanduva-Taió (S. Catarina)	26°23′	50°08′	270–350		Sandstones and Limestones	23–28	Hamza *et al.* (1976a) Vitorello *et al.* (1977)
22. Figueira (Paraná)	24°00′	50°25′	350–400		Sedimentary	22–28	Vitorello *et al.* (1977)
Alkaline intrusions							
23. Poços de Caldas (M. Gerais)	21°55′	46°25′	60– 80	Up to 400	Alkaline rocks	25–35	Araujo *et al.* (1976)
24. Piqueri (R. G. Sul)	30°11′	52°55′	<180	120–700	Alkaline rocks	24–26	present work
Shield and platform areas							
25. São Paulo (S. Paulo)	23°36′	46°34′	550–900	130–230	Granite	21–23	Hamza *et al.* (1976b) Vitorello *et al.* (1977)
26. Cabo Verde (M. Gerais)	21°28′	46°23′	550–900	50– 65	Granitoides	10.0	Present work
27. Botelhos (M. Gerais)	21°38′	46°24′	550–900	40– 60	Granitoides	9.2–22.5	,,

(BA = Bauru; BT = Botucatu; SG = Serra Geral; PR = Piramboia; EN = Estrada Nova).
– – Indicates low or near zero gradients.

Table 2—(*continued*)

Number and locality	Latitude	Longitude	Tectonic Age (m. ys)	Depth Interval (m)	Rock type	Temperature gradient (°C/km)	Reference
28. Cassiterita (M. Gerais)	21°07′	44°28′	550– 900	30– 65	Brasilides	12.0–31.8	Present work
29. Igarapé (M. Gerais)	20°06′	44°18′	900–1300	45– 80	PC-Undivided	8.9	,,
30. Piranga (M. Gerais)	20°41′	43°19′	>900	35– 85	,,	10.4	,,
31. S. D. Prata (M. Gerais)	19°52′	42°58′	>900	35– 65	,,	14.7	,,
32. C. Fabriciano (M. Gerais)	19°32′	42°37′	>900	40– 55	,,	8.7	,,
33. Currais Novos (R.G.N.)	6°20′	36°35′	550– 900		Paragneiss	19–25	VITORELLO *et al.* (1976, 1977)
34. Caraiba – Poço de Fora-(Bahia)	10°20′	40°10′	1800–2600		Mafics and Ultramafics	12–21	,,
35. Jacobina (Bahia)	11°15′	40°30′	1800–2600		Quartzites	6–10	,,
36. Arraial (Bahia)	12°32′	42°50′	1800–2600		Biotite Schist	11.5–13	,,
37. Cana Brava (Goiás)	13°30′	48°15′	550– 900		Serpentinites	16–18	VITORELLO *et al.* (1976, 1977)
38. Niquelândia (Goiás)	14°13′	48°18′	550– 900		Ultramafics	17–21	,,
39. Americano (Goiás)	16°25′	50°00′	550– 900		Mafics and Ultramafics	12–16	,,
40. Morro Agudo (M. Gerais)	17°30′	46°50′	550– 900		Dolomites	10–13	,,
41. Vazante (M. Gerais)	18°00′	46°45′	550– 900		Dolomites	11–12	,,
42. Nova Lima (M. Gerais)	20°00′	43°50′	1300–1800		Schists	14–16	,,
43. Bico de Pedra (M. Gerais)	20°26′	43°36′	1300–1800		Metasedimentary	7– 8	,,
44. Cachoeira do Itapemirim (E. Santo)	20°51′	41°06′	550– 900		Dolomites	11–12	,,

(BA = Bauru; BT = Botucatu; SG = Serra Geral; PR = Piramboia; EN = Estrada Nova).
– – Indicates low or near zero gradients.

and ages of tectonic units. Thermal conductivity measurements are being initiated and it is hoped that heat flow values will soon be available for most of the sites listed in Table 2. But for the present purpose we assume that the distribution of geothermal gradient values is indicative of the general pattern of terrestrial heat flow.

It is obvious from the results given in Tables 1 and 2 that regions of very high geothermal gradients are absent in Brazil. This is a natural consequence of the fact that there are no areas of young volcanism or active tectonics. The youngest volcanic activity has been 80–120 million years ago in the Parana Basin and manifested itself in the form of widespread eruption of tholeiitic flood basalts. Thus active geothermal systems such as vapour dominated systems are absent and one expects to encounter only low enthalpy hot water systems.

An interesting feature that can be noticed from Table 2 is an inverse correlation between geothermal gradient and geologic age (HAMZA *et al.*, 1976a, b; VITORELLO *et al.*, 1977). The shield areas are characterized by temperature gradients in the range 6 to 20°C/km while the Parana Basin is characterized by temperature gradients in the range 20 to 35°C/km. There are some areas on the eastern limits of the Parana Basin where the temperature gradients are in the range 10 to 20°C/km. This, however, simply reflects the transition from shield geotherm to the Basin geotherm. This transition is illustrated in Fig. 2, where geothermal gradients in compact sections of

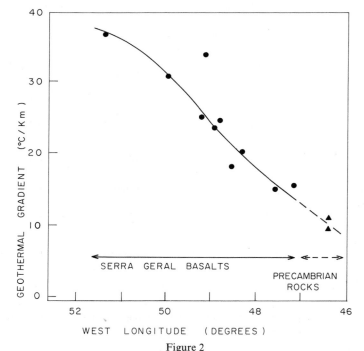

Figure 2

East–West variation of geothermal gradients in unfractured basaltic sections in the northeastern part of the Parana Basin.

basaltic flows are plotted against the west longitude for northeastern parts of the Parana Basin. The trend of increasing temperature gradients towards the central parts of the basin is in agreement with the preliminary results of MEISTER (1973).

Since geothermal gradient is directly proportional to terrestrial heat flow for constant thermal conductivity, a decrease of gradient with age suggests a similar decrease of heat flow with age. Available heat flow and geologic age data from continental regions have been interpreted by POLYAK and SMIRNOV (1968), HAMZA and VERMA (1969), VERMA et al. (1970), SCLATER and FRANCHETEAU (1970) and HAMZA (1976) as indicating a decrease of heat flow with age. Hence the gradient age relation suggests the important point that relatively high temperature gradients may be expected in younger sedimentary basins as for example in the Parana Basin. In addition, favourable hydrological conditions such as recharge and deep circulation exist in most of the Brazilian sedimentary basins. These two factors together therefore suggest that the suitable regions for geothermal energy exploitation in Brazil are younger sedimentary basins. Parana Basin in particular offers at present the best conditions for geothermal energy exploitation in Brazil because of its relatively high temperature gradients, highly permeable aquifers, impermeable cap rocks and suitable recharge zones.

Parana basin as a low temperature geothermal system

The Parana Basin is an intracratonic sedimentary basin that evolved during middle Paleozoic to early Cenozoic. It is approximately elliptical in shape and covers an area of about 1.6 million square kilometers. The maximum depth of the basin is estimated to be about 7800 m. Detailed descriptions of the geology of the basin have been published in the form of a large number of papers, short notes and review articles (NORTHFLEET et al., 1969; RAMOS, 1970; FULFARO, 1971; MÜHLMANN et al., 1974; RUEGG, 1975; SOARES, 1975 among others). The major rock types encountered in the basin are sandstones, siltstones, shales and tholeiitic flood basalts. Extensive hydrological studies of the basin have been carried out (HAUSMAN, 1966; LEINZ and SALLENTIEN, 1962; MEZZALIRA, 1967; ARID, 1970; PEREZ and HOLTZ, 1970; DAEE, 1974; REBOUÇAS, 1976 among others). The major aquifer in this basin is the Botucatu sandstone formation overlying in general Paleozoic Sedimentary formations and in some regions Piramboia formation of Triassic age. It is a highly permeable eolian sandstone of Triassic age with thicknesses of up to 500 m. It out-crops extensively along the eastern limits of the basin which are also the main recharge zones. The formation gradually dips down to about 2000 m in the central parts of the basin. Overlying the Botucatu formation is the basaltic lava flows of the Serra Geral Formation. The thickness of the basaltic flows vary considerably and there is a general increase in thickness towards the central parts of the basin. The average thickness has been estimated to be about 650 m (LEINZ et al., 1966), but thicknesses of up to 1500 m have been observed in some oil wells. The basalts, when not intensely fractured, have much less permeability than the Botucatu sandstone formation, thus acting as an effective cap rock for the Botucatu aquifer. Overlying the Serra Geral Formation are the Cenozoic sedimentary formations, principally the Bauru Sandstone.

In the southern parts of the Parana Basin temperature measurements were carried out in exploratory boreholes drilled for coal and uranium while in northeastern parts measurements were made in wells drilled for groundwater as well as in a few abandoned oil wells.

Figures 3 and 4, show some of the typical temperature distributions that were obtained in the major formations Bauru, Serra Geral and Botucatu. The most notable feature is the presence of undisturbed temperature gradients encountered in wells passing through basaltic formations (Fig. 3) in contrast to the absence of any observable gradient within aquiferous regions (Fig. 4). The temperature gradients in basalts are generally in the range 20 to 30°C/km; when the basalts are highly

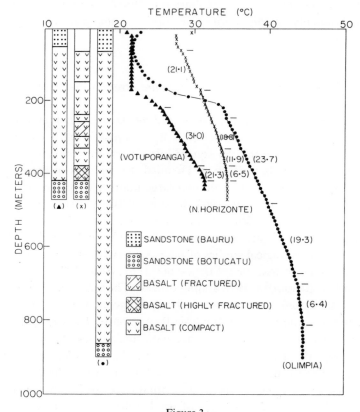

Figure 3

Example of temperature distributions in wells passing through basalts. (The numbers in brackets are gradients in °C/km.)

fractured however, the gradients are much lower as in the case of the bottom part of the well at Novo Horizonte in Fig. 3. Hence much of the variations in temperature gradients in basalts are most probably due to groundwater movement through fracture zones. On the other hand extremely low gradients were observed in aquiferous zones of the Bauru, Botucatu and Piramboia formations as for example in the case of the wells shown in Fig. 4. The well at Bauru is particularly interesting as it shows the absence of any temperature gradient in the Bauru and Botucatu aquifers and the presence of a small gradient of less than 5°C/km in the Piramboia aquifer. The aquifers seem to be separated by thin impermeable layers with the result that each aquifer has a characteristic temperature of its own for a specific depth. Of these the Botucatu sandstone (the main aquifer of the region) has a porosity of more than 20 percent. The average thickness of this aquifer is about 300 m and measurements in several wells have indicated (see Table 2) lack of any temperature gradient at least through the topmost 100 m of the formation. This indicates that the water layer in the Botucatu aquifer is convecting. In order to have convection in a porous

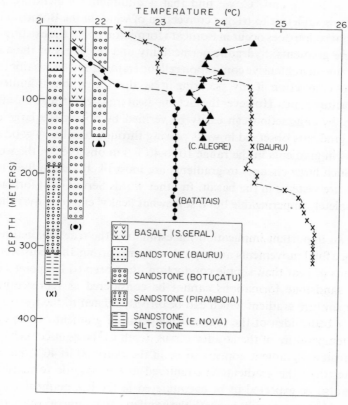

Figure 4
Examples of temperature distributions in wells passing through aquifers.

layer it is necessary that the following relation be satisfied (HORTON and ROGERS, 1945):

$$Kgh^2 \left(\frac{\partial T}{\partial Z}\right)\left(\frac{\partial \rho}{\partial T}\right) > 4\pi^2 D\eta$$

where K = Hydraulic permeability
         g = Acceleration due to gravity
         h = Thickness of the convective system
         D = Thermal diffusivity
         η = Viscosity of water
         $\dfrac{\partial T}{\partial Z}$ = The excess temperature gradient over the adiabatic gradient (the latter can be considered negligible for water)
         $\dfrac{\partial \rho}{\partial T}$ = Density variation with temperature

By assuming $D = 0.01$ cm^2/s, $\eta = 0.7 \times 10^{-2}$ poise, $K = 3 \times 10^{-8}$ cm^2 and $\partial\rho/\partial T = 0.5 \times 10^{-3}$ g/cm^3 °C, we find that a minimum temperature gradient of 15 to 35°C/km is sufficient to trigger convection of water in the Botucatu formation. Thus while convection does occur in Botucatu formation under the existing conditions of temperature gradients, hydraulic permeability and thickness of the aquifer, it is unlikely that any such extensive convection cells operate within the basaltic formations. It appears that convection, if any, occuring in basaltic formations is limited to highly permeable fracture zones. However the fact that heat transfer through basaltic sections is not wholly by conduction can easily be verified by noting the large changes in temperature gradients observed in wells passing through basalts. A good example is the variations in gradients in the range 10 to 40°C/km observed in the well at Novo Horizonte. Such large changes in gradients are most likely due to fluid movements through fracture systems in the basalt. In other words Serra Geral formation serves not as a completely impermeable but a somewhat 'leaky' cap rock over the Botucatu aquifer.

This has an important implication for estimating the true regional temperature gradient. Since fluid movements are capable of transporting and redistributing much larger amounts of heat than simple conduction, the observed temperature gradients in basalt or sandstone formations cannot be considered as representing the true regional temperature gradient which can be directly related to the true regional heat flux. In fact a better idea of the regional temperature gradient can be obtained by plotting the temperature of the aquifer versus depth to the aquifer such as shown in Fig. 5. The regional gradient appears to be in the range 30 to 40°C/km and is thus slightly higher than the gradients encountered in the basaltic formations. Such a gradient can also be expected to be encountered in the less permeable sedimentary formations and basement rocks beneath the aquifers, assuming of course that thermal conductivity contrasts are a minimum.

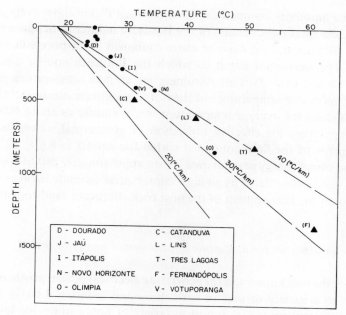

Figure 5

Water temperature in the aquifer versus depth to the top of the Botucatu aquifer. (Closed circles are measurements made in wells from which water is not being pumped. Triangles are measurements of water temperature at the mouth of the well during flow.)

Thus, in short, the vertical distribution of temperature gradients in the Parana Basin can be described in the following way:

Basaltic rocks of the Serra Geral (Leaky cap rock)	10 to 30°C/km
Bauru, Botucatu and Piramboia (Aquifers)	0 to 5°C/km
Paleozoic Sedimentary Formations and Basement Rocks (Regional geothermal gradient)	30 to 40°C/km

Such a distribution of temperature gradients can be considered as indicative of the existence of a low enthalpy geothermal system.

Estimates of energy reserves

An approximate estimate of the total energy contained in the geothermal waters in the Botucatu aquifer can be obtained based on certain simplifying assumptions. The area of the Parana Basin where the Botucatu aquifer is dipping to depths of more than 500 m is estimated to be roughly about 100 000 km². If the average thickness of the aquifer is assumed as 350 m (DAEE, 1972) and the average porosity as 20 percent, the total pore volume of the Botucatu aquifer at depths greater than 500 m is 7×10^{12} m^3 and the mass of water contained in the pore volume is about 7×10^{15} kg. The maximum depth to which the Botucatu aquifer dips down is estimated to be less than 2500 m. Assuming an average temperature gradient of 35°C/km, the maximum temperature of the Botucatu aquifer can be estimated to be about 90°C. Taking the average temperature of the aquifer as about 60°C and the minimum temperature for effective utilization of geothermal waters as 30°C the useful heat energy of the fluid contained within the aquifer is 8.8×10^{20} J (8.3×10^{17} Btu). The stored energy per unit area is then approximately 280 MW-years/km^2. These figures can be considered as giving a conservative estimate because it does not take into account the heat content of the host rock (Botucatu sandstone formation).

Possibilities for utilization of geothermal energy in Brazil

Apart from the well known use for generating electrical power, geothermal energy has been put to a variety of uses in agriculture and industry. LINDAL (1973) and KUNZE and RICHARDSON (1975) report a variety of applications for low enthalpy geothermal waters. The main uses for geothermal waters with temperatures in the range 40 to 100°C appear to be in space heating and in agriculture, for regions where relatively long and cold winter seasons prevail.

The availability of hot water from the Botucatu aquifer with temperatures in the range 40 to 90°C is limited principally to the western parts of Sao Paulo, Parana and Rio Grande do Sul states. The eastern parts of the state of Mato Grosso can also possibly be included but we have no measurements at present west of the Parana river. These are in general main agricultural areas with large plantations of coffee, sugarcane, soya beans and other common vegetables. Thus possible uses of geothermal waters in agriculture in these areas is worth investigating. However these regions have warm climates with short and relatively mild winter seasons and hence routine applications such as soil warming using geothermal waters are not appealing and economical. The damaging effects of very cold weather excursions on these agricultural plantations can in principle be minimised by the use of geothermal waters. However because of the large aerial extent an uneconomically big and extensive production and distribution system would be needed. In addition the cold weather excursions capable of producing damaging effects are of short duration, aperiodic and random in nature which makes countermeasures difficult. Applications involving space heating are also not very appealing because of the existing favorable climatic conditions. Conversion of geothermal energy into electrical energy is also very likely to be uneconomical because of the extremely low efficiency of the conversion process at low temperatures of 40 to 90°C. Thus the Parana Basin geothermal system cannot at present be considered as an independent economically exploitable energy resource.

There are however other possible secondary uses. POWELL *et al.* (1976) discussed a high-efficiency power cycle using molecular hydrogen as a working fluid. This process operates between a high temperature source (for example a nuclear reactor) and a low temperature source. The low temperature source as POWELL *et al.* (1976) pointed out could be a low enthalpy geothermal system. In view of the existing plans for generations of electrical power from nuclear fuels we believe a proper integration of nuclear power generating systems with low enthalpy geothermal sources would be suitable to the existing conditions in Brazil.

Another possible use of geothermal waters could be in the production of methane from waste fluids and biodegradable waste materials. The waste material needs to be heated to a temperature of 30 to 50°C for the bacteria to generate methane. Usually up to 30 percent of the gas produced in such a system is used for heating the fluids. Use of geothermal waters for heating could result in an improvement in the overall efficiency of such gas producing systems. Many of the industrial and other applications mentioned by LINDAL (1973) and KUNZE and RICHARDSON (1975) require temperatures higher than 100°C. This is slightly higher than the temperature range encountered for geothermal waters in the Botucatu aquifer. But significant savings can be realized through the use of a combined system of geothermal and conventional heating systems in which the latter is employed to raise the temperature of the geothermal waters to the desired level.

Conclusions

Analysis of geothermal investigations carried out so far suggest that younger sedimentary basins with favorable hydrological conditions are the best sites for the extraction of geothermal energy in Brazil. The Parana Basin appears at present to be the best region. Our measurements give a regional geothermal gradient of 30 to 40°C/km for the Parana Basin in contrast to gradients of less than 20°C/km for the Brazilian shield. Preliminary interpretation of the temperature distributions in the Botucatu formation suggests that this aquifer contains substantial quantities of waters in the temperature range 40 to 90°C and can be considered as a low enthalpy geothermal system.

The total energy content of this low enthalpy geothermal system is substantial and the cost of extraction is comparable to the normal costs encountered in drilling wells for groundwater. Several uses for these geothermal waters exist in agriculture and industry. But overall considerations of the economy suggest that the Parana Basin geothermal system cannot by itself be considered an economically exploitable independent energy resource. Nevertheless this geothermal system can make a significant contribution when properly integrated into another economically viable system where it can increase the overall efficiency of the latter and thus lead to substantial savings.

ACKNOWLEDGEMENTS

The present work was supported by Instituto Astronômico e Geofísico da Universidade de São Paulo, Instituto de Pesquisas Tecnológicas and Conselho Nacional de Pesquisas (operating research grant 12787/74). One of us (R.L.C. Araujo) gratefully acknowledges the receipt of a research scholarship from Fundação de Amparo à Pesquisa do Estado de São Paulo.

We are thankful to Departamento de Água e Energia Elétrica, São Paulo (DAEE), Companhia de Saneamento de Minas Gerais (COPASA), Departamento Nacional de Produção Mineral (DNPM), Companhia de Pesquisa de Recursos Minerais (CPRM) and Empresas Nucleares Brasileiras S.A. (NUCLEBRAS) for their active cooperation and assistance during the execution of this work.

We are indebted to Dr. A. E. Beck for donating part of the equipment used for temperature loggings and for the considerable help and advice that we received during the initial execution of the project. We are also thankful to our colleagues at IAG for interesting and stimulating discussions on this topic.

REFERENCES

ARAUJO, R. L. C., HAMZA, V. M., VITORELLO, I. e POLLACK, H. N. (1976), *Resultados do gradiente geotérmico obtidos na Chaminé Alcalina de Poços de Caldas*, 29º Congresso Brasileiro de Geologia, Resumo dos Trabalhos, p. 97.
ARID, P. M. (1970), *Estudos hidrogeológicos no município de São José do Rio Preto*, São Paulo, Bol. Soc. Bras. Geol., *19*, 43–69.
DAEE (1972), *Estudo de Águas Subterrâneas*, Avaliação Prelimiar, Report prepared for Departamento de Águas e Energia Elétrica by Geopesquisadora Brasileira Ltda. and Tahal Consulting Engineers Ltd.
DAEE (1974), *Estudo de Águas Subterrâneas*, Região Administrativa 6, Ribeirão Preto, Vol. 1–4. Report prepared for Departamento de Águas e Energia Elétrica by Geopesquisadora Brasileira Ltda. and Tahal Consulting Engineers Ltd.
FULFARO, N. J. (1971). *A evolução tectonica e paleogeografica da bacia sedimentar do Paraná pelo Trend Surface Analysis*, Esc. Eng. São Carlos, Geol. 14, pp. 112, Tese de Livre Docência.
HAMZA, V. M., ARAUJO, R. L. C., VITORELLO, I. and POLLACK, H. N. (1976a), *Gradientes térmicos na bacia do Paraná*, Ciência e Cultura (Suplemento), *28*, 207.
HAMZA, V. M., VITORELLO, I. and POLLACK, H. N. (1976b), *Estado atual das investigações geotérmicas no Brasil*, 29º Congresso Brasileiro de Geologia, Resumo dos Trabalhos, p. 96.
HAMZA, V. M. and VERMA, R. K. (1969), *The relationship of heat flow with age of basement rocks*, Bull. Volcanologique, *33*, 123–152.
HAMZA, V. M. (1976), *Possible Extension of Oceanic Heat Flow-Age Relation to Continental Regions and the Thermal Structure of Continental Margins*, Anais de Acad. Bras. Ciências, *48*, 121–131.
HAUSMAN, A. (1966), *Comportamento do freatico nas áreas basálticas do Rio Grande do Sul*, Bol. Paranaense de Geografia, *18–20*, 177–213.
HORTON, C. W. and ROGERS, F. T. (1945), *Convection currents in a porous medium*, J. Appl. Phys., *16*, 367–370.
KUNZE, J. F. and RICHARDSON, A. S. (1975), *National program definition study for the non-electric utilization of geothermal energy*, U.S. ERDA Report ANCR – 1214.
LEINZ, V., BARTOSELLI, A., SADOWSKI, G. R. and ISOTA, C. A. L. (1966), *Sobre o comportamento espacial do trapp basáltico da Bacia do Paraná*, Bol. Soc. Bras. Geol., *15*, 79–91.

LEINZ, V. and SALLENTIEN, B. (1962), *Água Subterrânea no Estado de São Paulo e regiões limítrofes*, Bol. Soc. Bras. Geol., *11*, 27–36.

LINDAL, B., *Industrial and other applications of Geothermal Energy*, in *Geothermal Energy: Review of Research and development* (ed. by H. C. H. Armstead) (UNESCO, Paris, 1973), 135–147.

MEISTER, E. (1973), *Gradientes Geotérmicos nas Bacias Sedimentares Brasileiras*, Boletim Técnico da Petrobrás, *16*, 221–232.

MEZZALIRA, S. (1967), *Atualização dos estudos e captação de água subterrânea feitos pelo IGG no estado de São Paulo, 1965–1967*; Bol. IGG, *19*, 83–91.

MÜHLMANN, H., SCHNEIDER, R. L., TOMMASI, E., MEDEIROS, R. A. and DAEMON, R. F. (1974), *Revisão estratigráfica da Bacia do Paraná*, 28º Congr. Bras. Geol., *1*. Resumo das Comunicações, 812–815.

NORTHFLEET, A. A., MEDEIROS, R. A. and HULMANN, H. (1969), *Reavaliação dos dados geológicos da Bacia do Paraná*, Bol. Tec. Petrobrás, *12*, 291–343.

PEREZ, H. H. and HOLTZ, A. C. (1970), *Bacia da Prata – Inventário e Análise crítica sobre os recursos naturais, mapa hidrogeológico*, OEA, Washington.

POLYAK, B. G. and SMIRNOV, Ya. B. (1968), *Relationship between terrestrial heat flow and the tectonics of the Continents*, Geotectonics, 205–213.

POWELL, J. R., SALZANO, F. J., WEN-SHI, Yu. and MILAU, J. S. (1976), *A high efficiency power cycle in which Hydrogen is compressed by absorption in metal hydrides*, Science, *193*, 314–317.

RAMOS, A. N. (1970), *Aspectos paleo-estruturais da Bacia do Paraná e sua influência na sedimentação*, Bol. Tec. Petrobrás, *13*, 85–93.

REBOUÇAS, A. C. (1976), *Recursos Hídricos Subterrâneos da Bacia do Paraná, análise de Pré-Viabilidade*, p. 143, Tese de Livre Docência, Universidade de São Paulo, Brasil.

RUEGG, N. R. (1975), *Modelos de variação química na província basáltica do Brasil meridional*, Tese de Livre Docência, Universidade de São Paulo, Brasil.

SCLATER, J. C. and FRANCHETEAU, J. (1970), *The implications of terrestrial heat flow observations on current tectonic and geochemical models of the crust and upper mantle of the earth*, Roy. Astr. Soc. Geophys. Jour. *20*, 509–542.

SOARES, P. C. (1975), *Divisão estratigráfica do Mesozóico no estado de São Paulo*, Rev. Bras. Geol. *5*, 229–251.

UYEDA, S. and WATANABE, T. (1970), *Preliminary report of terrestrial heat flow study in the south American continent; Distribution of geothermal gradients*, Tectonophysics, *10*, 235–242.

VERMA, R. K., HAMZA, V. M. and PANDA, P. K. (1970), *Further study on the correlation of heat flow with the age of basement rocks*, Tectonophysics, *10*, 301–320.

VITORELLO, I., HAMZA, V. M., POLLACK, H. N. and ARAUJO, R. (1977), *Geothermal Investigations in Brasil*, Submitted for Publication in Revista Brasileira de Geociências.

VITORELLO, I., POLLACK, H. N., ARAUJO, R. and HAMZA, V. M. (1976), *Resultados preliminares das investigações geotérmicas no escudo cristalino no território brasileiro*, Ciência e Cultura, (suplemento), *28*, 207.

(Received 7th November 1977; revised 16th May 1978)

Pageoph, Vol. 117 (1978/79), Birkhäuser Verlag, Basel

Investigation for Geothermal Energy in Sweden

By K. G. Eriksson, K. Ahlbom, O. Landström, S. Å. Larson, G. Lind and D. Malmqvist[1])

Abstract – Preliminary investigations of the geothermal energy potential in Sweden are being carried out in crystalline rocks of Precambrian age, as well as in the Triassic Buntsandstone. The geothermal potential of fracture zones is also being investigated. Different methods for prospecting have been tried and compared.

Key words: Precambrian granites; Fracture zones; Warm water aquifers.

Introduction

The bedrock of Sweden is part of the Baltic shield and is composed of Precambrian crystalline rocks, except for Paleozoic and Mesozoic sedimentary rocks in southern Sweden (Fig. 1). Cambrian-Silurian rocks are present as small outliers in several areas within southern Sweden, and in the Caledonides; however, the sedimentary rocks of the Caledonides are more metamorphosed than the sedimentary rocks in southern Sweden.

Recent volcanism as well as hydrothermal activity is lacking and the development of geothermal energy is therefore directed towards

- sedimentary rocks with aquifers containing warm water; mainly in southern Sweden, and
- crystalline rocks with higher than normal heat-flow.

The sedimentary rocks

In the southwesternmost part of Sweden, a sedimentary rock sequence of Paleozoic, Mesozoic and Cenozoic age is present. The total thickness is about 2.600 m (Fig. 2). Exploration for salt deposits during World War II resulted in the discovery of warm water aquifers in the Cenomanian sandstone (Cretaceous), and in the

---

[1]) Department of Geology, Chalmers University of Technology and University of Göteborg, S-402 20 Göteborg, Sweden.

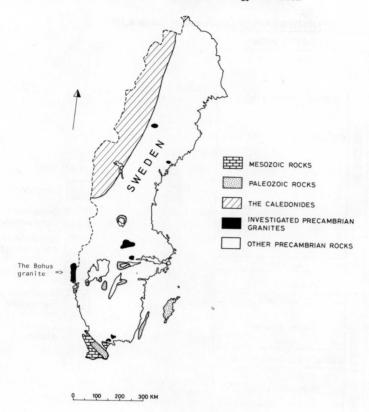

MESOZOIC ROCKS

PALEOZOIC ROCKS

THE CALEDONIDES

INVESTIGATED PRECAMBRIAN
GRANITES

OTHER PRECAMBRIAN ROCKS

The Bohus
granite =>

0   100   200   300 KM

Figure 1
Simplified petrological map of Sweden.

Buntsandstone (Triassic). Figure 3 shows a temperature–depth diagram from one of these boreholes, Ljunghusen 1, giving a mean temperature gradient of 35°C/km (Arbetsgruppen för geotermisk energi, 1977). The temperature of the water in the Cenemonian sandstone reaches only 40°C due to its shallow depth, whereas in the Buntsandstone, water temperature of 50–60°C has been measured.

The salt content of the water is fairly high, in some cases up to approximately 25 per cent (BROTZEN, 1956).

The investigation is divided into two parts:

– A regional study using geophysical and borehole logging data, to provide a hydrogeological model.
– Redrilling of an old borehole (Höllviksnäs 1) and a pumping test of the Buntsandstone aquifer. This sandstone is at a depth of 1.860–2.060 m and consists of a sequence of sandstones intermixed with thin layers of conglomerate, clay- and siltstones. The first drilling indicated high hydraulic conductivity.

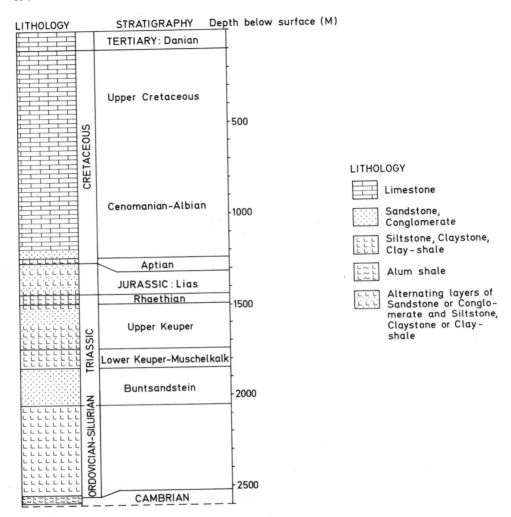

Figure 2

Stratigraphical profile through the sedimentary rock sequence at Höllviksnäs, southern Sweden (WESSLÉN, 1976).

The above mentioned projects will be carried out during 1978. The well site will serve as an experimental plant, where the use of warm saline water for domestic heating will be evaluated.

Crystalline rocks

Crystalline Precambrian rocks constitute the main part of the rocks in Sweden (Fig. 1). The search for 'hot regions' in Precambrian crystalline rocks having sufficiently high heat-flow and temperature gradients to be of interest for geothermal

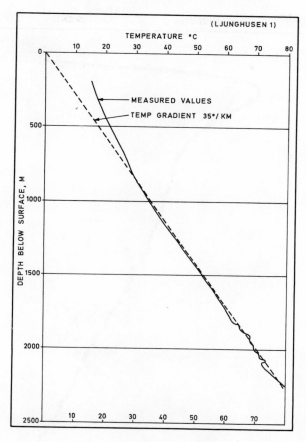

Figure 3

Temperature–depth diagram from borehole Ljunghusen 1, southern Sweden. The deviation of the measured temperature above 800 metres from the 35°C/km gradient can be explained by a higher thermal conductivity or/and the temperature rise at the end of the last ice age.

exploration started in 1975. In some areas a linear correlation exists between heat-flow and the heat generated from radioactive elements in crystalline rocks (BIRCH et al., 1968). Accordingly interest has been focused on rocks, rich in potassium, thorium and uranium.

From regional mapping of the radioactivity of different rocks in Sweden, which was carried out in the late fifties, as well as from the results of more recent prospecting for uranium, areas with abnormal radioactivity were selected for more detailed investigations. These areas are connected to anorogenic and serorogenic granites (LANDSTRÖM et al., in preparation). Fairly extensive geothermal investigations have been carried out in the Bohus granite, which outcrops on the Swedish west coast (LANDSTRÖM et al., in preparation). This granite is about 900 m.y. old (SKIÖLD, 1976).

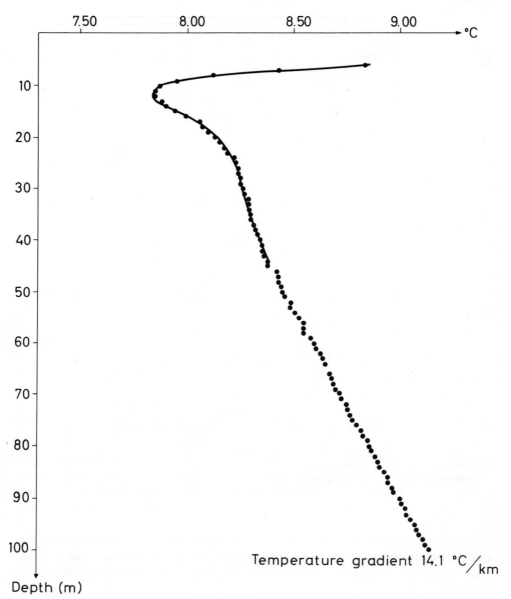

Figure 4
A typical temperature–depth curve in the Bohus granite area. The upper part of the curve shows the annual
temperature wave caused by the surface temperature variations down to approximately 50 metres below
the surface. Solid line: calculated model curve (from LANDSTRÖM *et al.*, in preparation).

Excellent exposures made the Bohus granite suitable for sites for exploration. Different geothermal parameters of the Bohus granite were determined including temperature, temperature gradient (Fig. 4), heat production, thermal conductivity, thermal diffusivity as well as heat capacity and heat-flow. Mean values of some of these parameters are listed in Table 1.

Table 1

Geothermal parameters of the Bohus granite

Th (ppm)	U (ppm)	K (%)	Heat generation ($\mu W\ m^{-3}$)	Thermal conductivity ($Wm^{-1}\ K^{-1}$)	Temperature gradient ($mK\ m^{-1}$)	Heat flow ($mW\ m^{-2}$)
44.3	9.7	4.3	6.0	4.43 ± 0.67	16.2 ± 3.3	72
Mean of 69 analyses				Mean of 33 measurements	Mean of average gradients in 17 drill-holes	

The geothermal prospecting is at present being continued in other granite areas in Sweden (Fig. 1).

Determinations of the radiogenic heat production are based on laboratory analysis of rock samples from outcrops and drillcores, and in addition on measurements *in situ* by gamma-spectrometric methods. The delayed neutron method (U and Th) and instrumental neutron activation analysis (Th and K) proved to be suitable techniques for the rock types involved.

Heterogeneous distribution of the trace elements Th and U and the leaching effect due to weathering render the collection of surface samples a somewhat difficult problem. Some advantage of *in situ* analyses has been manifested in comparative experiments: It is possible to make a great number of surface analyses, and the volume sampled is considerable. In core-free drilled boreholes the depth distribution of the specific heat generation can be studied by gamma spectrometric methods. This is of importance as core drilled boreholes are usually rare in non-mineralized areas. Both NaI and Ge(Li) detectors have been tested.

Thermal conductivity has been measured *in situ* by using a heating rod and measuring the temperature, assuming a homogeneous cylindrical configuration of the heat distribution. The advantage of this *in situ* method is that the natural moisture in the rock is preserved and a large rock volume is measured (for further details see LANDSTRÖM *et al.*, in preparation).

In Sweden, few heat-flow values have been determined. However, other heat-flow values from the Baltic shield have been recorded in Norway and Finland. Most of the heat-flow values in Norway and Sweden have been determined from measurements in mines. Figure 5 shows a compilation of these results including a value from the sedimentary rocks in southern Sweden. Due to the unknown thermal conductivity in these rocks, an approximate conductivity value of 5.5 mcal/cm, sec, °C (2.3 W/m, K)

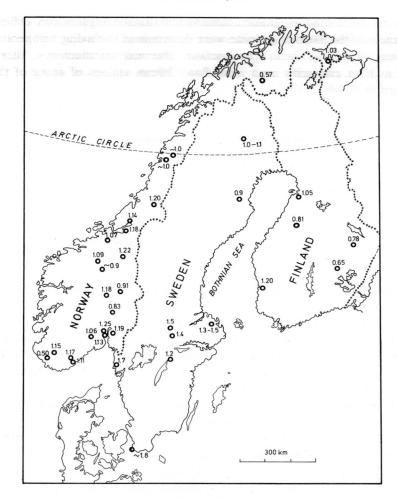

Figure 5

Heat flow values in Sweden (from LANDSTRÖM *et al.*, in preparation), Norway (GRØNLIE *et al.*, 1977) and Finland (PURANEN, 1968). Figures give heat-flow in HFU units. 1 HFU = 41.8 mW/m². Concerning heat-flow in Denmark, see BALLING and SAXOV (1978) in this volume.

has been chosen. It is also evident from Fig. 5 that the Bohus granite shows the highest known heat-flow in Precambrian crystalline rocks in the Baltic shield.

Fracture zones

The Swedish geothermal program also includes studies on the extraction of geo-thermal energy from fracture zones. Due to the comparatively low heat-flow present in the Baltic shield, a great amount of warm water has to be extracted. It is therefore necessary to find fractures that could provide a large contact area between the fluid and the rock. A combination of a rock with high heat production and a fracture zone through this rock could provide a geothermal system that could be run economically. In summer 1977, an investigation started of a topographical lineament, which can be traced for 80 kilometres within Precambrian gneisses along the Bullaren lakes, just east of the Bohus granite. Geological mapping, seismic, electric and gravity measurements were used to study the character of this lineament (AHLBOM et al., in preparation). The interpretation of the gravity measurements show that the Bohus granite dips gently beneath the gneisses and is also present beneath the lineament (LIND, 1967). Furthermore, gravity and seismic refraction measurements show the thickness of the Quaternary cover in the valley along the lineament. The seismic velocity is also being used to obtain an indication of the fracturing of the rock. Another method that will be used to determine the degree of fracturing of the rock is magnetic and electromagnetic airborne mapping. From the geological mapping it is possible to get information about the petrology and the tectonic history of the area. If the result of these investigations indicate that the lineament is a suitable fracture zone, the next proposed step will be shallow drilling to investigate the hydraulic parameters of the fracture zone.

Conclusions

Prospection for geothermal energy in Sweden is at present in an initial phase. However, there is good hope to be able to use aquifers within sedimentary rocks in southernmost Sweden for domestic heating. In the future, an exploitation of crystalline 'hot' dry rock is also possible, especially considering the techniques already developed at Los Alamos Scientific Laboratory, USA. Investigations already carried out in Sweden show that heat-flow even in Precambrian shield areas locally can be comparatively high. Fracture zones crossing anomalous heat generating rocks could be favourable as sites for geothermal plants.

REFERENCES

Arbetsgruppen för geotermisk energi, LUND, L. T. H., *Geotermisk energiutvinning i Skåne*. Delrapport 1. *Nämnden för energiproduktionsforskning* (Stockholm, 1977).

AHLBOM, K., LARSON, S. Å. and LIND, G., in preparation, *Geofysisk-geologisk undersökning av en sprickzon i östra Bohuslän*.

BIRCH, F., ROY, R. F. and DECKER, E. R. *Heat flow and thermal history in New York and New England*, in *Studies of Appalachian geology: Northern and Maritime* (eds.: Zen, E., White, W. S., Haddly, J. B. and Thomas, J. B.) (John Wiley & Sons Inc., 1968).

BROTZEN, F. (1956), *Sveriges geologiska undersöknings borrningar efter salt, gas och olja i sydvästra Skåne*, in *En redogörelse för arbeten utförda under åren 1937–1955*, Internal report, S.G.U. (1956).

GRØNLIE, G., HEIER, K. S. and SWANBERG, C. A. (1977), *Terrestrial heat-flow determinations from Norway*, Norsk Geologisk Tidskrift, *57*, 153–162.

LANDSTRÖM, O. and MALMQVIST, D. (in preparation). *Prospecting for heat resources in Precambrian shield areas.*

LANDSTRÖM, O., LARSON, S. Å., LIND, G. and MALMQVIST, D. (in preparation), *Geothermal investigations in the Bohus granite area in southwestern Sweden.*

LIND, G. (1967), *Gravity measurements over the Bohus granite in Sweden.* Geol. Fören. Stockholm Förh. *88*, 542–548.

PURANEN, H., JÄRVIMÄKI, P., HÄMÄLÄINEN, U. and LEHTINEN, S. (1968), *Terrestrial heat flow in Finland* Geoexploration, *6*, 151–162.

SKIÖLD, T. (1976), *The interpretation of the Rb–Sr and K–Ar ages of late Precambrian rocks in southwestern Sweden,* Geol. Fören. Stockholm Förh. *98*, 3–29.

WESSLÉN, A. *Förstudie för utvinning av geotermisk energi. Slutrapport. Nämnden för energiproduktions-forskning* (Stockholm, 1976).

(Received 7th November 1977, revised 3rd April 1978)

Pageoph, Vol. 117 (1978/79), Birkhäuser Verlag, Basel

Low Enthalpy Geothermal Energy Resources in Denmark

By Niels Balling and Svend Saxov*)

Abstract – The deep oil exploration drillings in Denmark have shown that especially the Danish Embayment contains low enthalpy geothermal resources associated with warm aquifers. The most promising reservoirs have been found in highly permeable Upper Triassic sand and sandstone beds, which cover at least 5000 km^2 at depths of 2000–3000 m and at temperatures of 60–100°C. The porosity of the main reservoir is of 15–25%, and the permeability is presumed to be approximately 1 darcy ($\simeq 10^{-12}$ m^2) or higher. A layer thickness of 30–60 m has been observed on a number of localities. Also the Middle Jurassic and the Lower Triassic contain reservoirs of interest. A major geothermal exploration work is planned with seismic investigations, drillings to depths of 2000–4000 m and probably establishment of pilot district heating plants.

Key words: Low enthalpy geothermal energy resources; Sedimentary basin; Economics of geothermal energy utilization; Denmark.

Introduction

About 50 years ago prospecting for oil was commenced in Denmark. Some 200 on shore drillings have been undertaken, but most of the wells were very shallow. About 50 wells were deeper than 1000 m. Due to the drillings (Sorgenfrei and Buch, 1964; Rasmussen, 1972 *et al.*) and to detailed gravimetric and seismic investigations the structural geological conditions are rather well known, and also some valuable information on temperature is available (Madsen, 1975).

In Fig. 1 we give a simplified structural geological map of Denmark. The area is part of the North Sea (or Northwest European) Sedimentary Basin and contains the following main structural units: The Fennoscandian Border Zone and the Ringkøbing-Fyn High, both with shallow Precambrian basement (depth 500–2000 m), the Danish Embayment and the northern part of the North German Basin with thick sedimentary sequences, which in the Danish Embayment may reach 8000–10 000 m. Here low enthalpy geothermal energy resources are associated with porous and permeable strata. We will concentrate on the Embayment and especially on the northwestern part where rather detailed information is available.

*) Laboratory of Geophysics, Aarhus University, Finlandsgade 6–8, DK-8200 Aarhus N, Denmark.

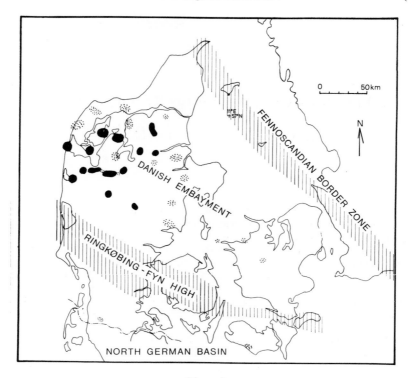

Figure 1
Simplified structural geological map of Denmark. Areas with shallow Precambrian basement (depths
500–2000 m) are hatched. Shallow salt structures (200–1000 m) are shown in black, and deeper seated
structures are indicated with dots.

Geology and temperatures

The main stratigraphic units are given in Fig. 2. The Haldager formation in the
Middle Jurassic, the Gassum formation mainly of Upper Triassic (Rhaetic) age and
the Lower Triassic (Buntsandstein) strata are of primary interest (HANDELS-
MINISTERIET, 1977). They all contain porous and permeable beds of sand and sandstone. In the deepest part of the Danish Embayment the permeable strata have been
found at depths between 1500 m and 5000 m. The average geothermal gradients in
this depth interval vary from 20 to 35°C km^{-1}. Regionally, the highest gradients are
found to the northwest in the Embayment, and the lowest values along the Fennoscandian Border Zone associated with a heat flow transition zone (BALLING, 1976a).
Locally, the salt structures due to their higher thermal conductivity cause both positive and negative temperature anomalies.

Buntsandstein rocks have been observed in 3 drillings with thicknesses from
700 m to more than 1200 m at depths from about 3000 m to about 5000 m. At these

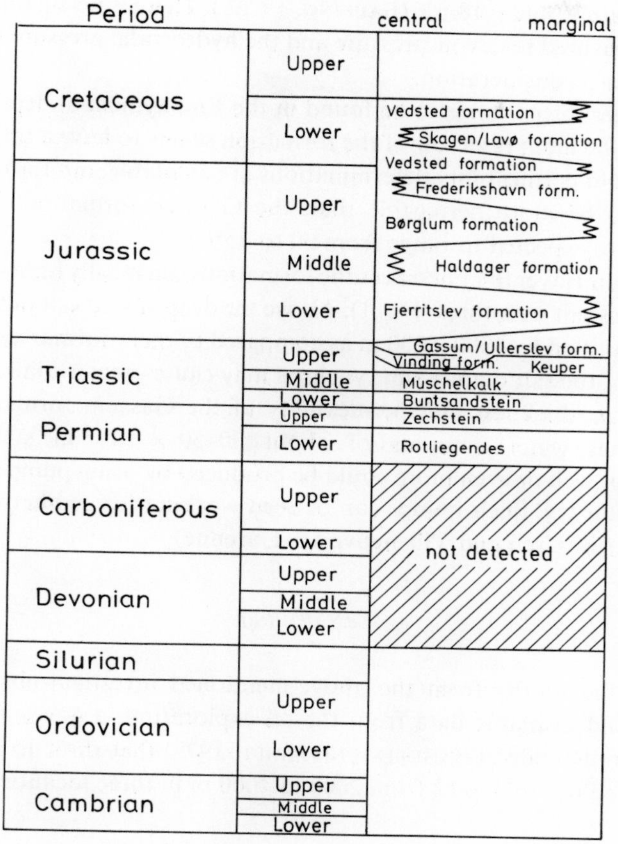

Period		central marginal
Cretaceous	Upper	
	Lower	Vedsted formation ⋛ ⋛ Skagen/Lavø formation
Jurassic	Upper	Vedsted formation ⋛ ⋛ Frederikshavn form. Børglum formation ⋛
	Middle	⋛ Haldager formation
	Lower	Fjerritslev formation ⋛
Triassic	Upper	Gassum/Ullerslev form. Vinding form. Keuper
	Middle	Muschelkalk
	Lower	Buntsandstein
Permian	Upper	Zechstein
	Lower	Rotliegendes
Carboniferous	Upper	
	Lower	not detected
Devonian	Upper	
	Middle	
	Lower	
Silurian		
Ordovician	Upper	
	Lower	
Cambrian	Upper	
	Middle	
	Lower	

Figure 2
Main stratigraphic units in the Danish Embayment.

depths the temperatures are estimated at 90–140°C. No permeability values are available.

The Gassum formation seems to offer the best exploitation possibilities. The main reservoir of the formation covers an area of at least 5000 km² situated at a depth of 2000 to 3000 m and with highly permeable beds of 30 to 60 m thick sand and sandstone. In general, the temperature at these depths will be between 60 and 100° C.

In 1976 the Laboratory of Geophysics, Aarhus University, had the opportunity to carry out some investigations including a drill stem test in the deep drilling Oddesund 1 in NW Jylland (Fig. 3). The main reservoir within the Gassum formation was tested, and the following data were obtained: depth 1949–1961 m, temperature 77°C, porosity 25%, permeability approximately 1 darcy ($\simeq 10^{-12}$ m²), formation water pressure $2.16 \pm 0.03 \times 10^7$ N m^{-2} (213 $\pm$ 3 atm), water density 1.11×10^3 kg m^{-3}, and salinity (mainly NaCl) 18.6% (BALLING, 1976b). There is no significant difference between the measured reservoir pressure and the hydrostatic pressure taking the high water density into consideration.

The Haldager formation is to be found in the Embayment at depths up to 1500–2000 m, where the main reservoir of the formation seems to have a thickness of up to 50 m or more. No permeability determinations are available, but formation intervals seem locally to be more permeable than the Gassum formation. The formation temperatures are expected to range from 50 to 75°C.

The depths and layer thicknesses of the formations can locally be heavily influenced by the numerous salt structures (Fig. 1). Above the deep-seated salt pillows the formations may be uplifted by 1000–2000 m as compared to their normal level, and around the periphery of the salt diapirs rim synclines may cause greater than normal depths.

The reservoir thickness and permeability of the Gassum formation justify the assumption that water amounts of about $30–50 \times 10^{-3}$ m^3 s^{-1} (about 110–180 m^3 h^{-1}) per well or even more could be produced by using pump systems known from the oil industry. Such pumps can be seen working for geothermal purposes in France near Paris (Creil and Villeneuve-La-Garenne).

Test drillings

Based on the results from the above-mentioned investigations in the drilling Oddesund 1 and available data from the oil exploration, a governmental working group has recommended (HANDELSMINISTERIET, 1977) that three to six test drillings are to be carried out to depths from 2000 to 4000 m in three localities (Fig. 3). If the

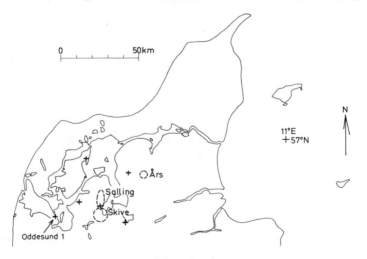

Figure 3
Three areas recommended for test drillings. Temperature data from six deep drillings in the area (marked with crosses) are shown in Fig. 4.

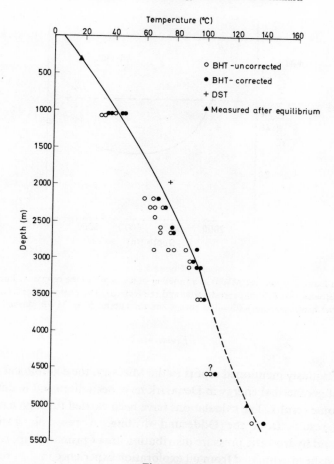

Figure 4

Estimated generalized temperature–depth relation of NW Jylland based on data from six deep wells (Fig. 3). Measured and calculated (corrected) 'bottom hole temperatures' (BHT), a drill stem test (DST) value, and two values measured after thermal equilibrium, are shown.

results are favourable, pilot plants should be established with combined production and reinjection, i.e. at least two drillings per plant. These plans are expected to be implemented in the period 1978–80. The test areas have been selected with a view to geology, temperature, and marketing conditions for heat energy.

Figure 4 gives the observed and calculated temperatures from six oil exploration drillings in the area in question. Based on this we may expect temperatures of approximately 40, 70, and 95°C at the depths 1000, 2000, and 3000 m, respectively, and less accurate data (one drill hole only) from greater depths indicate about 100°C at 4000 m and about 130°C at 5000 m. Above the deep-seated salt structures, e.g. near Skive (Figs. 1 and 3), positive temperature anomalies of 5–15°C should be present.

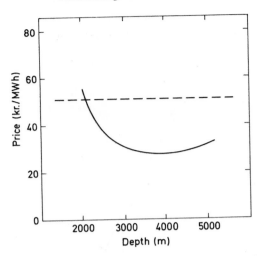

Figure 5

Energy price (in Danish kroner per MWh) as a function of the depth to the reservoir. The calculations are based on a plant with two drillings (production and reinjection). The energy price from typical Danish district heating plants with oil as energy source is about 52 kr/MWh (dashed line).

Economy

In the previously mentioned report to the Ministry, the economical aspects of the utilization of geothermal energy in Denmark have been discussed in detail. Figures 5 and 6 give some results. The calculations have been carried through a.o. on the basis of the test results from the Oddesund drilling. A reservoir transmissivity of 30 darcy m and hydrostatic pressure distribution are of primary importance. Drilling expenses have been estimated from oil exploration experience in the area in question; e.g. the expenses per drilling to the depths 2000 m, 3000 m, and 4500 m are estimated at 3–5, 6–8, and 12–14 mill kr, respectively. A 20 years' depreciation, and an inflation-proofed interest of 4% p.a. have been used. The stated energy prices are the prices at the production plant which consists of two drillings (production and re-injection), heat exchangers and pump systems. The cost estimates are based on 1976-prices.

The general relation in Fig. 5 is based on a constant production rate of 150 $m^3 h^{-1}$ and sale of the total produced energy at temperatures above 55°C (injection tempera-ture). A variation of the geothermal gradient with depth (Fig. 4) has been considered. In Fig. 6 the temperature has been kept constant at 100°C (depth about 3000 m), and an energy amount of maximum 4600 MWh per year is expected to be consumed. These constraints have been chosen with a view to test locality Års (Fig. 3).

As very few data are available it is difficult to evaluate the size of the salinity in the geothermal reservoirs, but the concentration at Oddesund is expected to be higher than normal. However, it is taken for granted that corrosion resistant casings and heat exchangers have to be used, which has also been considered in the economic

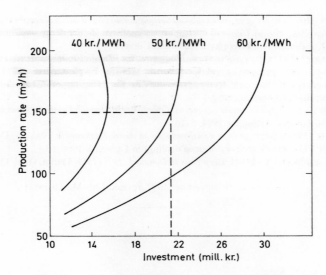

Figure 6

Relations between production rate, investments and energy price with special reference to the conditions at test locality Års (Fig. 3). Here a local oil-fired district heating plant is operating at a cost of 50 kr/MWh. If a geothermal plant has a production rate of e.g. 150 m³/h and investments below 21 mill kr it should be able to produce cheaper energy.

calculations. As the temperature of the return water is typically around 50°C in Danish district heating plants, most calculations are based on an injection temperature of 55°C. A further cooling could improve the economy, if this does not cause problems of chemical precipitation. These conditions cannot be further evaluated until final measurement results from the test drillings are available. However, it appears that favourable possibilities exist for geothermal energy production at competitive prices.

Conclusion

The Danish Embayment contains highly permeable aquifers of Upper Triassic age at temperatures between 60 and 100°C, and at depths of 2000 to 3000 m. It should be possible to exploit the associated geothermal resources, e.g., for district heating at prices lower than or of the same size as for heat energy produced with oil as energy source. Also Middle Jurassic and Lower Triassic strata contain warm aquifers of interest. The planned comprehensive investigations including deep drillings should give information on the actual exploitation possibilities.

REFERENCES

BALLING, N. (1976a), *Geothermal models of the crust and uppermost mantle of the Fennoscandian Shield in South Norway and the Danish Embayment*, J. Geophys. *42*, 237–256.

BALLING, N. (1976b), *Rapport over geotermiske energiundersøgelser i dybdeboringen Oddesund 1, Laboratoriet for Geofysik, Aarhus Universitet.* (Report on geothermal energy investigations in the deep well Oddesund 1, Laboratory of Geophysics, Aarhus University.)

HANDELSMINISTERIET (1977), *Udnyttelse af geotermisk energi i Danmark*, Rapport fra Handelsministeriets arbejdsgruppe vedr. geotermisk energi. (Danish Ministry of Commerce (1977), Exploitation of geothermal energy in Denmark, Report from the working group appointed by the Ministry of Commerce concerning geothermal energy).

MADSEN, L. (1975) *et al., Approximate geothermal gradients in Denmark and the Danish North Sea sector*, Danm. Geol. Unders. Årbog for 1974, 5–16.

RASMUSSEN, L. B. (1972), *Oversigt over dybdeboringer på dansk landområde 1965–68*, Dansk Geol. Foren. Årsskrift for 1971, 41–48. (Survey of deep drillings in Denmark 1965–68.)

SORGENFREI, T. and BUCH, A. (1964), *Deep tests in Denmark 1935–1959*, Danm. Geol. Unders. III raek. 36.

(Received 7th November 1977, revised 13th March 1978)

Pageoph, Vol. 117 (1978/79), Birkhäuser Verlag, Basel

Heat Flow and the Geothermal Potential of Egypt

By Paul Morgan and Chandler A. Swanberg[1])

Abstract – Preliminary heat flow values ranging from 42 to 175 mW m^{-2} have been estimated for Egypt from numerous geothermal gradient determinations with a reasonably good geographical distribution, and a limited number of thermal conductivity determinations. For northern Egypt and the Gulf of Suez, gradients were calculated from oil well bottom hole temperature data; east of the Nile, and at three sites west of the Nile, gradients were calculated from detailed temperature logs in shallow boreholes. With one exception, the heat flow west of the Nile and in northern Egypt is estimated to be low, 40–45 mW m^{-2}, typical of a Precambrian Platform province. A local high, 175 mW m^{-2}, is probably due to local oxidational heating or water movement associated with a phosphate mineralized zone. East of the Nile, however, including the Gulf of Suez, elevated heat flow is indicated at several sites, with a high of 175 mW m^{-2} measured in a Precambrian granitic gneiss approximately 2 km from the Red Sea coast. These data indicate potential for development of geothermal resources along the Red Sea and Gulf of Suez coasts. Water geochemistry data confirm the high heat flow, but do not indicate any deep hot aquifers. Microearthquake monitoring and gravity data indicate that the high heat flow is associated with the opening of the Red Sea.

Key words: Heat flow; Temperature gradients; Silica geothermometer; Sodium–potassium–calcium geothermometer; Precambrian rocks; Egypt; Red Sea; Gulf of Suez.

1. Introduction

There is a clear correlation between the principal areas of current geothermal development (e.g. see Muffler, 1976; United Nations, 1976) and the seismically active boundaries of the moving segments of lithosphere defined by the plate tectonic models of the earth (e.g. see Le Pichon *et al.*, 1973, p. 83). It would therefore seem logical to concentrate the search for new geothermal resources along the plate boundaries. The Arab Republic of Egypt lies in the north-eastern corner of the African plate, bounded to the north by a zone of compression in the north-eastern Mediterranean and Anatolia (McKenzie, 1972), and to the east by what has been interpreted as a median spreading center in the Red Sea (McKenzie *et al.*, 1970). There is a complex transition between these two types of plate boundaries (Ben-Menahem *et al.*, 1976). The potential for geothermal resources in Egypt is good therefore if the geothermal anomalies associated with the plate boundaries are broad enough to extend into the coastal regions.

[1]) Departments of Earth Sciences and Physics, Box 3D, New Mexico State University, Las Cruces, New Mexico 88003, USA.

The plate margin to the north of Egypt appears to be too distant to result in any geothermal anomalies in northern Egypt. Several areas of high geothermal potential have been located in Turkey (ALPIN, 1976), but the Mediterranean to the south, including the island of Cyprus, is characterised by low heat flow, 30–45 mW m^{-2} (ERICKSON, 1970; RYAN *et al.*, 1970; MORGAN, 1973; ČERMÁK *et al.*, 1977).

To the east, however, the data are more encouraging: heat flow measurements ranging from 50 to 170 km from the axial trough of the Red Sea have a mean of 111 mW m^{-2} or about twice the world mean (GIRDLER and EVANS, 1977), and temperature gradient data from oil wells within a few kilometers of the Sudan coast are high and do not define a lateral limit to the elevated heat flow (EVANS and TAMMEMAGI, 1974). Until the present study, however, no heat flow data were available from the continental regions bordering the Red Sea from which a preliminary assessment of geothermal potential of these regions could be made. During 1976 and 1977 heat flow and other geophysical data have been collected from Egypt as part of an integrated geophysical study of the area, and these data indicate that the geothermal anomaly associated with the Red Sea is broader than the Red Sea itself.

2. New geothermal data from Egypt

New geophysical data from Egypt related to geothermal studies include temperature gradient and heat flow data, microearthquake studies and regional gravity profiles, which have been supplemented by a water geothermometry study. The temperature gradient and heat flow data have been compiled from existing subsurface temperature data, the main source being bottom hole temperatures from oil wells, mostly in northern Egypt and the Gulf of Suez, together with new measurements in water, mineral exploration and specially drilled heat flow boreholes.

Heat flow

Two hundred and forty-eight bottom hole temperatures from 128 oil wells in northern Egypt and the Nile Delta between longitudes 25°E and 33°E and latitudes 27°N and 32°N have yielded a mean temperature gradient of 20.6 ± 2.0 mK m^{-1} (°C/km), assuming a mean annual surface temperature of 26.7°C. There is a systematic geographical distribution to the gradients with lower values north of 31°N, south of 29°N, and in the Nile Delta. This gradient variation is thought to be due to the effects of sedimentation to the north and in the Delta, and due to thermal conductivity contrasts, and is not thought to reflect a variation in heat flow. The heat flow is estimated to be 42–47 mW m^{-2} and extends the low heat flow province of the eastern Mediterranean at least as far south as 29°N (MORGAN *et al.*, 1977; ČERMÁK *et al.*, 1977).

From the Gulf of Suez, one hundred and ten bottom hole temperatures from seventy-eight oil wells gave a mean gradient of 26.7 ± 5.5 mK m^{-1}. This value agrees well with an average gradient of 26.1 mK m^{-1} given by TEWFIC (1975) using data from three hundred and twenty-one Gulf of Suez oil wells in an unpublished report. The higher gradients in the Gulf of Suez are not thought to be a function of uniform heat flow with a lower conductivity section, and it is hoped to confirm this if lithologic logs for the wells can be obtained for thermal conductivity analysis. It is certain, however, that the heat flow in the Gulf of Suez is more than 30% higher than the northern Egypt heat flow, as the conductivity will be higher due to the presence of high thermal conductivity evaporites in the section. The Evaporite Series attains variable thicknessess in different parts of the Gulf, perhaps explaining some of the scatter in the temperature data, and may reach several thousand meters in thickness (SAID, 1972, pp. 182–183). A minimum estimate for the Gulf of Suez heat flow is 61 mW m^{-2} but it is probably higher, 80 or even 100 mW m^{-2} (MORGAN et al., 1977). This higher heat flow is more consistent with the mean value of 116 mW m^{-2} reported by HAENEL (1972) for the Red Sea.

The geographical range of the heat flow estimates from the oil wells has been extended south and into the principal area of interest, the Red Sea coastal region, by more conventional heat flow measurements in mineral exploration, water, and specially drilled heat flow boreholes. Boreholes at two sites in the Western Desert, west of the Nile, have been temperature logged, and measurements have been made in boreholes at thirteen sites between the Nile and the Red Sea in the Eastern Desert (Fig. 1). The temperature gradient results from fifty-two temperature logs are summarized in Table 1.

With one notable exception, the data from the Western Desert (Abu Tartur and West Kharga) indicate low gradients (15–19 mK m^{-1}), extending the low heat flow province of the eastern Mediterranean south to 26°N. In eight boreholes at Abu Tartur (Phosphate in Table 1), however, the gradients averaged 74 mK m^{-1}, although these holes were only a few kilometers from holes having more normal gradients. The high gradients were measured over a proven phosphate deposit at 80–150 m depth, and the temperature logs showed the deposit to be essentially isothermal. Although temperature measurements were not possible below the phosphate bed, it is thought that similar gradients would have been measured to those outside the deposit. The gradients above the phosphate are believed to be enhanced by oxidative heating of pyrite in the phosphate (cf. LOVERING, 1948), and/or by mass transfer of heat by water in the phosphate. The Western Desert therefore appears to have a fairly uniform upper crustal heat flow in the range 42–47 mW m^{-2}, similar to the mean value of 44 mW m^{-2} given by POLYAK and SMIRNOV (1968) for other Precambrian platform provinces. Samples have been collected from Abu Tartur and West Kharga for which thermal conductivity measurements will be made to improve the quality of the heat flow data from these sites.

Twelve of the thirteen heat flow sites in the Eastern Desert are in Precambrian rocks, and although thermal conductivity measurements have not yet been made on

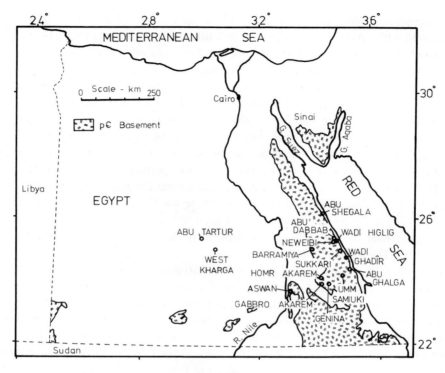

Figure 1
Locations of borehole temperature logging sites in Egypt excluding oil well data sites.

the samples collected at most of the sites, the heat flow can be estimated for the nine sites in granitic rocks using preliminary conductivity measurements of ninety-six samples from the Abu Dabbab and Neweibi sites. These heat flow estimates are summarized in Table 2. Several sites show significantly high heat flow values even if extremely large error limits are placed on the conductivity estimates. The temperature data are considered to be reliable, as shown by the examples in Fig. 2. A significant heat flow anomaly has therefore been discovered in the Precambrian of eastern Egypt.

From the locations of the heat flow sites shown in Fig. 1, it can be seen that the high heat flow values appear to be concentrated in the coastal zone of the Red Sea. This is further illustrated by a plot of heat flow against distance from the Red Sea coastline shown in Fig. 3. Estimated conductivities for the non-granitic heat flow sites in the Eastern Desert give heat flow values consistent with the pattern of a high heat flow anomaly restricted to the coastal zone: Abu Shegala and Abu Ghalga are predicted to be high heat flow, Umm Samiuki and Gabbro Akarem will be normal to low. All the data at present therefore indicate a high heat flow anomaly, up to four times normal Precambrian values, restricted to a zone extending inland from the Red Sea coast approximately 30 km.

Table 1
Egypt geothermal gradient data.

Location	Lithology	Gradient mK m^{-1} (n)	Remarks
A. Northern and Western Egypt			
Northern Egypt oil wells	Precambrian to Recent sediments	20.6 ± 2.0 (128)	248 bottom hole temperatures
Abu Tartur	Palaeocene carbonate beds + clastics	18.7 ± 1.0 (4)	
Abu Tartur phosphate	Palaeocene carbonate beds + clastics	74.4 ± 6.0 (8)	
West Kharga	Cretaceous sandstone	15.2 (1)	
B. Eastern Egypt			
Gulf of Suez oil wells	Cenozoic sediments	26.7 ± 5.5 (78)	110 bottom hole temperatures
Abu Shegala	Tertiary sediments	30–50 (1)	Two distinct linear gradients (probably due to conductivity contrast)
Abu Dabbab	Precambrian granite	28.9 ± 2.9 (8)	
Wadi Higlig	Precambrian granite	23.4 (1)	
Neweibi	Precambrian granite	20.3 ± 2.6 (10)	
Sukkari	Precambrian granite	18.9 (3)	
Wadi Ghadîr	Precambrian granite gneiss	54.0 (1)	
Abu Ghalga	Precambrian gabbro	18.8 (5)	
Umm Samiuki	Precambrian volcanics	19.1 (2)	
Barramîya	Precambrian granite	16.7 (1)	
Genina	Precambrian granite	12.0 (1)	
Homr Akarem	Precambrian granite	17.6 (2)	
Gabbro Akarem	Precambrian gabbro	8.2 (3)	
Aswan	Precambrian granite	13.9 (1)	

(n) = number of boreholes at each site.

Water geothermometry

A water geothermometry study was carried out in Egypt to investigate the possible effects of deep water circulation on the surface heat flow and to extend the range of the heat flow measurements using geothermal indicators. Eighty-four water samples were collected from nearly all parts of Egypt and quantitatively chemically analysed for sodium, potassium, calcium, and silica. These samples represent thermal and non-thermal wells and springs, salt lakes, the Red Sea, and the River Nile (Fig. 4). The surface temperatures of nineteen of the samples exceeded 35°C and can thus be considered thermal. The remaining samples, together with data from the literature, establish the background chemistry.

Several artesian wells from the principal oases in the Western Desert (Dakhla, Kharga and Bahariya, Fig. 4) have temperatures in the range 35 to 43°C. Most of

Table 2
Preliminary Eastern Egypt heat flow values.

Data number	Descriptive code (1)	Site name (2)	Latitude	Longitude	Elevation m (3)	Gradients (mK m^{-1})	Conductivity (W m^{-1} K^{-1})	Heat flow (mW m^{-2})	No. of boreholes
EG01	Abaahb	Abu Dabb	25° 20'N	34° 33'E	250	28.9	3.20*)	92	8
EG02	Abafhb	Wadi Hig	25° 14'N	34° 41'E	200	23.4	3.25	76	1
EG03	Abaahb	Neweibi	25° 13'N	34° 31'E	500	20.3	3.30*)	67	10
EG04	Abafhb	Sukkari	24° 57'N	34° 42'E	400	18.9	3.25	61	3
EG05	Abafhb	Wadi Gha	24° 49'N	34° 58'E	80	54.0	3.25	175	1
EG06	Abafhb	Barramiy	25° 06'N	33° 47'E	400	16.7	3.25	54	1
EG07	Abafhb	Gemina	24° 04'N	34° 15'E	400	12.0	3.25	39	1
EG08	Abafhb	Homr Aka	24° 13'N	34° 03'E	400	17.6	3.25	57	2
EG09	Abafhb	Aswan	24° 00'N	33° 00'E	250	13.9	3.25	45	1

(1) Codes in accordance with international standard for descriptive heat flow codes.
(2) Abbreviated site names: see Table 1 for full names.
(3) Approximate elevations taken from 1:1000 000 topographic map.
*) Mean conductivities from measurements on 48 samples from each site. An estimated conductivity value is used at other sites.

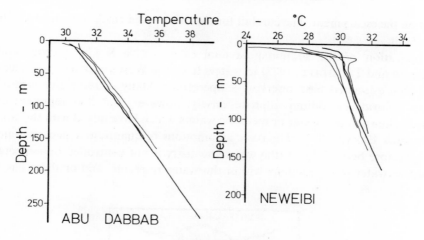

Figure 2

Examples of the borehole temperature data from the Abu Dabbab (*left*) and Neweibi (*right*) sites. The Neweibi temperature data show greater divergence at shallow depths as these holes were drilled from the surface whereas the Abu Dabbab holes were drilled from an exploratory adit.

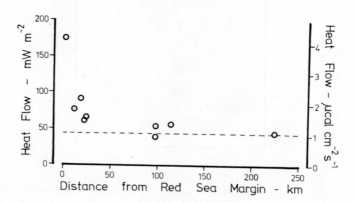

Figure 3

Heat flow data from Table 2 plotted as a function of distance from the Red Sea coast.

these surface temperatures can be explained simply by the depths of the wells and a normal to low geothermal gradient, however, on the order of 16.5 mK m^{-1}. The majority of discordant data are from Dakhla Oasis and are concentrated towards an escarpment which forms the northern boundary of the oasis. These data are most easily reconciled by assuming that water, heated by a normal to low geothermal gradient, is ascending along conduits at the north end of the oasis and migrating south through the principle aquifers before being tapped by the wells (SWANBERG, *et al.*, 1977). A reconnaissance microearthquake survey in this area subsequent to the discovery of the anomalous water temperatures recorded a single event with an epicenter on the escarpment. The conduit for the rising water therefore appears to be an active fault.

Application of the sodium–potassium–calcium (Na–K–Ca) geothermometer (FOURNIER and TRUESDELL, 1973) to waters from the Western Desert gave inconsistently high estimated base reservoir temperatures. Many of the waters are of the calcium chloride or sodium sulphate variety, however, and thus represent either original saline connate water or meteoric waters which have mixed with the connate waters (ALY EZZAT, 1974). The basic assumptions of quantitative geothermometry have therefore been violated (the water chemistry is not controlled by temperature dependent water-rock reactions within the water reservoir, and/or there has been

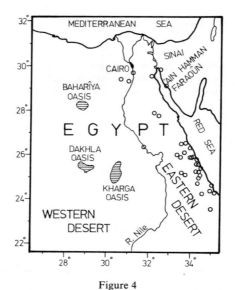

Figure 4

Map of water sample locations for the water geothermometry study. Open circles indicate single or multiple sample sites. The ruled areas indicate major oases from which multiple samples were collected.

significant mixing of waters), and consequently the Na–K–Ca geothermometer is not a viable tool for these waters (SWANBERG *et al.*, 1977).

The silica geothermometer (FOURNIER and ROWE, 1966) applied to the Western Desert waters indicates low base reservoir temperatures (range 44–61°C, average 53°C). This base temperature is similar to the average silica geotemperature of 55°C obtained for waters from the Colorado Plateau and the Eastern United States (SWANBERG and STOYER, 1976; SWANBERG and MORGAN, 1977), two tectonic provinces having a fairly low mean heat flow of 50 mW m^{-2} (BLACKWELL, 1971), and is further evidence for low heat flow throughout the Western Desert (SWANBERG *et al.*, 1977; see also below).

Most of the samples from the southern part of the Eastern Desert were collected from wells dug to a depth of a few meters into valley-fill sediments. None of these wells, or the five springs sampled show evidence of thermal activity. A single pumped well, Umm Khariga, is thermal (surface temperature 35.8°C), but this can probably be explained by the well depth and the local geothermal gradient. The average silica content of these waters is double the average for the Western Desert waters, however, giving a mean silica geotemperature of 78°C (range 44–107°C). By comparison, the Basin and Range and Rio Grande Rift provinces in the U.S. yield silica geotemperatures averaging 80°C (SWANBERG and STOYER, 1976; SWANBERG and MORGAN, 1977, also this volume), with a mean heat flow of 80 mW m^{-2} (BLACKWELL, 1971). This is further evidence for high heat flow in the Eastern Desert, and the silica analyses indicate that the elevated heat flow may extend as far west as Cairo in the north (SWANBERG et al., 1977).

The hottest springs in Egypt are located in the Gulf of Suez area. Unfortunately the thermal waters from this region show the same range of silica geotemperatures as the non-thermal waters to the south. Ain Hamman Faraoûn (Fig. 4), the hottest spring in Egypt at 75°C (EL RAMLY, 1969), gives an Na–K–Ca geotemperature of 129°C, but this value is suspect due to the probable mixing of geothermal fluids with sea water (SWANBERG et al., 1977).

Gravity and microseismics

Gravity measurements have been made at 1 km intervals along the three main hard-top roads between the Red Sea and the Nile. These data are not yet tied into an absolute base station, but several relative small Bouguer gravity anomalies (up to 40 milligals) can be identified on the profiles. Most of the relatively negative gravity anomalies can be correlated with surface outcrops of Precambrian granitic rocks shown on the geologic map of the Basement rocks of the Eastern Desert (EL RAMLY, 1972). At the eastern (Red Sea) end of each line, however, the data do show a relatively positive Bouguer anomaly.

From published data there appears to be very little seismic activity in the northern Red Sea (MCKENZIE et al., 1970; FAIRHEAD and GIRDLER, 1970), which is rather surprising as it is generally accepted that an extensional plate boundary passes through the center of the Red Sea, bifurcating around the Sinai Peninsula into the Gulfs of Suez and Aqaba (e.g. see LE PICHON et al., 1973, p. 99). Preliminary microearthquake studies indicate a high level of microseismicity, however, localized in two areas, one at the southern end of the Gulf of Suez, and the other in the Precambrian crust close to Abu Dabbab (Fig. 1) and the location of a single magnitude 6.0 earthquake in 1955 reported by FAIRHEAD and GIRDLER (1970). Minor microseismic activity has been recorded at other locations, for example at the northern end of the Gulf of Suez, but no clearly defined pattern has emerged from the microseismicity.

3. Interpretation of the geothermal data

The new heat flow data indicate low heat flow 42–47 mW m^{-2} in western Egypt consistent with a Precambrian platform tectonic setting (mean 44 mW m^{-2}, POLYAK and SMIRNOV, 1968). This is also in good agreement with the mean of 46 mW m^{-2} obtained from heat flow measurements in the South African shield (BULLARD, 1939; CARTE, 1954; GOUGH, 1963). To the east of the Nile, however, significantly high heat flow values have been determined, up to 175 mW m^{-2}, the high values being restricted to a 30 km coastal zone, and seemingly having a negative correlation with distance from the coast.

A recent compilation of Red Sea heat flow data confirms that the whole of the Red Sea is associated with high heat flow (GIRDLER and EVANS, 1977). Observations within 5 km distance of the deepest water of the axial trough have a mean of 467 mW m^{-2}, and observations in the range 50 to 170 km from the axial deep water have a mean of 111 mW m^{-2} (*idem.*). The high heat flow values measured in the Precambrian crust therefore appear to be continuous with the main Red Sea heat flow anomaly.

The heat flow anomaly appears to tail off to more normal values 30–40 km from the coast, indicating a relatively shallow (crustal) origin for the anomaly. Preliminary gravity data suggest that a relatively positive Bouguer gravity anomaly may be associated with the heat flow anomaly which is consistent with a rapid thinning of the crust as has been detected by gravity and seismic methods in the Gulf of California (HARRISON and MATHUR, 1964; PHILLIPS, 1964), an area which also exhibits high heat flow (VON HERZEN, 1963). There are currently three views on the amount of oceanic crust beneath the Red Sea: (1) only the axial trough is underlain by oceanic crust (GIRDLER, 1958; DRAKE and GIRDLER, 1964; HUTCHINSON and ENGLES, 1970, 1972; LOWELL and GENIK, 1972; ROSS and SCHLEE, 1973); (2) the whole of the Red Sea was formed by sea-floor spreading (MCKENZIE *et al.*, 1970; R. W. GIRDLER, personal communication, 1977); and (3) a compromise view that considerably more than the axial trough, but not the whole of the Red Sea is underlain by oceanic crust (GIRDLER and WHITMARSH, 1974). The new heat flow and gravity data are most easily reconciled by the second view, and even if the whole of the Red Sea is not underlain by oceanic crust, the data suggest a significant thinning of the crust very close to the coastline. A complementary thinning of the lithosphere is also predicted by the high heat flow in and along the margin of the Red Sea (see CHAPMAN and POLLACK, 1977; POLLACK and CHAPMAN, 1977).

Oil well temperature data suggest that the high heat flow may extend north into the Gulf of Suez. A study of two of the large earthquakes near the mouth of the Gulf of Suez suggest that there may be low Q material beneath the Gulf (R. W. GIRDLER, personal communication, 1977), and the occurrence of thermal springs on both sides of the Gulf all agree with the conclusion of high heat flow from the oil well data. The silica geothermometry data suggest that the high heat flow may extend as far west as Cairo in the north.

No well defined plate boundary can be defined from the new seismic data, although the new data do indicate a high level of seismicity associated with active tectonics. The relationship between the active seismicity and the heat flow anomaly is not clear, but it is thought that they are both related to extensional tectonics in the northern Red Sea.

4. The geothermal potential of Egypt

The heat flow data indicate that the most promising areas for geothermal development in Egypt are along the Red Sea coast. Temperature gradients of up to 54 mK m^{-1} have been measured, and it is quite possible that higher gradients may exist, as the present survey has made no attempt to investigate specific local geothermal areas. The oil well temperature data and thermal springs also indicate good prospects for geothermal development along the Gulf of Suez coast.

Water geothermometry data from eastern Egypt suggest that there is little or no deep circulation of the surface waters which indicates dry geothermal resources rather than fluid dominated systems. In view of this lack of water in the geothermal systems, and the generally arid nature of the region, probably the greatest use for geothermal resources in this area would be in use for desalination of sea water to provide potable water and water for irrigation.

5. Conclusions

Temperature gradient and heat flow data indicate a low heat flow 42–47 mW m^{-2} in western Egypt, consistent with the Precambrian platform tectonic setting, and extending the low heat flow province of the eastern Mediterranean south into north-eastern Africa. High heat flow values, up to 175 mW m^{-2}, approximately four times normal, have been measured in eastern Egypt, and the heat flow appears to increase towards the Red Sea coast. This pattern of normal low heat flow in western Egypt and high heat flow in eastern Egypt is confirmed by water geothermometry data, although the same data do not indicate any deep circulation of hot water in the east. A high level of microseismicity measured close to the Red Sea coast indicates active tectonics in eastern Egypt, but this activity cannot be simply related to a median spreading ridge in the northern Red Sea. A relatively positive Bouguer gravity anomaly close to the Red Sea coast, together with the heat flow data, suggest that the eastern Egypt heat flow anomaly is associated with the opening of the Red Sea.

The greatest areas of geothermal potential in Egypt are close to the coasts of the Red Sea and Gulf of Suez. These potential resources are probably dry, and their greatest application may be in the desalination of sea water to provide fresh waters for this arid region.

Acknowledgements

Dr Rushdi Said, Chairman of the Egyptian Geological Survey and Mining Authority, is thanked for his sincere help in making the facilities for this study possible. Senior U.S. scientists participating in the project with the authors are Drs David D. Blackwell and Eugene Herrin. Field operations are being directed by Dr Fouad K. Boulos of the Egyptian Survey. Personnel from the Egyptian Geological Survey, Southern Methodist University and New Mexico State University have participated in the collection, reduction and interpretation of the geophysical data, and special thanks are expressed to Mr Paul H. Daggett for his work with these data. Ms Marilyn K. Martinez assisted in the preparation of this manuscript. The study has been financed by grant no. DES75-21851 from the Earth Science Program of the National Science Foundation, Public Law Fund 480 grant no. 01P76-02108, administered by the National Science Foundation, and minigrant award 3103-102 from the New Mexico State University Arts & Science Research Center.

REFERENCES

ALPIN, S. *Geothermal energy exploration in Turkey*, in *Proceedings, Second United Nations Symposium on the Development and Use of Geothermal Resources* (U.S. Govt. Printing Office, Washington, D.C., 1976), vol. 1, pp. 25–28.

ALY EZZAT, M. Exploitation of ground water in El-wadi El-gedid project area, Part 1, Regional hydrogeologic conditions, *Groundwater Series in A.R.E.* (Cairo, 1974).

BEN-MENAHEM, A., NUR, A. and VERED, M. (1976), *Tectonics, seismicity and structure of the Afro-Eurasian junction – the breaking of an incoherent plate*, Physics Earth Planet. Int. 12, 1–50.

BLACKWELL, D. D. *The thermal structure of the continental crust*, in *The structure and physical properties of the earth's crust* (ed. J. G. Heacock), Geophysical Monograph Series, No. 14, 169–184 (Am. Geophys. Union, Washington, D.C., 1971), 316 pp.

BULLARD, E. C. (1939), *Heat flow in South Africa*, Proc. R. Soc. Lond., A *173*, 474–502.

CARTE, A. E. (1954), *Heat flow in the Transvaal and Orange Free State*, Proc. Phys. Soc. Lond., B *67*, 664–672.

ČERMÁK, V., HURTIG, E., KUTAS, R. I., LODO, M., LUBIMOVA, E. A., MONGELLI, F., MORGAN, P., SMIRNOV, YA. B. and TEZCAN, A. K. (1977), *Heat flow map of Europe – southern and Mediterranean Region*, in *Proceedings, International Congress on Thermal Waters, Geothermal Energy and Vulcanism of the Mediterranean Area* (National Technical Univ. of Athens, Greece, 1977), vol. 1, pp. 149–168.

CHAPMAN, D. S. and POLLACK, H. N. (1977), *Regional geotherms and lithospheric thickness*, Geology 5, 265–268.

DRAKE, C. L. and GIRDLER, R. W. (1964), *A geophysical study of the Red Sea*, Geophys. J. R. astr. Soc. 8, 473–495.

EL RAMLY, I. M. (1969), *Recent review of investigations on the thermal and mineral springs in the U.A.R.*, XXIII Int. Geol. Cong. *19*, 201–213.

EL RAMLY, M. F. (1972), *A new geological map for the basement rocks in the Eastern and South-western Deserts of Egypt*, Annals Geol. Surv. Egypt 2, 1–18.

ERICKSON, A. J. (1970), 'The measurement and interpretation of heat flow in the Mediterranean and Black Sea', Ph.D. thesis, Massachusetts Institute of Technology, Cambridge, Massachusetts 02139, USA.

EVANS, T. R. and TAMMEMAGI, H. Y. (1974), *Heat flow and heat production in north-east Africa*, Earth Planet. Sci. Letts. *23*, 349–356.

FAIRHEAD, J. D. and GIRDLER, R. W. (1970), *The seismicity of the Red Sea, Gulf of Aden and Afar triangle*, Phil. Trans. Roy. Soc. Lond. A *267*, 49–74.

FOURNIER, R. O. and ROWE, J. J. (1966), *Estimation of underground temperatures from the silica content of water from hot springs and steam wells*, Amer. J. Sci. *264*, 685–697.

FOURNIER, R. O. and TRUESDELL, A. H. (1973), *An empirical Na–K–Ca geothermometer for natural waters*, Geochim. Cosmochim. Acta *37*, 1255–1275.

GIRDLER, R. W. (1958), *The relationship of the Red Sea to the East African Rift System*, Q. Jl. Geol. Soc. Lond. *114*, 79–105.

GIRDLER, R. W. and EVANS, T. R. (in press, 1977), *Red Sea Heat Flow*, Geophys. J. R. astr. Soc.

GIRDLER, R. W. and WHITMARSH, R. B. *Miocene evaporites in Red Sea cores, their relevance to the problem of the width and age of oceanic crust beneath the Red Sea*, in R. B. Whitmarsh, O. E. Weser, D. A. Ross, *et al.*, *Initial Reports of the Deep Sea Drilling Project*, vol. 23, pp. 913–921 (U.S. Government Printing Office, Washington, D.C., 1974), 1180 pp.

GOUGH, D. I. (1963), *Heat flow in the Southern Karroo*, Proc. R. Soc. Lond., A *272*, 207–230.

HAENEL, R. (1972), Heat flow measurements in the Red Sea and the Gulf of Aden, Zeit. Geophys. *38*, 1032–1047.

HARRISON, J. C. and MATHUR, S. P. (1964), *Gravity anomalies in Gulf of California*, Am. Assoc. Petrol. Geol., Mem. *3*, 76–89.

HUTCHINSON, R. W. and ENGELS, G. G. (1970), *Tectonic significance of regional geology and evapourite lithofacies in northeastern Ethiopia*, Phil. Trans. R. Soc. Lond., A *267*, 313–329.

HUTCHINSON, R. W. and ENGELS, G. G. (1972), *Tectonic evolution in the southern Red Sea and its possible significance to older rifted continental margins*, Bull. Geol. Soc. Am. *83*, 2989–3002.

LE PICHON, X., FRANCHETEAU, J. and BONNIN, J. *Plate Tectonics* (Elsevier, Amsterdam, 1973), 300 pp.

LOVERING, T. S. (1948), *Geothermal gradients, recent climatic changes, and rate of sulphide oxidation in the San Manuel district, Arizona*, Econ. Geol. *43*, 1–20.

LOWELL, J. D. and GENIK, G. J. (1972), *Sea-floor spreading and structural evolution of the southern Red Sea*, Bull. Am. Assoc. Petrol. Geol. *56*, 247–259.

MCKENZIE, D. P. (1972), *Active tectonics of the Mediterranean region*, Geophys. J. R. astr. Soc. *30*, 109–185.

MCKENZIE, D. P., DAVIES, D. and MOLNAR, P. (1970), *Plate tectonics of the Red Sea and East Africa*, Nature, *226*, 243–248.

MORGAN, P. (1973), 'Terrestrial heat flow studies in Cyprus and Kenya', Ph.D. thesis, London University, England.

MORGAN, P., BLACKWELL, D. D., FARRIS, J. C., BOULOS, F. K. and SALIB, P. G. (1977), *Preliminary geothermal gradient and heat flow values for northern Egypt and the Gulf of Suez from oil well data*, in *Proceedings, International Congress on Thermal Waters, Geothermal Energy and Vulcanism of the Mediterranean Area* (National Technical Univ. of Athens, Greece, 1977), vol. 1, pp. 424–438.

MUFFLER, L. J. P. *Summary of section 1, present status of resources development*, in *Proceedings, Second United Nations Symposium on the Development and Use of Geothermal Resources* (U.S. Govt. Printing Office, Washington, D.C., 1976), vol. 1, xxxiii–xliv.

PHILLIPS, R. P. (1964), *Seismic refraction studies in Gulf of California*, Am. Assoc. Petrol. Geol. Mem. *3*, 90–121.

POLLACK, H. N. and CHAPMAN, D. S. (1977), *On the regional variation of heat flow, geotherms, and lithospheric thickness*, Tectonophysics *38*, 279–296.

POLYAK, B. G. and SMIRNOV, YA. B. (1968), *Relationship between terrestrial heat flow and the tectonics of continents*, Geotectonics *4*, 205–213.

ROSS, D. A. and SCHLEE, J. (1973), *Shallow structure and geologic development of the southern Red Sea*, Bull. geol. Soc. Am. *84*, 3827–3848.

RYAN, W. B. F., STANLEY, D. J., HERSEY, J. B., FAHLQUIST, D. A. and ALLAN, T. D. *The tectonics and geology of the Mediterranean Sea*, in *The Sea* (ed. A. E. Maxwell), vol. 4, pt. II (Wiley-Interscience, New York, 1970), 387–492.

226 Paul Morgan and Chandler A. Swanberg

SAID, R., *The Geology of Egypt* (Elsevier, Amsterdam, 1962), 377 pp.

SWANBERG, C. A. and MORGAN, P. *The correlation among the silica content of groundwaters, regional heat flow, and the geothermal potential of the Western United States*, in *Abstracts with Program* (Joint General Assemblies, IASPEI/IAVCEI, Durham, England, 1977), p. 38.

SWANBERG, C. A., and STOYER, C. H. (1976), *The correlation among water chemistry, regional heat flow, and the geothermal potential of Arizona*, in *S.E.G., Proc. 46th Ann. Mtg.*

SWANBERG, C. A., MORGAN, P., HENNIN, S. F., DAGGETT, P., MELIC, Y. S. and EL-SHERIF, A. A. *Preliminary report on the thermal springs of Egypt*, in *Proceedings, International Congress on Thermal Waters, Geothermal Energy and Vulcanism of the Mediterranean Area* (National Technical Univ. of Athens, Greece, 1977), vol. 2, pp. 540–554.

TEWFIC, R. (1975), *Geothermal gradients in the Gulf of Suez*, in *Gulf of Suez Petroleum Co., Exploration Report No. 212*, 8 pp. + figures.

UNITED NATIONS, *Present status of world geothermal development*, in *Proceedings, Second United Nations Symposium of the Development and Use of Geothermal Resources* (U.S. Govt. Printing Office, Washington, D.C., 1976), vol. 1, pp. 3–9.

VON HERZEN, R. P. (1963), *Geothermal heat flow in the Gulfs of California and Aden*, Science *140*, 1207–1208.

(Received 15th November 1977, revised 28th March 1978)

Pageoph, Vol. 117 (1978/79), Birkhäuser Verlag, Basel

The Linear Relation Between Temperatures Based on the Silica Content of Groundwater and Regional Heat Flow: A New Heat Flow Map of the United States

By Chandler A. Swanberg and Paul Morgan[1])

Abstract – Application of the silica geothermometer to over 70,000 non-thermal groundwaters from the United States has shown that there is a correlation between the average silica geotemperatures for a region (T SiO_2 in °C) and the known regional heat flow (q in mW m^{-2}) of the form:

$$T\,SiO_2 = mq + b, \qquad (1)$$

where m and b are constants determined to be 0.67°C m^2 mW^{-1} and 13.2°C respectively. The physical significance of 'b' is the mean annual air temperature. The slope 'm' is related to the minimum average depth to which the groundwaters may circulate. This minimum depth is estimated to be between 1.4 and 2.0 km depending on the rock type. A preliminary heat flow map based on equation (1) is presented using the $T\,SiO_2$ for new estimates of regional heat flow where conventional data are lacking. Anomalously high local $T\,SiO_2$ values indicate potential geothermal areas.

Key words: Silica geothermometer; Regional heat flow; Geothermal energy; Ground water circulation; United States.

Introduction

The water resources division of the U.S. Geological Survey maintains a centralized water quality data bank called WATSTORE. This computerized file represents both surface and groundwaters (wells and springs) for the entire United States, and includes such information as chemical analysis, well depth, temperature, date of sampling, location of the sample in latitude and longitude, and various other parameters. The entire file represents over a million individual waters and perhaps 20 million individual pieces of information. The present study, however, includes only the several hundred thousand samples representing groundwaters. Of these, between 70,000 and 100,000 have reasonably complete chemical analyses and are included in the present study.

Since this large amount of chemical data exists in a computerized data bank, and since the chemical quality of groundwaters is such an important tool in the search for and evaluation of potentially valuable geothermal resources, it was decided to search the WATSTORE file for groundwaters whose chemical nature seemed to suggest that

---

[1]) Departments of Earth Sciences and Physics, Box 3D, New Mexico State University, Las Cruces, New Mexico 88003, USA.

they might have originated within an active geothermal system. A summary of the chemical indicators of geothermal activity (geothermometers), the assumptions upon which they are based, and their applications to geothermal studies is given by MARINER and WILLEY (1976), FOURNIER et al. (1974), and TRUESDELL (1976).

Our initial search of WATSTORE for waters possibly associated with geothermal activity revealed regional trends in geothermal indicators (SWANBERG and STOYER, 1976) which made it very difficult to determine whether or not an individual sample was actually anomalous. Thus our initial effort in utilizing WATSTORE for geothermal studies has been to define 'normal' against which anomalously high geotemperatures can be recognized and associated with geothermal activity. This has been completed for the Na–K–Ca and silica geothermometers and for in situ temperature. We report here our results for the silica geothermometer.

The Silica Geothermometer

The silica geothermometer is based on the temperature dependence of quartz solubility in water (FOURNIER and ROWE, 1966) and is usually applied to thermal waters to estimate reservoir base temperatures of geothermal systems. This geothermometer can be quantitatively expressed according to the conductive cooling equation of TRUESDELL (1976):

$$T\,SiO_2 = \frac{1315}{5.205 - \log_{10} SiO_2} - 273.15 \qquad (2)$$

where $T\,SiO_2$ is the silica geotemperature in degrees Celcius and dissolved silica is expressed in parts per million. The quantitative use of this geothermometer requires that water-rock equilibrium exist within the geothermal system and that there is neither silica precipitation, continued water-rock reactions, or mixing with non-thermal groundwaters as the water migrates from the geothermal reservoir to the supplying point. These assumptions are supported by the general reluctance of quartz to precipitate from super saturated solutions (TRUESDELL, 1976) and the slow water-rock reactions at low temperatures (RIMSTIDT, 1977). In the event mixing has occurred, more accurate estimates of subsurface temperatures can usually be obtained by applying the methods of TRUESDELL and FOURNIER (1977).

In the following sections, we examine the nationwide distribution of silica geotemperatures as calculated by applying equation (2) to the silica concentrations of groundwaters. The geotemperatures are used rather than actual silica concentrations as the geotemperatures are linearly related to heat flow (equation 1) and normal distributions about the mean are obtained.

Histograms of silica geotemperatures

The first step in correlating silica geotemperatures with heat flow is to examine the distribution of silica within the various heat flow provinces of the United States. This

Table 1

Silica geotemperature data for the United States

Province	Mean $T\,SiO_2$ (°C)	Modal $T\,SiO_2$ (°C)	Standard deviation (°C)	Samples	Silica equivalent of mean (ppm)
Atlantic Coast Province	36.2	35	26.8	6186	9.0
Eastern United States (North)	41.3	*)	19.7	4285	10.6
Eastern United States (Central)	42.6	35	20.6	1920	11.0
Wyoming Basin	42.9	35	22.4	455	11.1
Middle Rocky Mountains (Colorado)	45.9	35	23.8	693	12.1
Colorado Plateau	53.4	55	20.2	2126	15.1
Southern Rocky Mountains	55.6	55	20.4	1110	16.0
Great Plains	60.5	*)	19.6	10550	18.4
Northern Rocky Mountains	67.3	*)	35.4	149	22.0
Great Valley (California)	70.5	75	19.3	322	23.9
Geopressured Area	71.7	*)	23.7	1250	24.6
Basin and range (Nevada–Utah)	72.8	75	22.8	5049	25.3
Basin and range (Arizona–New Mexico)	78.9	75	19.0	2737	29.5
California coast ranges	80.8	85	18.4	1312	30.9
Rio Grande Rift	82.7	75	22.0	1366	32.3
Cascade Mountains	83.8	*)	30.4	105	33.2
Nebraska	85.5	105	20.2	1821	34.6
Battle Mountain High (Nevada–Idaho)	95.3	105	31.3	795	43.3

*) Bimodal-mean not a reliable estimate of surface heat flow.

has been done for the major physiographic provinces of the United States and the results are presented in histogram form in Figs 1 to 9 and summarized in Table 1. The province boundaries are taken from BLACKWELL (1971) and modified slightly to simplify the data retrieval and to avoid including unrepresentative values located near province boundaries. The mean silica geotemperature for each province is then calculated and is considered to be representative of the entire province if the standard deviation is less than 25°C and if the mean value is approximately equal to the modal value.

Mean silica geotemperature – heat flow plot

Figure 10 shows the correlation of mean silica geotemperature with heat flow for the United States. The silica data are taken from the histograms (Figs. 1–9) and are plotted only if the standard deviations are less than 25°C and if the mean value is in good agreement with the modal value. The heat flow values are averaged from the

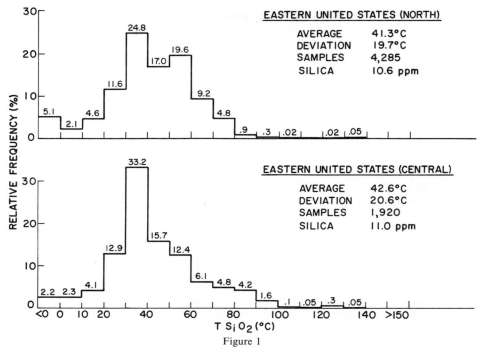

Figure 1

Histogram of $T\,SiO_2$ for the United States east of Meridian 96°W. Plot (a) includes data north of parallel 40°N and plot (b) includes data between parallels 34°N and 40°N but not the Atlantic Coast Province. Values of $T\,SiO_2$ in the 50–60°C range in plot (a) come mostly from Minnesota.

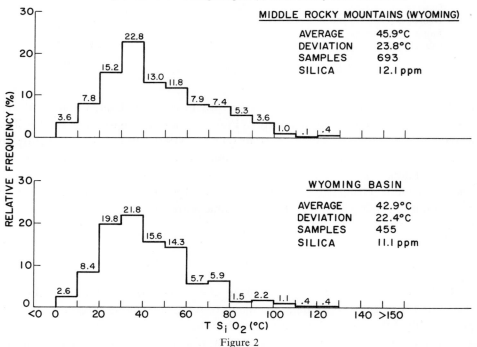

Figure 2

Histogram of $T\,SiO_2$ for the Middle Rocky Mountains of Wyoming (plot a) and for the Wyoming Basin (plot b).

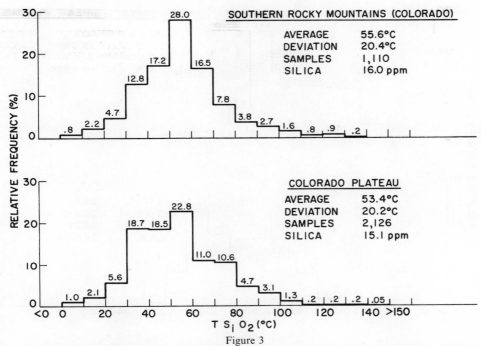

Figure 3

Histogram of $T\,SiO_2$ for the Southern Rocky Mountains (plot a) and for the Colorado Plateau (plot b).

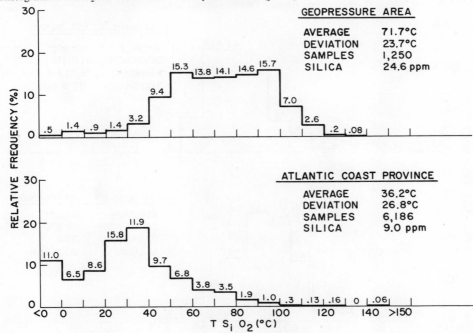

Figure 4

Histogram of $T\,SiO_2$ for the Southern Mississippi Embayment – Geopressured area of Louisiana and East Texas (plot a) and for the Atlantic Coast Province (plot b). Data west of meridian 85°W are omitted from plot (b).

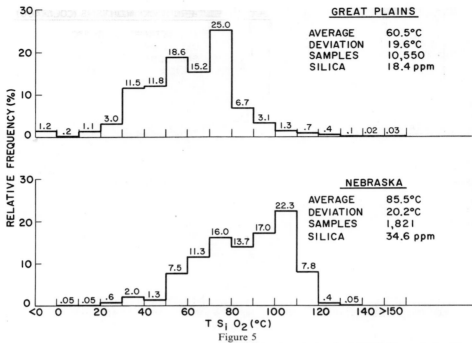

Figure 5

Histogram of $T \, SiO_2$ for the Great Plains (plot a) and for Western Nebraska (plot b). Note a lack of data from Kansas, Oklahoma, and Texas.

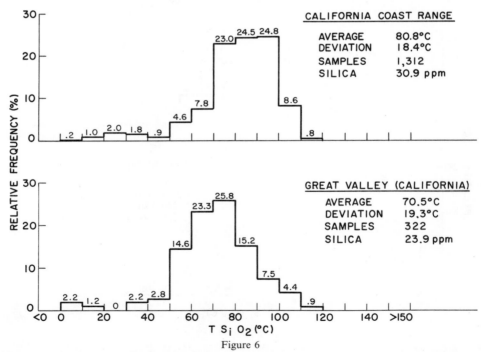

Figure 6

Histogram of $T \, SiO_2$ for the California Coast Ranges (plot a) and for the Great Valley of California (plot b).

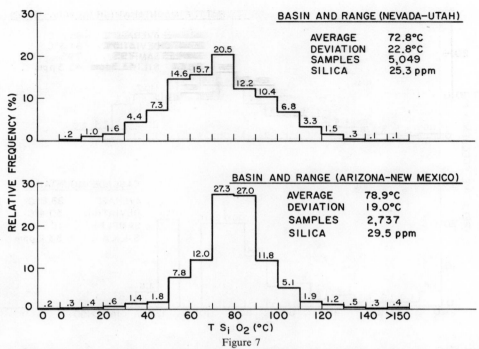

Figure 7

Histogram of T SiO$_2$ for the Basin and Range Province of Nevada and Utah (plot a) and for the Basin and Range Province of Arizona and New Mexico (plot b).

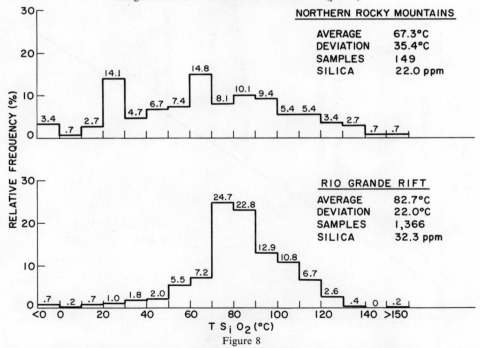

Figure 8

Histogram of T SiO$_2$ for the northern Rocky Mountains (plot a) and for the Rio Grande Rift of southern New Mexico and West Texas (plot b). Note the poor data set for plot (a).

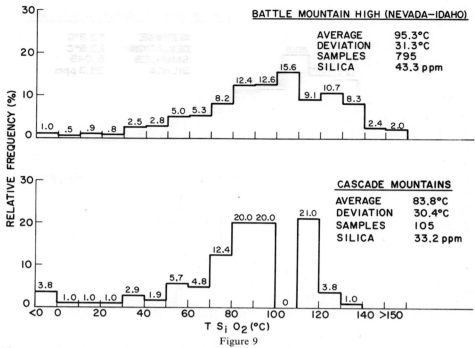

Figure 9

Histogram of $T\,SiO_2$ for the Battle Mountain High region of Nevada and Idaho (plot a) and for the Cascade Mountains of Washington and Oregon (plot b).

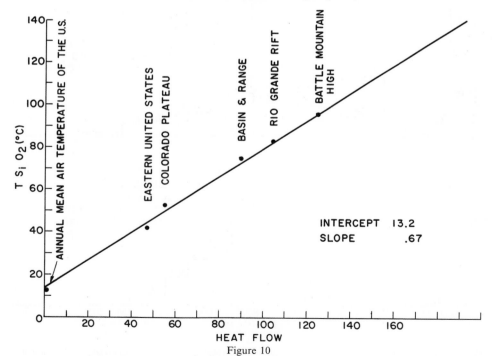

Figure 10

Plot of mean $T\,SiO_2$ versus heat fow for provinces where both parameters are well defined. Heat flow in $mW\ m^{-2}$.

compilation data of SASS *et al.* (1976) with one exception: the high heat flow values from New Hampshire which are associated with extremely radioactive plutons (i.e. the White Mountain magma series) have been omitted as these values are not representative of the entire eastern United States.

The linear correlation between mean silica geotemperature and heat flow can be represented by:

$$T\,SiO_2 = mq + b \qquad\qquad (3)$$

where $T\,SiO_2$ is mean silica geotemperature in degrees Celcius (°C), q is heat flow in milliwatts per square meter (mW m^{-2}) and m and b are constants determined to be 0.67°C m^2 mW^{-1} and 13.2°C by the method of least squares. The physical significance of the intercept 'b' is that this value should represent the national mean annual air temperature since the absence of heat flow implies that all thermal energy be solar in nature. A national mean annual air temperature has been calculated by averaging the mean air temperatures for 84 cities located throughout the conterminous United States. The resulting value of 12.7°C has been included in the least squares analysis shown in Fig. 10 in order to give control to low end of the plot. The slope 'm' multiplied by thermal conductivity (in W m^{-1} K) has the units of depth and reflects the mean depth to which groundwaters may circulate: a minimum depth if the circulating groundwaters contribute a convection component to the total heat flow or if silica precipitation has occurred. Assuming conductivity values of 3.0 and 2.1 W m^{-1} k for crystalline rocks and sediments respectively, depths of 1.4 and 2.0 km are implied, and although these values cannot be confirmed or rejected they are not unrealistic.

Regional distributions of $T\,SiO_2$

Figures 11 and 12 show respectively the distribution of silica geotemperatures and heat flow throughout the United States. The similarity of the two figures both in general and in detail underscores the potential value of using silica geotemperatures as a means of estimating regional heat flow. Figure 11 was prepared by averaging silica geotemperatures over a 1° × 1° grid and then contouring the results. The data are presented to outline areas of normal heat flow ($T\,SiO_2 < 60°C$; $q < 70$ mW m^{-2}) and very high heat flow ($T\,SiO_2 > 90°C$; $q > 115$ mW m^{-2}). In addition, contours of 25, 70, and 75°C have been added in some places to reveal more detailed heat flow patterns in areas where silica data are abundant and their distribution reveals the finer structure of surficial heat flow variations throughout the United States. Several of the more interesting areas are discussed below.

Atlantic Coast Province – Many of the high silica geotemperatures in the eastern United States are located in the Atlantic Coast Province (Fig. 11), yet the histogram of this province (Fig. 4b) is not appreciably different from the histograms representing the remainder of the eastern United States (Figs. 1a, 1b), all having mean values of about 40°C ± 20°C (s.d.). We interpret these data to mean that the background heat flow in

Figure 11
Contoured map of T SiO$_2$ values as averaged over a 1° × 1° grid. The dots refer to grid blocks having silica data.

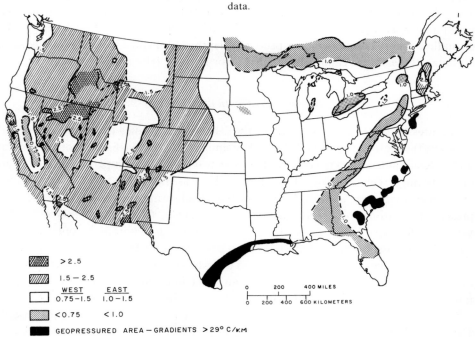

Figure 12
Heat flow in the United States from SASS *et al.* (1976) modified to show areas of abnormally high gradients in the Atlantic coast province (COSTAIN *et al.*, 1977) and in the geopressured region of the Southern Mississippi Embayment (WHITE and WILLIAMS, 1975).

the Atlantic Coast Province is typical of the eastern United States, but that numerous isolated areas exist which have higher than normal heat flows. A similar conclusion has been reached by COSTAIN et al. (1977) who note elevated heat flow associated with highly radioactive plutons. The spatial relation among radioactive plutons, elevated heat flow, and high silica geotemperatures is generally good (J. COSTAIN, personal communication, 1977) and central Georgia is a notable example (Figs. 11, 12).

Geopressured Area – High silica geotemperatures are also located along the Texas–Louisiana Gulf coast and their geographic distribution appears to reflect the presence of geopressured-geothermal resources. Unlike the Atlantic Coast Province, the histogram of this area (Fig. 4a) implies high regional heat flow and not normal heat flow with superimposed anomalies. Traditional heat flow data from this area are lacking (SASS et al., 1976) but ample evidence of high regional heat flow can be found in the deep oil well data (AAPG-USGS, 1976) and from the temperature gradients calculated using the geopressured temperature–depth data of WHITE and WILLIAMS (1975). Additional groundwater data obtained from the Texas Natural Resources Information System (TNRIS) indicate that the highest heat flow region is southern Texas in the area where the hottest geopressured regions are located nearest the surface (regions AT_1, BT_1, CT_1: WHITE and WILLIAMS, 1975).

Eastern Tectonic Areas – The remainder of the eastern United States is characterized by low silica geotemperatures. Notable exceptions occur, however, in tectonically active areas such as New Madrid (Missouri), northern New York, and South Carolina. The presence of active faults may be responsible for deeply circulating groundwaters with greater quantities of dissolved silica.

Great Plains – Figure 11 shows the presence of a major silica anomaly extending north from western Nebraska to the eastern Dakotas. Additional data from the TNRIS show the anomaly to also exist throughout much of the West Texas Panhandle region north of parallel 32°N. Thus, the anomaly covers a sizeable portion of the central United States. It seems premature to speculate the existence of a major heat flow anomaly of such a large areal extent since the highest silica waters generally come from Pliocene to Recent continental deposits and may reflect the presence of amorphous silica originating as air fall ash from silicic volcanic centers such as Yellowstone, Bishop and the Jemez Mountains to the west. Still, all six heat flow values from this area are high (up to 90 mW m^{-2}: SASS et al., 1976) and the AAPG-USGS (1976) bottom hole temperatures are also elevated throughout much of the region.

Rio Grande Rift – A recognizable silica anomaly can be traced over the entire extent of the Rio Grande Rift and the southern Rocky Mountains as far north as southern Wyoming, and coincides with the heat flow anomaly of REITER et al. (1975). Silica data are uniformly high throughout the entire Rio Grande Rift, but this is not apparent from Fig. 11 because the Rift becomes narrow in northern New Mexico and the 1° × 1° grid blocks used in compiling Fig. 11 have included lower silica values from the Colorado Plateau and the Great Plains either side of the Rift. A 75°C contour has been added to more clearly delineate the heat flow anomaly associated with the Rio Grande

Rift in West Texas and southern New Mexico. This contour has also been extended as far as southern Utah to demonstrate the sharp transition between the Colorado Plateau and the Basin and Range-Rio Grande Rift provinces.

Western United States – The highest silica geotemperatures and the highest heat flow in the United States are located in the west, primarily in Nevada and Idaho. However, we are unable to predict heat flow in this area with the same confidence as the remainder of the country for the following reasons. First, much of the region is sparsely populated and the silica data available for study are both fewer in number and distributed much more unevenly than the remainder of the country. This means that municipal water supplies, developed in search of high quality water, may represent a disproportionately large percentage of the total data set. If these waters are low in silica because they represent surface waters circulating in shallow aquifers (i.e., in arroyos, next to lakes and rivers, etc.) their inclusion may bias the average silica geotempera- tures towards the low side. Second, the area in question includes a majority of the nation's hydrothermal convection systems (see WHITE and WILLIAMS, 1975, Figs. 1 and 3). Many of these hydrothermal systems have been studied in connection with geothermal resource studies (e.g., the data of YOUNG and MITCHELL, 1973 are included in WATSTORE) and the inclusion of these data may bias the average silica geotemperatures toward the high side. Finally, the western United States contains both the nation's highest heat flow region (Battle Mountain High) and lowest heat flow region (Sierra Nevada) along with numerous isolated hot and cold spots (SASS *et al.*, 1976).

Despite these potential shortcomings, a few general conclusions can be drawn. The highest average silica geotemperatures seem to be associated with the Battle Mountain High, a region in northern Nevada and southern Idaho having anomalously high heat flow (SASS *et al.*, 1976) and containing numerous hydrothermal convective systems (WHITE and WILLIAMS, 1975). Other areas of high silica geotemperatures are located along both flanks of the Basin and Range province, near the Sierra Nevada province of California and near the Colorado Plateau province in south central Utah, southeast Arizona and southwest New Mexico. The Cascade Mountains of Washington and Oregon also have very high silica geotemperatures, although there is very little data available from these regions (Figs. 9b, 11).

The portion of the United States having the lowest average silica geotemperatures is a narrow band which includes the Sierra Nevada and Klamath Mountains, the Oregon Coast Ranges, and the Olympic Mountains of Washington. This area also includes the lowest heat flow values in the United States (SASS *et al.*, 1976). This area is dominated by extremely low silica geotemperatures (36% have values less than 10°C) which probably represent shallow circulating meteoric groundwaters which have never reached equilibrium with the host rock. It is perhaps more realistic to characterize this area by its lack of normal or high silica values rather than by its low average. A supplemental 25° contour has been added to Fig. 11 to outline the regions of lowest silica geotemperatures.

Geothermal applications

To this point we have been mainly concerned with mean silica geotemperatures and their correlation with regional heat flow. However, the economic implications of the silica data should not be overlooked. For example, waters giving high silica geotemperatures may reflect waters which are of thermal origin or which have a thermal water component, and the geographic distribution of these waters may bear significantly upon the search for and evaluation of economic geothermal fields. The approach is basically qualitative because there is no guarantee that the assumption necessary for quantitative use of the silica geothermometer will be met when applied randomly to non-thermal waters. Still waters of thermal origin are likely to have more dissolved silica than non-thermal waters, and once the mean silica concentration for waters from a given province has been determined, the waters with abnormally high concentrations of silica can be recognized and correlated with geothermal activity.

The first step in delineating potential thermal waters is to select a cut-off geotemperature and assign geothermal significance to geotemperatures above the cut-off value. Figures 1–9 present the data necessary to assign cut-off temperatures for the various physiographic provinces in the United States. In general, the selection of a high cut-off temperature will delineate the best geothermal prospects, but may overlook many other promising areas. The selection of a lower cut-off temperature will delineate a greater number of prospects, but may also include waters of non-thermal origin. Our experience in the southwest United States has shown that a cut-off temperature of one standard deviation above the mean is a workable compromise. Application of this criterion to waters from Arizona, New Mexico and southern California (SWANBERG et al., 1977; SWANBERG and ALEXANDER, in press) has led to the rediscovery of most known geothermal prospects and the delineation of an equal number of promising new areas, without being overwhelmed with a vast quantity of prospects of marginal geothermal significance. Of course, the most reliable use of the silica geothermometer in geothermal exploration occurs when it is used in conjunction with other quantitative and qualitative geothermometers such as the Na–K–Ca geothermometer, in situ temperature, and temperature gradients. This procedure is discussed more fully by SWANBERG and ALEXANDER (in press).

Conclusions

The silica geothermometer has been applied to over 70,000 non-thermal groundwaters from the United States. Analysis of these data lead to the following conclusions:

1. There is a linear correlation between average silica geotemperatures ($T \, SiO_2$ in °C) for a region and regional heat flow (q in mW m^{-2}) of the form:

$$T \, SiO_2 = mq + b \qquad (4)$$

where m and b are constants determined to be $0.67°C\,m^2\,mW^{-1}$ and $13.2°C$ respectively.

2. The slope 'm' in equation (4) is related to the minimum average depth to which groundwaters may circulate. This minimum depth is estimated to be between 1.4 and 2.0 km depending upon rock type.

3. Using equation (4) a new heat flow map of the United States has been prepared. This map is in excellent agreement with published heat flow values and also provides new insight into regional heat flow patterns where traditional data are lacking.

4. Individual waters or groups of waters giving silica geotemperatures greater than one standard deviation above the mean value for the province in which they are located are likely to be of thermal origin or to have a component of thermal water. The geographic distribution of such waters bears significantly upon the search for and evaluation of geothermal prospects.

Acknowledgments

We would like to thank the U.S. Geological Survey, Water Resources Division for making available the WATSTORE file for our use. We also wish to thank C. H. Stoyer and Bruce Stewart for their help with the data retrieval, and Shari Alexander for her help with the data reduction. The work was funded in part by U.S. Geological Survey grant 14-08-0001-G-406.

REFERENCES

AAPG-USGS, *Subsurface Temperature Map of North America* (U.S. Geol. Survey, Arlington, Virginia, 1976).

BLACKWELL, D. D., *The thermal structure of the continental crust*, in *The Structure and Physical Properties of the Earth's Crust*, Geophys. Monogr. ser., vol. 14 (ed. J. G. Heacock) (Am. Geophys. Un., Washington, D.C., 1971), pp. 169–184.

COSTAIN, J. K., GLOVER III, L. and SINHA, A. K., *Evaluation and targeting of geothermal energy resources in the southeastern United States*, in *Vir. Poly. Tech. Prog. Rept. VPI-SU-5103-3* (Blacksburg, Virginia, 1977).

FOURNIER, R. O. and ROWE, J. J. (1966), *Estimation of underground temperatures from the silica content of water from hot springs and wet steam wells*, Amer. Jour. Sci. *264*, 685–697.

FOURNIER, R. O., WHITE, D. E. and TRUESDELL, A. H. (1974), *Geochemical indicators of subsurface temperature, 1, Basic Assumptions:* Jour. Res. U.S. Geol. Sur. *2*, 259–262.

MARINER, R. H. and WILLEY, L. M. (1976), *Geochemistry of thermal waters in Long Valley, Mono County, California*, Jour. Geophys. Res. *81*, 792–800.

REITER, M., EDWARDS, C. L., HARTMAN, H. and WEIDMAN, C. (1975), *Terrestrial heat flow along the Rio Grande Rift, New Mexico and Southern Colorado*, Geol. Soc. Amer. Bull. *86*, 811–818.

RIMSTIDT, J. D. (1977), *Kinetic evaluation of the quartz geothermometer*, abs. Geol. Soc. Am. Ann. Mtg., pp. 1142–1143.

SASS, J. H., DIMENT, W. H., LACHENBRUCH, A. H., MARSHALL, B. V., MUNROE, R. J., MOSES, T. H., JR and URBAN, T. C., *A new heat flow contour map of the conterminous United States*, in *U.S. Geol. Survey, open-file Rpt. 76–756* (Menlo Park, California, 1976), 24 pp.

SWANBERG, C. A., and ALEXANDER, S., *The use of the random groundwater data from WATSTORE in geothermal exploration: An example from the Imperial Valley, California* (in press).

SWANBERG, C. A. and STOYER, C. H. (1976), *The correlation among water chemistry, regional heat flow, and the geothermal potential of Arizona*, abs. Soc. Exp. Geophys., Ann. Mtg., Oct. 23–28, p. 93.

SWANBERG, C. A., MORGAN, P., STOYER, C. H. and WITCHER, J. C. (1977), *An appraisal study of the geothermal resources of Arizona and adjacent areas in New Mexico and Utah and their value for desalination and other uses*, in *New Mex. Energy Inst. Rpt. 006* (Las Cruces, New Mexico, 1977), 76 pp.

TRUESDELL, A. H., *Summary of Section III, Geochemical techniques in exploration;* in *Proc. 2nd U.N. Symp. on the Development and Use of Geothermal Resources, San Francisco, 1975*, vol. 1 (1976), pp. liii–lxxix.

TRUESDELL, A. H. and FOURNIER, R. O. (1977), *Procedure for estimating the temperature of a hot water component in a mixed water using a plot of dissolved silica vs. enthalpy*, Jour. Res. U.S. Geol. Survey, *5*, 49–52.

WHITE, D. F. and WILLIAMS, D. L., *Assessment of Geothermal Resources of the United States – 1975*, U.S. Geol. Survey, Circ., 725 (1975), 155 pp.

YOUNG, H. W. and MITCHELL, J. C. (1973), *Geothermal investigations in Idaho, part I, Geochemistry and geologic setting of selected thermal waters*, Idaho Dept. Water Admin., Water Info., Bull. *30*, 43 pp.

(Received 15th November 1977, revised 28th March 1978)

Pageoph, Vol. 117 (1978/79), Birkhäuser Verlag, Basel

Applied Volcanology in Geothermal
Exploration in Iceland

By Ingvar B. Fridleifsson[1]

Abstract – In an active spreading area like Iceland, where the regional geothermal gradient is in the range 50–150°C/km, it is normally not a problem to find high enough temperature with deep drilling, but the difficulties arise with finding permeable layers at depth within the strata. Various volcanological methods can be applied in the search for aquifers and geothermal reservoir rocks. The flow pattern (as deduced from deuterium studies) indicates that the thermal water flows preferentially along high porosity stratiform horizons and dyke swarms from the recharge areas in the highlands to the hot spring areas in the lowlands. The primary porosity of the volcanic strata is dependent on the chemical composition and the mode of eruption of the volcanic units. Both the reservoir rocks and the flow channels forming the geothermal plumbing system are thought to vary from the Tertiary to the Quaternary provinces due to environmental conditions at the eruptive sites.

Key words: Porosity of volcanic rocks; Pillow lava; Hyaloclastite, Chemical composition of pillows; Aquifers in volcanic rocks; Iceland.

1. Introduction

The utilization of geothermal energy is very important for the national economy of Iceland. During the last three decades geothermal energy has been harnessed on an increasingly large scale, mainly in the form of space heating (Pálmason *et al.*, 1975). When the energy crisis came on at the beginning of this decade, Iceland like most other countries tried to increase the use of its natural energy sources, hydropower and geothermal energy, in order to cut down the use of fossil fuel, which has to be imported. This effort has been fairly successful, as can be seen in Fig. 1. The total net energy consumption has increased by about 23 % from 1972 to 1976. This increase has been met entirely by hydropower and geothermal energy. Between 60 and 70 % of buildings in Iceland are now centrally heated by thermal water. By using geothermal energy the nation is saving close to 2 metric tons of fossil fuel per inhabitant per year.

In geothermal research different scientific disciplines are involved. In Iceland various geological, geophysical, and geochemical methods are applied in order to locate, and develop new geothermal fields (e.g., Arnórsson *et al.*, 1975; Stefánsson and Arnórsson, 1975; Tómasson *et al.*, 1975). The present paper deals mainly with the application of one of these disciplines, volcanological research, in geothermal exploration in Iceland.

---

[1] National Energy Authority, Laugavegur 116, Reykjavík, Iceland.

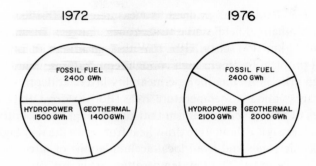

Figure 1
A comparison of the net energy consumption of Iceland in 1972 and 1976.

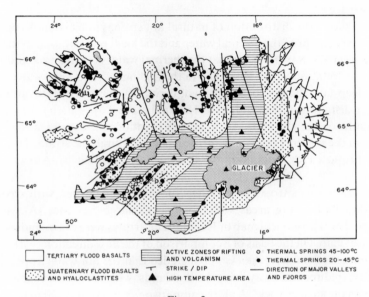

Figure 2
Geological map of Iceland (simplified from SAEMUNDSSON, 1973) showing the distribution of natural geothermal activity (compiled by JÓNSSON, 1967) and the direction of major erosional features (valleys and fjords). The volcanic strata dip towards but age away from the active volcanic zones. High temperature areas (with temperatures above 200°C in the uppermost 1 km) are confined to the active zones of rifting and volcanism. The low temperature activity is most intense in areas where the major erosional directions are approximately parallel with the geological strike.

2. Distribution of geothermal activity in Iceland

Iceland lies astride the Mid-Atlantic Ridge, and like other segments of an active ocean ridge Iceland is characterized by very high heat flow. The regional heat flow varies from about 80 mW/m² furthest away from the active volcanic zones crossing the country to about 300 mW/m² in some regions at the margins of the longest lived volcanic zone. The geothermal gradient as measured in over 100 m deep drillholes outside known geothermal fields and outside zones of active volcanism, ranges from 37°C/km to 165°C/km (PÁLMASON, 1973). With such a high thermal gradient it is normally not a problem to obtain high enough temperatures with deep drilling, but difficulties often arise in finding high permeability layers at depth within the strata.

Active volcanoes and high-temperature areas (BÖDVARSSON, 1961) are confined to the active zones of rifting and volcanism that run through the country (Fig. 2). The high-temperature areas are thought to draw heat from both the very high regional heat flow in the volcanic zones and from local accumulations of intrusions cooling at a shallow depth in the crust. The low temperature areas are, on the other hand, in Quaternary and Tertiary volcanics, and are thought to withdraw heat from the regional heat flow. This paper will mainly deal with geothermal exploration in the low temperature areas.

Figure 2 shows the distribution of natural hot springs in Iceland. A comparison of the deuterium content of the thermal water and the local precipitation in the individual areas has shown (ÁRNASON, 1976) that the thermal water is of meteoric origin. In most cases it is precipitation which has fallen in the highlands. There the water manages to percolate deep into the bedrock and then, driven by the hydrostatic gradient, flows laterally for distances of tens and up to 150 km before it appears on the surface along dykes or faults on the lowlands. As previously mentioned the water is believed to take up heat from the regional heat flow during its flow through the strata. This model was originally proposed by EINARSSON (1942), but later verified by ÁRNASON's isotope studies.

Figure 3 shows ÁRNASON's flow pattern for the thermal water, with arrows joining the individual hot spring areas with possible recharge localities (ÁRNASON, 1976) superimposed on a geological map of Iceland. The arrows were drawn independently of the geological map. On the basis of a comparison of the flow directions with hydrostatic pressure isolines of country, ÁRNASON concluded that the hot water appeared to 'flow equally in every direction away from the highlands, without regard to the direction of the fissures in the surface rock'. A close comparison of the geological map and the flow pattern arrows, however, shows a remarkably good correlation between the flow directions and the geological strike. This suggests that the water may flow along the same high porosity layers in the strata and/or dykes and faults along the strike all the way from the highlands to the lowlands. A detailed comparison of the flow pattern with unpublished data on dyke swarms in some Tertiary provinces indicates that there the dykes are more important as controlling factors for the regional flow than the stratiform horizons (SAEMUNDSSON, personal communication).

Further indication of the preferential flow of the water along stratiform horizons and/or dyke swarms (which generally have a direction deviating 0–30° from the strike) is seen in the distribution of hot springs with regard to erosional features. Although hot springs are very widely distributed in Iceland (Fig. 2) there are certain areas, particularly in the eastern part of the country, that are almost devoid of thermal

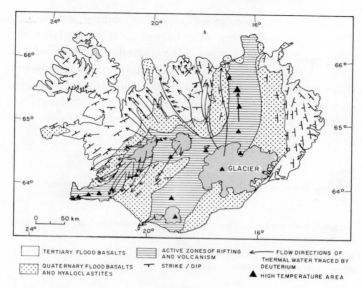

Figure 3

The general flow pattern of thermal groundwater systems in Iceland according to deuterium measurements (ÁRNASON, 1976) superimposed on the geological map of Iceland (simplified from SAEMUNDSSON, 1973). The arrows join the thermal areas in the lowlands with possible recharge areas in the highlands. The arrows were modified such that they are almost perpendicular to the isolines of average topographic heights based on rectangular areas of 520 km² (ÁRNASON, 1976). The arrows were drawn independently of the geological map. Note how closely the flow direction arrows generally fall with the geological strike.

activity. A comparison of the distribution of hot springs and geological strike with the direction of major erosional features, such as fjords and valleys, shows that all the major thermal areas of the country are characterized by the erosional directions being approximately parallel to the strike. This implies that water can flow undisturbed along the same permeable horizons from the recharge areas in the mountains to the outflow areas in the lowlands. The regions that are devoid of hot springs, such as the eastern fjords, are on the other hand characterized by the erosional directions being nearly perpendicular to the strike directions. This can be interpreted to indicate that water seeping into the bedrock in the mountains in these areas cannot flow for but a few kilometers along strike, as the permeable horizons and the dyke swarms are intersected by erosional features (valleys and fjords) at relatively short intervals. As the flow distance is so short the water does not get the same opportunity to withdraw heat from the regional heat flow as water that flows undisturbed for tens of kilometers. Indeed the few hot spring localities in eastern Iceland are in areas where the erosional directions are nearly parallel to the strike of the bedrock (Fig. 2).

3. Geothermal flow channels and reservoir rocks

The crustal thickness of Iceland varies from 8 to 15 km and the crust is formed almost entirely of igneous rocks (e.g., PÁLMASON and SAEMUNDSSON, 1974). The uppermost 2 to 3 km are composed of subaerial lavas and much subordinate airborne tuffs in the Tertiary provinces, but of subaerial lavas intercalated with morainic horizons and thick piles of hyaloclastites in the Quaternary provinces, which flank the active zones (Fig. 2). Each eruptive unit is fed by a dyke, and consequently the dyke intensity increases with depth in the crust. Below 3 to 4 km the crust probably consists mostly of very low porosity impermeable intrusions (e.g., FRIDLEIFSSON, 1977). This layer (the oceanic layer, $V_p = 6.5$ km/s) may form the base to water circulation in the crust (BÖDVARSSON, 1961).

Table 1

Potential geothermal reservoir rocks in Iceland (from FRIDLEIFSSON, 1975)

Tertiary

  Group 1   Stratiform horizons of pyroclastics, ignimbrites, sediments, and olivine tholeiite compound lava shields.

  Group 2   Local accumulations in central volcanoes of high porosity lavas, pyroclastics, agglomerates, and hyaloclastites*) (in caldera lakes).

Quaternary

  Group 3   Same as Group 1 plus primary and reworked subglacial hyaloclastites.*) Hyaloclastite*) horizons reach maximum thickness over the eruptive sites.

  Group 4   Local accumulations in central volcanoes of hyaloclastites*) and subaerial eruptives same as in Group 2.

*) The term hyaloclastite is here used in a collective sense for all subaquatic volcanic products, thus comprising pillow lavas, pillow breccias and tuffs.

On basis of the structure and the relative primary porosity of the various rock types, FRIDLEIFSSON (1975) proposed that potential geothermal reservoir rocks in the Tertiary strata could be divided into two groups (Table 1). The flow channels or aquifers forming the plumbing system from the recharge areas in the highlands to the hot spring areas in the lowlands are probably the thin, high porosity stratiform horizons of Group 1 (Table 1) as well as dykes and permeable faults (BÖDVARSSON, 1961). The geothermal reservoirs, which are fed by the stratiform aquifers and which supply the water to the major hot springs, however belong probably to both Group 1 and Group 2.

Since 3 m.y. ago (SAEMUNDSSON, 1974) there have been over twenty glaciations with intermittent warmer periods in the Iceland region. The continuous volcanic activity during this period is reflected in strata characterized by successions of subaerial lavas intercalated, at intervals corresponding to glaciations, with morainic horizons and thick (tens to hundreds of meters) and commonly elongated piles of subglacial volcanics. FRIDLEIFSSON (1975) divided the potential geothermal reservoir rocks of the Quaternary strata into two groups (Table 1). The flow channels from the recharge areas to the lowlands are thought to belong mainly to Group 3 (Table 1) as well as dykes and faults, but the reservoirs of the main hydrothermal systems are thought to belong to both Group 3 and Group 4.

The division between geothermal aquifers and geothermal reservoirs must clearly be a matter of debate, and a clear definition of the terms will not be attempted here. Low temperature thermal areas, where the free flow from wells can be multiplied several times by lowering the water table some tens of meters by pumping can, however, be regarded as supplied by reservoirs. The thermal areas, where the free flow from wells can only be doubled or so by pumping and a similar drawdown, can similarly be regarded as being fed by individual, narrow aquifers. Both types of thermal areas are presently harnessed in Iceland.

4. Application of volcanological studies in the search for aquifers and geothermal reservoir rocks

A. Geological mapping

It is evident from the assumed nature of the flow channels and the geothermal reservoir rocks that detailed geological mapping on a regional scale is of a primary importance in the geothermal exploration. The volcanics are divided into mappable units on basis of lithology and paleomagnetic polarity directions. The average thickness and general characteristics of lava flows in individual units as well as the thickness variation and grain size distribution of volcanoclastics and sediments are studied. Dykes and faults are traced and the displacement on the latter measured when possible. After an area has been mapped in detail the potential aquifer horizons and reservoir rocks can be identified and subsequently tested by drilling.

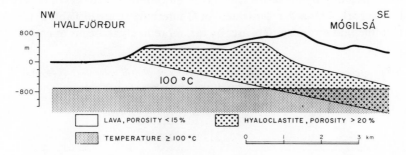

Figure 4

Schematic cross section of Quaternary volcanic strata in SW-Iceland. The lavas are of a relatively low porosity compared with the subglacially erupted hyaloclastites, which include large proportions of pillow lavas and pillow breccias. The 100°C isoline is drawn on the basis of a 240 m well (with a linear thermal gradient) in the left-hand side of the picture. An exploration well to test for hydrothermal convection would be sited in the right-hand side of the diagram, where the high porosity rocks are at a depth great enough so that the water in pores and fractures is hot enough for commercial use. In fact the thermal areas of Reykir and Reykjavík (TÓMASSON et al., 1975) would project on the right-hand side of this section. The thermal areas are characterized by hydrothermal convection, and the best wells (1–2 km deep) yield over 70 l/s of 80–90°C water by pumping with a drawdown of 30 m or less (THORSTEINSSON, 1975).

B. Siting of boreholes

A simplified cross section of Quaternary volcanic strata in SW-Iceland is shown in Fig. 4 to demonstrate the siting of a borehole. The rocks are divided into two groups; lavas with relatively low primary porosity and hyaloclastites (including pillow lavas and pillow breccias) of a much higher primary porosity. In this particular area the regional geothermal gradient is 165°C/km as measured in a 240 m deep drillhole which had no aquifers. The country rock can therefore be assumed to be at a temperature of 100°C or higher below a depth of 800 m or so, if there is no hydrothermal convection in the strata. It is clear that an exploration well should be sited where the high-porosity rocks are found below the 100°C isotherm in the diagram. The primary aim of the exploration well would be to test whether there was active hydrothermal convection in the strata. Prior to the exact siting of the well, detailed geological and geophysical analyses would be made of the potential drilling area in order to locate and trace structural irregularities such as dykes and faults that might have enhanced the permeability of the strata locally.

C. Aquifers in volcanic strata

It should be stressed here that the model shown in Fig. 4 is greatly simplified, as high primary porosity of a rock does not necessarily mean high permeability. In a subglacially erupted hyaloclastite sequence for instance field analyses indicate that the pillow lava portion is relatively highly permeable, whereas the tuffaceous part is commonly impermeable.

Experience from deep drilling in Iceland has shown that aquifers in volcanic strata are most likely to occur at the contacts of lithological units, such as interbeds in a lava sequence, contacts of hyaloclastites and lavas, and contacts of dykes or other intrusives with the host rock. This is demonstrated for Quaternary strata in Table 2; it is apparent from the table that large aquifers are, by far, more likely to occur at contacts than in

Table 2

Occurrence of aquifers in the different rock types of 29 drill holes (800–2045 m) in the Reykir thermal area (from TÓMASSON et al., 1975)

Rock types	Aquifers			Total number
	≤ 2 l/s	> 2–20 l/s	> 20 l/s	
Lavas	44	27	2	73
Hyaloclastites*)	29	12	4	45
Dolerites		1	1	2
Lavas + hyaloclastites*)	53	38	20	111
Lavas + dolerites	13	1	3	17
Hyaloclastites*) + dolerites	5	2	1	8

*) The term hyaloclastite is here used in a collective sense for all subaquatic volcanic products, thus comprising pillow lavas, pillow breccias, and tuffs. Included in this group are also reworked hyaloclastites and detrital beds.

lavas alone or hyaloclastites alone. This should be taken into account in well siting. For example, in Quaternary volcanic strata large hyaloclastite mountains are sometimes literally buried in subsequent subaerial lavas that bank up against the hyaloclastite slopes. During intervals between lava eruptions aprons of sediments spread out over the lava plains at the base of the easily eroded hyaloclastite mountains. The overall effect of this is a 'cedar-tree' structure with a bulky 'stem' formed of mainly primary hyaloclastites (including pillow lavas and pillow breccias) with thin (tens of cm to tens of m), wedge shaped 'branches' of resedimented hyaloclastite, intercalated in the lavas submerging the 'stem' (FRIDLEIFSSON, 1975). In order to drill through as many contacts between lithological units as possible it is clearly advisable to site a well at the margin of the bulky high porosity hyaloclastite body rather than into its center.

Detailed mapping of dykes and faults is very important, as these cause structural irregularities in the strata and commonly create significant secondary permeability. Natural hot springs in Iceland are most commonly located along dykes and faults, and especially in the Tertiary strata the main upflow zones are generally controlled by such structures. However, in Quaternary strata characterized by good reservoir properties and horizontal aquifers, such as in the Reykjavík and Reykir thermal areas, dyke swarms and large faults respectively, appear to act as vertical impermeable barriers that separate harnessable thermal systems (THORSTEINSSON, 1975; TÓMASSON et al., 1975).

D. Primary porosity of volcanics – petrochemical and lithological control

In volcanic areas both the environmental conditions at the eruptive sites and the chemical composition of the volcanics affect the primary porosity of the strata. Lithological and petrochemical research is therefore of importance in the search for potential aquifers and geothermal reservoirs. To demonstrate this the discussion will be devoted firstly to subaerial lavas and secondly to the much more diverse subaquatic volcanics.

The primary porosity of the central part of basaltic lavas of the tholeiitic suite depends more on their thickness than their chemical composition (FRIDLEIFSSON, 1975). The average thickness of the various lava types is, however, related both to their chemical composition and the mode of the eruption. By simple calculations, assuming that the vesicular top and base of each basaltic lava has a porosity of 25% but the massive central part of each flow a porosity of 5%, FRIDLEIFSSON (1975) showed that a suite of thin flank tholeiites, typical for the central volcanoes of Iceland (WALKER, 1963), had about 50% higher primary pore volume than a suite of floodbasalt tholeiites. Similarly he showed that lava shields (olivine tholeiite compound lavas) could be expected to have about twice the primary pore volume of a normal flood-basalt sequence in Iceland.

The term hyaloclastite is commonly used in the Icelandic geological literature in a collective sense for all subaquatic volcanic products. The subglacially erupted hyaloclastites in Iceland are generally formed in water shallower (commonly much

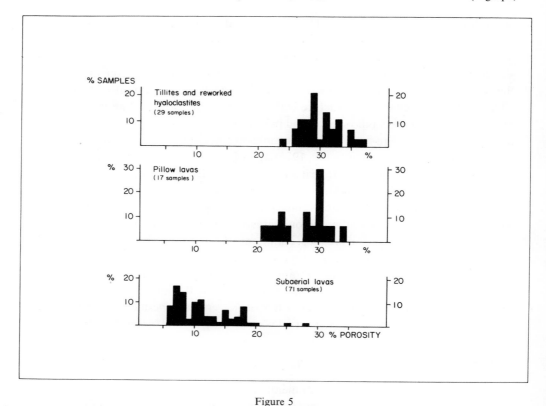

Figure 5
The percentage distribution of porosity (core samples from drill holes crushed to ≤1 mm grain size) of fresh subaerial lavas, subaquatic pillow lavas, tillites, and reworked hyaloclastites (from FRIDLEIFSSON, 1975). The volcanics are of olivine tholeiite composition.

shallower) than 1 km. Due to the shallow depth the pillows may be highly vesicular (e.g. MOORE, 1965), and the pillow breccia and tuff fraction is commonly very large. On basis of a comparison of the porosity (Fig. 5) of fresh core samples of subaerial lavas and pillow lavas, both of olivine tholeiite composition, as well as reworked hyaloclastites and tillites, FRIDLEIFSSON (1975) concluded that the primary porosity of a typical subglacial hyaloclastite sequence appeared to be at least twice that of a subaerial lava sequence.

Experience from drilling indicates that pillow lavas generally speaking have higher effective permeability than other major rock types in Icelandic geothermal areas. During a glaciation, fissure eruptions (which are the most common volcanic type in Iceland) can produce series of parallel hyaloclastite ridges 1–5 km broad, a few hundred meters thick, but up to tens of kilometers long. In subsequent subaerial eruptions lava flows may bank up against the ridges and eventually bury them. The core axis of the ridges is formed of pillow lavas, but this is coated on the sides and above by a thick layer of pillow breccias, and further out by tuffaceous hyaloclastite. The

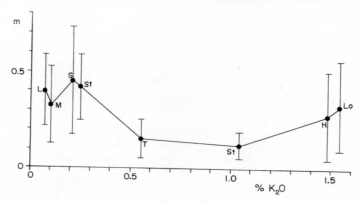

Figure 6

Variation diagram of the diameter (average of the vertical and horizontal axes) of subglacially erupted pillows versus the $\%K_2O$ composition of the pillows (FRIDLEIFSSON *et al.*, in preparation). The standard deviation is shown by bars. The composition of the rocks and the number of measurements at each locality is the following; olivine tholeiite L(100), M(256), S(100), St(49); tholeiite T(140); basaltic andesite Sf(66); andesite H(40); Lo(67). The rock classification is according to CARMICHAEL (1964).

permeability of the core can be expected to be high, but the outermost glassy hyaloclastite layer can be expected to have very low permeability and thus serve as an aquiclude. Thus a monogenetic eruptive unit may form both the geothermal reservoir rock and the cap rock. TÓMASSON *et al.* (1975) suggested that such hyaloclastite ridges served as flow channels for the water from the central highlands to the thermal areas of Reykir and Reykjavík in SW-Iceland.

The primary porosity of a pillow lava sequence is of two types. Firstly there is the interspace between individual pillows, which is large because of the geometric shape of the pillows. Secondly there is the pore space of the shallow water pillow bodies themselves. Both factors are affected by the chemical composition of the pillows. Figure 6 shows the variation of the average size of pillows with a change in the chemical composition (FRIDLEIFSSON *et al.*, in preparation). Considering secondary mineralization processes it is likely that a close-packed pillow sequence will be less effectively sealed by mineralization the larger the diameter of the pillow is. This is because of the inverse relation between the average diameter of the pillows and the surface area of the rock that is in contact with the water. Hence a sequence of large pillows can be expected to have higher effective porosity than a sequence of small pillows. Systematic porosity measurements of pillows of different compositions are in progress, but field inspection and preliminary laboratory results indicate that there is a very marked drop in porosity from Mg-rich olivine tholeiites to FeTi-rich quartz normative tholeiites. This effect is complementary with the decrease in the average diameter of pillows from olivine tholeiites to tholeiites. One implication of this is that effective porosity can be expected to be higher in the olivine tholeiite pillow sequences in the volcanic zone outside central volcanoes than in central volcanic sequences, which are very often dominated by quartz normative tholeiites.

Acknowledgements

Sincere thanks are due to Drs. S. Arnórsson, K. Saemundsson, V. Stefánsson, and J. Tómasson for constructive criticism of the manuscript. Mr. R. Halldórsson is thanked for calculating the data for Fig. 1 from the records of the National Energy Authority, Reykjavík.

REFERENCES

ÁRNASON, B. (1976), *Groundwater systems in Iceland traced by deuterium*, Societas Scientarium Islandica *42*, 236 pp.

ARNÓRSSON, S., BJÖRNSSON, A., GÍSLASON, G. and GUDMUNDSSON, G., *Systematic exploration of the Krísuvík high-temperature area, Reykjanes Peninsula, Iceland*, in *Second UN Geothermal Symposium Proceedings* (Lawrence Berkeley Laboratory, Univ. of California, 1975), pp. 853–864.

BÖDVARSSON, G. (1961), *Physical characteristics of natural heat resources in Iceland*, UN Conference on New Sources of Energy, Rome, Proceedings, Geothermal Energy *2*, 82.

CARMICHAEL, I. S. E. (1964), *The petrology of Thingmuli, a tertiary volcano in eastern Iceland*, Jour. Petrology *5*, 435–460.

EINARSSON, T. (1942), *Über das Wesen der heissen Quellen Islands*, Societas Scientarium Islandica *26*, 91 pp.

FRIDLEIFSSON, I. B., *Lithology and structure of geothermal reservoir rocks in Iceland*, in *Second UN Geothermal Symposium Proceedings* (Lawrence Berkeley Laboratory, Univ. of California, 1975), pp. 371–376.

FRIDLEIFSSON, I. B. (1977), *Distribution of large basaltic intrusions in the Icelandic crust and the nature of the layer 2–layer 3 boundary*, Geol. Soc. Am. Bull. *88*, 1689–1693.

FRIDLEIFSSON, I. B., FURNES, H., ATKINS, F. B., SAEMUNDSSON, K., and GRÖNVOLD, K. (in preparation), *Subglacial volcanics, Part I. On the control of magma chemistry on pillow dimensions*.

JÓNSSON, J., *Map of the Distribution of Hot Springs in Iceland* (National Energy Authority, Reykjavík, 1967).

MOORE, J. G. (1965), *Petrology of deep-sea basalt near Hawaii*, Am. Jour. Sci. *263*, 40–53.

PÁLMASON, G. (1973), *Kinematics and heat flow in a volcanic rift zone, with application to Iceland*, Geophys. J. Roy. Astron. Soc. *33*, 451–481.

PÁLMASON, G. and SAEMUNDSSON, K. (1974), *Iceland in relation to the Mid Atlantic Ridge*, Ann. Rev. Earth Plan. Sci. *2*, 25–50.

PÁLMASON, G., RAGNARS, K. and ZOEGA, J. *Geothermal energy developments in Iceland 1970–1974*, in *Second UN Geothermal Symposium Proceedings* (Lawrence Berkeley Laboratory, Univ. of California, 1975), 213–217.

SAEMUNDSSON, K., *Geological map of Iceland* (National Energy Authority, Reykjavík, 1973).

SAEMUNDSSON, K. (1974), *Evolution of the axial rifting zone in northern Iceland and the Tjörnes fracture zone*, Geol. Soc. Am. Bull. *85*, 495–504.

STEFÁNSSON, V. and ARNÓRSSON, S., *A comparative study of hot-water chemistry and bedrock resistivity in the southern lowlands of Iceland*, in *Second UN Geothermal Symposium Proceedings* (Lawrence Berkeley Laboratory, Univ. of California, 1975), pp. 1207–1216.

TÓMASSON, J., FRIDLEIFSSON, I. B. and STEFÁNSSON, V., *A hydrological model for the flow of thermal water in southwestern Iceland with special reference to the Reykir and Reykjavík thermal areas*, in *Second UN Geothermal Symposium Proceedings* (Lawrence Berkeley Laboratory, Univ. of California, 1975), pp. 643–648.

THORSTEINSSON, T., *Redevelopment of the Reykir hydrothermal system in southwestern Iceland*, in *Second UN Geothermal Symposium Proceedings* (Lawrence Berkeley Laboratory, Univ. of California, 1975), pp. 2173–2180.

WALKER, G. P. L. (1963), *The Breiddalur central volcano, eastern Iceland*, Quart. J. Geol. Soc. Lond. *119*, 29–63.

(Received 2nd November 1977, revised 28th March 1978)

Pageoph, Vol. 117 (1978/79), Birkhäuser Verlag, Basel

Determination of Characteristics of Steam Reservoirs by Radon-222 Measurements in Geothermal Fluids

By F. D'Amore[1]), J. C. Sabroux[2]) and P. Zettwoog[3])

Abstract – The authors conducted a Rn^{222} survey in wells of the Larderello geothermal field (Italy) and observed considerable variations in concentrations. Simple models show that flow-rate plays an important part in the Rn^{222} content of each well, as it directly affects the fluid transit time in the reservoirs. Rn^{222} has been sampled from two wells of the Serrazzano area during flow-rate drawdown tests. The apparent volume of the steam reservoir of each of these two wells has been estimated from the Rn^{222} concentration versus flow-rate curves.

Key words: Radon concentrations of steam; Radon emanating power of rocks; Flow rate of producing wells; Steam reservoirs; Fluid transit time; Larderello, Italy.

List of symbols

Q Flow-rate (kg h^{-1}).

λ Decay constant of Rn^{222} ($\lambda = 7.553 \times 10^{-3}$ h^{-1}).

ϕ Porosity of the reservoir (volume of fluid/volume of rock).

ρ_1 Density of the fluid in the reservoir (kg m^{-3}).

ρ_2 Density of the rock in the reservoir (kg m^{-3}).

M Stationary mass of fluid filling the reservoir (kg).

E Emanating power of the rock in the reservoir (nCi kg$_{\text{rock}}^{-1}$ h^{-1}).

P Production rate of Rn^{222} in the reservoir: number of atoms of Rn^{222} (divided by 1.764×10^{7}) transferred by the rock to the mass unit of fluid per unit time (nCi kg$_{\text{fluid}}^{-1}$ h^{-1}).

N Specific concentration of Rn^{222} in the fluid (nCi kg^{-1}).

τ Characteristic time of the steam reservoir at maximum flow-rate ($\tau = M/Q$).

1. Radon in geothermal fields

Rn^{222} is naturally occurring radioactive gas present in almost all geofluids in contact with uranium (and therefore radium) bearing rocks. The quantity of Rn^{222} which emanates from a parent rock per unit of time depends on various parameters, the most important among these appear to be the following:

---

[1]) Istituto Internazionale per le Ricerche Geotermiche, C.N.R., Via del Buongusto, 1, 56100 Pisa, Italy.

[2]) Laboratoire de Volcanologie du C.N.R.S., Centre des Faibles Radioactivités, B.P. N° 1, 91190 Gif sur Yvette, France.

[3]) Section Technique d'Etudes de la Pollution dans l'Atmosphère et dans les Mines, C.E.A. B.P. N° 6, 92260 Fontenay aux Roses, France.

(a) the total quantity of radium in the rock;

(b) the distribution of radium within the rock;

(c) the effective escape area, dependent on imperfections in the radium bearing crystals, the nature of the mineral grains' boundaries, the structure of the glassy matrix, the porosity, etc.;

(d) the stress field in the parent rock;

(e) the temperature; and

(f) the physico-chemical characteristics of the fluid filling the rock pores.

In spite of the work of BARRETO *et al.* (1974) who examined the physical characteristics of the emanation of Rn^{222} from rocks, soils and minerals under laboratory conditions, the escaping fraction of Rn^{222} produced within a crystal under likely geothermal conditions is still not sufficiently well known.

It has already been demonstrated that Rn^{222} cannot be carried from great depths up to the surface solely by diffusion (TANNER, 1964a). However, if convective motion is established in the carrier fluid, then Rn^{222} can bring information on geological environments situated at great depths.

Exploited geothermal fields are precisely hydrothermal systems where natural fluids (mixtures of steam, gas and liquid water) are forced convectively towards the surface. This feature suggested the use of Rn^{222} as a natural tracer of the subsurface travel conditions of the geothermal fluids (STOKER and KRUGER, 1975). Moreover, like every natural or artificial radioactive tracer, Rn^{222} gathers information on processes whose durations are of the same order of magnitude as its half life, namely 3.835 days. It is thought that this figure approximates the transit time of geothermal fluids in small secondary reservoirs.

Two types of studies were carried out from March to June 1976 on the Larderello geothermal field:

(a) the Rn^{222} survey of the whole geothermal area (193 wells have been sampled); it was a part of the programme for studying the present geochemical state of the field; and

(b) the exploration of the possible use of Rn^{222} for the determination of time characteristics of steam reservoirs.

2. The Rn^{222} survey of the Larderello geothermal field

From a geological point of view, the Larderello geothermal field can be placed in a simple scheme as follows:

(a) a basal complex which represents the potential geothermal reservoir and is made up of a schistose-quartzitic series (Paleozoic-Triassic), overlain by a series made up of an anhydritic-dolomitic formation in its lower part, mainly limestone in the middle and sandstone in its upper part (Upper Triassic to Oligocene); and

(b) a cap-rock complex of prevalently shaly terms in flysch facies (Upper Jurassic to Cretaceous) lying in structural discordance over the basal complex.

This discordance between the mainly shaly complex and the underlying terrains is the result of prevalently horizontal movements caused by compressive phenomena during the Alpine orogenesis. After this compressive phase the region was affected by distensive movements which produced horst-and-graben tectonics.

The reservoir rocks are characterized by high permeability in the fractured zones only, having a very low matrix permeability. Information obtained from drilling shows that the highest permeability is in the upper part of the basal complex, and particularly where this is made up of carbonate layers.

The results of the whole Rn²²² survey have been presented elsewhere (D'AMORE, 1975). We shall only consider here the results along a NNW–SSE cross section of the Larderello geothermal field (Fig. 1).

Generally, as noticed by MAZOR (1977) and by TRUESDELL and NEHRING (1977): the (i) depth of wells, (ii) flow-rate, (iii) flowing temperature and pressure, (iv) gas/steam ratio, (v) hydrogen content and (vi) relative $\delta^{18}O$ versus SMOW decrease from the newly exploited Gabbro zone in the north, south-south-eastward toward the formerly

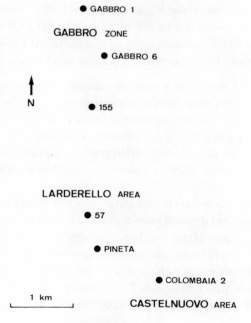

Figure 1

Relative locations of the steam wells corresponding to the cross section studied. The Serrazzano area is located 10 km to the south-west of the Larderello–Castelnuovo zone depicted here.

Table 1

The NNW–SSE section across the Larderello Geothermal Field: specific concentration of Rn^{222} *compared with some physical parameters of the steam wells.*

	Well	Flow-rate t/h	kg_{gas}/kg_{fluid} %	Radon content nCi/kg_{fluid}
NNW				
	Gabbro 1	43.0	8.69	39.2
	Gabbro 6	53.2	11.44	62.1
	155	17.9	5.80	64.6
4.5 km	57	9.7	6.52	75.1
	Pineta	8.5	2.39	161.0
	Colombaia 2	8.6	0.06	7.8
SSE				

drilled Castelnuovo area. All these observations suggest the existence of a deep circulation of the fluids in the Gabbro zone, and of a mixing process with shallower circulating waters in the Castelnuovo area (D'AMORE *et al.*, 1976b).

On the other hand the Rn^{222} content tends to increase along the same NNW–SSE cross section (see Table 1). Tritium, missing in the northernmost wells (unpublished data), seems to follow the same pattern as Rn^{222}.

In order to relate the Rn^{222} variations over the field to the hydrothermal convection systems, one must first explain the highest values of Rn^{222} concentration. Indeed the Rn^{222} content in some wells (up to 500 nCi kg^{-1}) is too high to be derived from a 'mean' thermal water (Rn^{222} concentrations ranging from 0.1 to 100 nCi kg^{-1}, quoted by STOKER and KRUGER, 1975). Several enrichment mechanisms might be considered:

(a) The emanating power of a rock in vacuum seems to be very low, possibly because Rn^{222} atoms recoiling from a mineral grain cannot be slowed down by an interstitial fluid and thus become embedded in an adjacent grain (LINDSTROM *et al.*, 1971) with no further possibilities of diffusion (LAMBERT *et al.*, 1972). While liquid water, owing to its higher density, may be supposed to extract Rn^{222} from a rock more efficiently than water vapour, there is nevertheless no strong experimental support for this assumption.

(b) At a given mass flow-rate liquid water does not transport Rn^{222} for as long a distance as steam does; but it is well known that Rn^{222} can be efficiently removed from a water solution if a gas phase bubbles through it (ZAGIN and SASKINA, 1966). Consequently, Rn^{222} tends to accumulate in the gaseous phase, more markedly if this gaseous phase crosses numerous water layers. If this process is effective, the abnormally low gas/steam ratio of well Colombaia 2 may account for the Rn^{222} concentration, surprisingly low with regard to the location of this well in the geothermal field, in its fluid.

(c) When flashing occurs, Ra^{226} can be deposited as a very thin layer of radium-bearing crystals whose emanating power will be very high. That the underground

Rn²²² distribution might largely be determined by the migration of its Ra²²⁶ precursor has been widely emphasized by TANNER (1964a and b). If uranium-radium minerals are deposited at the boiling level by processes of hydrothermal mineralization, the closer the boiling level is to the well bottom, the higher should the Rn²²² concentration be in the discharged fluid. These considerations lead one to conclude that high concentrations of Rn²²² in a geothermal fluid would generally indicate circulation of liquid water and boiling level close to the well bottom.

(d) Another mechanism which can be accounted for would lead to somewhat different conclusions. Let us consider a homogeneous porous reservoir (porosity ϕ) with an emanating power independent of the nature of the pore fluid (i.e. the first mechanism of enrichment described above is ineffective). Let us define P (nCi kg_{fluid}^{-1} h^{-1}) as the number of atoms[4]) of Rn²²² transferred by the rock to the mass unit of fluid per unit time and M (kg) as the stationary mass of fluid filling the reservoir. It can be easily shown that the Rn²²² specific concentration N (nCi kg^{-1}) in the stationary pore fluid is equal to P/λ ($\lambda = 7.553 \times 10^{-3}$ h^{-1}, decay constant of Rn²²²) as long as the flow-rate of the well tapping the reservoir is negligible in comparison with the product λM (D'AMORE and SABROUX, 1976a). If E (nCi kg_{rock}^{-1} h^{-1}) is the emanating power of the rock in the reservoir, then:

$$P = \frac{\rho_2 E}{\rho_1 \phi} \tag{1}$$

where ρ_1 and ρ_2 (in consistent units) are the densities of the fluid and rock respectively.

Hence, under the stationary conditions described above, the specific concentration of Rn²²² in the fluid is:

$$N = \frac{\rho_2 E}{\rho_1 \phi \lambda}, \tag{2}$$

i.e., inversely proportional to the density of the pore fluid.

Other things being equal, the specific concentration of Rn²²² in the fluid of a steam-filled reservoir can be up to a hundred times higher than in the fluid of a liquid water-filled reservoir; much higher Rn²²² specific concentrations should be generally expected in geothermal fluids from vapour-dominated systems than from water-dominated systems. This assertion is consistent with the data obtained by KRUGER et al. (1976).

3. Reservoir time characteristics

For a given productive geothermal well the flow-rate is the dominant parameter governing the Rn²²² content of the discharged fluid, and is the only one which can be

---

[4]) 1 nCi = 1.764×10^7 atoms of Rn²²².

modified (reduced) as long as stimulation techniques (water reinjection, hydraulic or explosive fracturing) are not applied.

The flow-rate determines the time available for Rn^{222} to accumulate in Ra^{226}-bearing media and to decay in Ra^{226}-free media. If relative emanating powers of the various geological layers forming the geothermal system could be measured on core-samples, simple models (D'AMORE and SABROUX, 1976a) would enable the volume of those different media to be deduced from the variation of the Rn^{222} concentration as a function of flow-rate.

Unfortunately, in a geothermal field generating electric energy, it is not possible to obtain experimental characteristic curves of Rn^{222} for a great number of wells, as this would involve a large waste of energy; the situation is similar for the flow-rate vs. wellhead-pressure curves (RUMI, 1972). No more than two drawdown tests have been performed during our survey of the field (see Fig. 2).

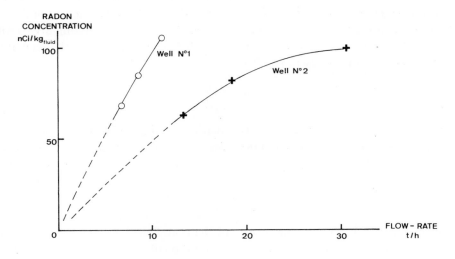

Figure 2
Specific concentration of Rn^{222} as a function of flow-rate for the wells Grottitana (no. 1) and Lustignano 5 (no. 2), Serrazzano area.

After each test, several days were allowed to pass in order to ensure an approach to radioactive equilibrium in the reservoir before sampling. While the transients are by themselves highly significant they have not been considered in this early study.

The overall shape of the characteristic curves suggests that the steam reservoirs are not of high emanating power; nevertheless, the extrapolation of the Rn^{222} concentration towards a zero value for lower flow-rates is purely speculative. It is a convenient assumption, for in that case, the Rn^{222} specific concentration does not directly depend on the density ρ_1 of the fluid in the reservoir. Indeed, measurements at highly reduced flow-rates require very long periods of stabilization.

We have considered two basic models of flow: pipe-flow and diffuse-flow. The main features of both models are:

(a) the existence of homogeneous reservoirs with regard to both the physical properties and the emanating power;
(b) the lack of any effect of flow-rate variations upon the pattern of the flow in the reservoirs; and
(c) the absence of large-scale phase changes and density variations in the geothermal fluid feeding each reservoir.

For each model there is a simple relation between the mass flow-rate and the Rn^{222} specific concentration in the fluid (D'AMORE and SABROUX, 1976a):

(a) Pipe-flow:

$$N = \frac{P}{\lambda} + \left(N_0 - \frac{P}{\lambda} \right) e^{-\lambda(M/Q)} \tag{3}$$

(b) Diffuse-flow:

$$N = \frac{QN_0 + PM}{Q + \lambda M} \tag{4}$$

N_0 is the specific concentration of Rn^{222} in the fluid for an infinite flow-rate (i.e. the concentration in a theoretical 'primary' reservoir whose Rn^{222} content would remain unaffected by any flow-rate variation in a tapping well).

Given two different values Q_1 and Q_2 of the flow-rate of a well, the working equations of the models for a Ra^{226}-free reservoir ($P = 0$) whose volume equals M/ρ_1 are:

$$M = \frac{1}{\lambda} \frac{Q_1 Q_2}{Q_1 - Q_2} \ln \frac{N_1}{N_2} \tag{5}$$

and

$$M = \frac{1}{\lambda} Q_1 Q_2 \frac{N_1 - N_2}{Q_1 N_2 - Q_2 N_1} \tag{6}$$

for pipe-flow and for diffuse-flow respectively. N_1 and N_2 are the Rn^{222} concentrations corresponding to the values Q_1 and Q_2 of the flow-rate of the well.

Table 2 displays the results of computation on the basis of the highest and lowest experimental values for each of the two wells studied. $\tau = M/Q$ is the characteristic time of the steam reservoir at maximum flow-rate.

These values must be interpreted very carefully: strictly speaking, M represents the mass of stationary fluid whose Rn^{222} content is affected by a flow-rate variation, and τ the time for this mass of geothermal fluid to be discharged by the well at the maximum flow-rate. Our models being over-simplified, these numerical values are only given for

Table 2

Reservoir time characteristics on the basis of the Rn^{222} *vs. flow-rate curve.*

Well	Flow-rate (t/h)	Mass of fluid M (t)	Characteristic time τ (days)
1 (Grottitana)	10.9	1051 (15444)	4
2 (Lustignano 5)	31.5	1449 (3334)	2

Values of M correspond to the pipe-flow model. Values corresponding to the diffuse-flow model are given in brackets; this last model fits very poorly and the corresponding values must not be considered. The accuracy is better for well no. 2. Both wells are situated in the Serrazzano area.

the sake of example. Owing to the scarcity of accurate data concerning the physical parameters of geothermal reservoirs, more sophisticated models, according to their authors (SAKAKURA *et al.*, 1959), would lead to unreliable numerical solutions.

It appears from the computations that the pipe-flow model gives a much better fit with the experimental results. In all likelihood, a network of macroscopic fractures represents the feeding system of the wells more realistically than does a homogeneous porous medium. This could be deduced from the low emanating power of the steam reservoir (doubtlessly due to a small exchange area) previously inferred from the shape of the Rn^{222} content characteristic curves.

4. Conclusions

Throughout our Rn^{222} survey, the method of sampling and analysis (described in D'AMORE, 1975 and in PRADEL and BILLARD, 1959 respectively) have revealed themselves particularly suited for geothermal studies. The wide variations in Rn^{222} concentrations over the whole geothermal field are related to other physical and geochemical parameters and possibly to pathways of fluid circulation. Long term variations (of several years) of the Rn^{222} content in a given well (D'AMORE *et al.*, 1976b) can then be interpreted in terms of an evolution of the related hydrothermal system.

Finally, it has been established that the Rn^{222} concentration vs. flow-rate curves could be an effective tool in computing reservoir time characteristics. The apparent capacity of the steam reservoir for each of the wells studied in detail is of the order of 10^5 m^3. Obviously, extensive additional work will be required to make the Rn^{222} measurements currently usable for the evaluation of other reservoir parameters.

Acknowledgements

The authors thank E.N.E.L. for all the facilities it provided for field and laboratory operations. The French C.E.A. technical assistance has been supported in part by the C.N.R.S.-A.T.P.: 'Transferts thermiques à travers l'écorce terrestre'.

REFERENCES

BARRETTO, P. M. C. CLARK, R. B. and ADAMS, J. A. S., *Physical characteristics of Radon-222 emanation from rocks, soils and minerals: its relation to temperature and alpha dose,* in *The Natural Radiation Environment II* (eds. J. A. S. Adams, W. M. Lowder and T. F. Gesell), (U.S.E.R.D.A., 1974), pp. 731–740.

D'AMORE, F. (1975), *Radon-222 Survey in Larderello geothermal field, Italy (Part 1),* Geothermics *4*, 96–108.

D'AMORE, F. and SABROUX, J. C. (1976a), *Signification de la présence de radon-222 dans les fluides géothermiques,* Bull. Volcanol. *40* (in press).

D'AMORE, F., FERRARA, G. C., NUTI, S. and SABROUX, J. C. (1976b), *Variations in Radon-222 content and Its implications in a geothermal field,* International Congress on thermal Waters, Geothermal Energy and Vulcanism of the Mediterranean Area, Athens.

KRUGER, P., STOKER, A. and UMAÑA, A. (1976), *Radon in geothermal reservoir engineering,* Geothermics *5*, 13–19.

LAMBERT, G., BRISTEAU, P. and POLIAN, G. (1972), *Mise en évidence de la faiblesse des migrations du radon à l'intérieur des grains de roche,* C.R. Acad. Sci., Ser. D, *274*, 3333–3336.

LINDSTROM, R. M., EVANS, J. C., FINKEL, R. C. and ARNOLD, J. R. (1971), *Radon emanation from the lunar surface,* Earth Planet. Sci. Lett. *11*, 254–256.

MAZOR, E. (1977), *Noble gases in a section across the vapor-dominated geothermal field of Lardello, Italy,* this volume.

PRADEL, J. and BILLARD, F. (1959), *Le thoron et les risques associés dans la manipulation du thorium,* Progress in Nuclear Energy Ser. XII, *1* (*Health Physics*), 239–261.

RUMI, O. (1972), *Some considerations on the flow-rate/pressure curve of the steam wells of Larderello,* Geothermics *1*, 13–23.

SAKAKURA, A. Y., LINDBERG, C. and FAUL, H. (1959), *Equation of continuity in geology with applications to the transport of radioactive gas,* U.S.G.S., Bull. 1052-I.

STOKER, A. K. and KRUGER, P., *Radon measurements in geothermal systems,* Stanford Geothermal Program (Stanford University, Stanford, California, 1975), 116 pp.

TANNER, A. B., *Radon migration in the ground: a review,* in *The Natural Radiation Environment* (eds. J. A. S. Adams and W. M. Lowder), (University of Chicago Press, 1964a), pp. 161–190.

TANNER, A. B., *Physical and chemical controls on distribution of Radium-226 and Radon-222 in ground water near Great Salt Lake, Utah,* in *The Natural Radiation Environment,* (eds. J. A. S. Adams and W. M. Lowder), (University of Chicago Press, 1964b), pp. 253–276.

TRUESDELL, A. H. and NEHRING, N. L. (1977), *Gases and water isotopes in a geochemical section across the Lardello, Italy, geothermal field,* this volume.

ZAGIN, B. P. and SASKINA, N. N. (1966), *Transfert du radon dans la chambre d'émanation à l'aide du gaz carbonique,* Radiochimie *8*, unabridged translation of Radiokhimiya, pp. 125–127.

(Received 8th December 1977, revised 14th April 1978)

Pageoph, Vol. 117 (1978/79), Birkhäuser Verlag, Basel

Noble Gases in a Section across the Vapor Dominated Geothermal Field of Larderello, Italy

By Emanuel Mazor[1]

Abstract – Noble gases were studied in six wells, located on a 4.5 km south to north section across the Larderello field. Depth of wells, flow and gas/steam ratios are known to increase from south to north. Exploitation progressed in the same direction. The following noble gas patterns are observable: (a) Atmospheric Ar, Kr and Xe reflect productions of gas-depleted water at Colombaia 2 and progressively more gas-enriched steam towards the Gabbro wells. (b) Radiogenic ^{4}He and ^{40}Ar are observed in increasing concentrations from south to north. (c) The radiogenic and atmospheric gases reveal a positive correlation, indicating that the recharging water enters deep into the system, and gets well mixed with the radiogenic gases prior to the steam separation. (d) Gas contents and relative abundances of radiogenic argon decrease with production, thus supplying markers for the degree of exploitation in a well and a guide for optimum well spacing. (e) Excess neon over argon is observed and discussed in terms of 'crustal' origin versus possible fractionation of atmospheric noble gases due to pertial steam separation.

Key words: Atmospheric Ar, Xe, Kr; Radiogenic ^{4}He, ^{40}Ar; Gas content of steam; Partition coefficients of noble gases; Recharging water; Production well spacing.

1. Introduction

1.1. Initial contents of atmospheric noble gases

The atmosphere is a well mixed reservoir of known contents of He, Ne, Ar, Kr and Xe, each being characterized by a known isotopic composition. Surface waters equilibrate with the air before their descent into the ground and are thus tagged by the dissolved atmospheric noble gases. The amounts of gas dissolved depend on the ambient temperature (Fig. 1), the altitude (barometric pressure) and salinity (Mazor, 1977). In the case of Larderello, recharge occurs near sea level and is of fresh water, hence the amount of noble gases contained in the recharge water may be directly deduced from Fig. 1. Recent work (Herzberg and Mazor, in preparation) has shown that recharge waters equilibrate with air along their path through the aerated zone, but as they reach the saturated zone no further exchange occurs (Mazor, 1972). The temperature at the base of the aerated zone is commonly close to the local mean annual temperature. In the case of Larderello one may assume that the recharge water equilibrated with air in the range of 5–15°C, or an average of 10°C.

---

[1] Geo-Isotope Group, Weizmann Institute of Science, Rehovot, Israel.

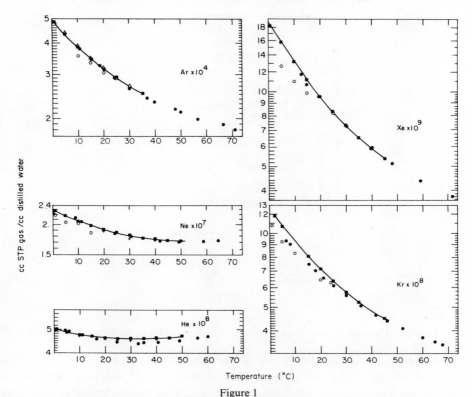

Figure 1

Solubility of atmospheric noble gases in fresh water. Sources of data: ○ – KÖNIG (1963), multiplying his seawater data by his given ratios for solubility in distilled water to solubility in sea water; △ – DOUGLAS (1964); ● – MORRISON and JOHNSTONE (1954); + – WEISS (1971) and the line is from BENSON and KRAUSE (1976). All data were multiplied by the atmospheric abundances. In the present work we applied the values of Benson and Krause's line.

1.2. Deep seated radiogenic noble gases

Radiogenic ^{4}He is commonly observed in thermal waters in amounts that are significantly higher than the value of about 5×10^{-8} cc STP/cc water that may be expected from air solution (Fig. 1). The excess is thus added at depth, commonly attributed to flushing of radiogenic He evolved in the aquifer rocks.

Occasionally also endogenic ^{40}Ar is observed. Its amount may be deduced via the ^{36}Ar concentration, knowing that in the atmosphere the ^{40}Ar/^{36}Ar ratio is 295.6. Any excess ^{40}Ar is, thus, of non-atmospheric origin and is commonly attributed to flushing of radiogenic ^{40}Ar evolved in the country rocks.

1.3. Techniques applied at Larderello

Samples were collected from six wells, constituting a section through the geo-thermal field (Fig. 2). The southern wells are shallower (less than 1000 m) and are

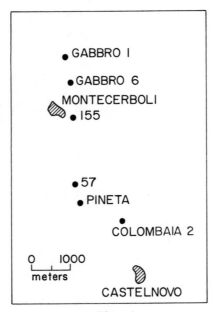

Figure 2
Location of wells sampled in a south–north section across the Larderello geothermal field.

older, whereas the northern ones are deeper (about 1500 m) and relatively newer. The steam was cooled in a condenser and collected in a 1 cc glass tube that had two high-vacuum spring-loaded stopcocks (see note, Table 1). The samples were glass sealed to an extraction line, cleaned with a titanium getter and measured in a Reynold's type 4.5″ 60° glass mass spectrometer (Nuclide Co.). Concentrations were calculated from peak heights of pipetted atmospheric noble gas standards, measured with each sample.

2. Results and discussion

The analytical data are summed up in Tables 1 and 2.

2.1. Dry gas/steam ratios

The ratio of dry gas to steam was measured with the aid of the condenser that was applied to collect the samples. The results reveal a geographical pattern, being low at Colombaia 2 and increasing towards the Gabbro 6 and 1 wells (Fig. 3).

2.2. Atmospheric noble gases

The isotopic composition of Ne, Kr and Xe was found to be atmospheric (the analytical error being estimated as up to $\pm 2\%$ of the ratio of the more abundant

Table 1
Noble gases, 10^{-8} cm^3 STP/cm^3 dry gas*).

Run no.	Well	°C	He	Ne	Ar	$\dfrac{(^{40}Ar/^{36}Ar)\ samp\dagger}{(^{40}Ar/^{36}Ar)\ air}$	^{40}Ar rad	Ar atm	Kr	Xe
384	Colombaia 2	229	600	2.55	2000	1.032	60	1940	–	–
393			585	5.29	5200	1.033	170	5030	0.80	0.14
415			600	3.33	2760	1.030	85	2680	0.60	0.16
Rep†			600	3.72	3300	1.032	105	3220	0.70	0.15
388	Pineta	200	470	2.14	3000	–	–	3000	–	–
390			55	1.17	3930	1.003	–	3930	0.97	0.19
414			240	3.68	3920	0.999	–	3920	1.06	–
Rep‡			470	2.33	3620	1.001	–	3617	1.00	0.19
389	57	214	1475	2.15	1920	1.040	80	1840	0.36	0.06
391			1350	1.28	1720	1.043	70	1650	0.30	0.05
412			1330	3.89	3530	1.037	130	3400	–	–
Rep‡			1470	2.44	2390	1.040	90	2300	0.33	0.06
386	155	222	2465	3.11	3030	1.077	230	2800	0.63	0.12
394			1470	3.56	3400	1.093	315	3085	0.62	0.14
413			2220	3.65	3460	1.041	140	3320	0.79	–
Rep‡			2460	3.44	3330	1.070	230	3070	0.67	0.13
387	Gabbro 6	242	1020	3.74	2400	1.062	150	2250	0.56	0.12
394			1110	2.85	2100	1.076	160	1940	0.54	0.12
416			1040	4.42	3940	1.043	170	3770	0.84	–
Rep‡			1110	3.67	2810	1.060	160	2655	0.65	0.12
392	Gabbro 1	240	1370	2.54	2850	1.075	215	2635	0.51	0.10
411			1810	3.70	3540	1.056	200	3340	0.78	–
Rep‡			1800	3.12	3200	1.066	205	2990	0.64	0.10

*) Estimated analytical error: ±4%. A larger error was, however, introduced in the sampling: the samples were collected in glass tubes closed by stopcocks. One end was connected to a condensator and the other one was dipped into water. The condensed steam was flushed through and the stopcocks were closed. This closing of stopcocks was not systematic, in times the one closer to the condensator was closed first and in times the other one. In the first cases, samples were collected at atmospheric pressure but in the second case slight over pressures could build up. This mechanism is called upon to explain the variations observed in the triplicate samples of Colombaia 2, well 57 and Gabbro 6. This over-pressure influenced the absolute abundances of the samples but not the relative patterns.

†) Estimated analytical error ±2% of the ratio.

‡) Representative values (Rep.) were calculated for each well by selecting the highest He value (assuming least leakage losses) and by calculating the average of the values for Ne, Ar, Kr and Xe.

isotopes of each element). The atmospheric Ar content were calculated from the ^{36}Ar contents.

The atmospheric noble gases may be expressed as parts of the dry gas fractions (Table 1) or as parts of the condensed steam (Table 2). The data expressed in the first mode revealed no elemental or geographic correlations, but the data expressed in the

Table 2
Noble gases, 10^{-8} cm^3 STP/cm^3 condensed steam*).

Well	$\dfrac{\text{cm}^3 \text{ STP dry gas\dagger)}}{\text{cm}^3 \text{ STP condensed steam}}$	He	Ne	Ar atm	Ar rad	Kr	Xe	He/Ar rad
Colombaia 2	3.3	1 900	12	10 500	340	2.4	0.5	5.7
Pineta	17.5	8 300	41	6 300	–	17	3.3	–
57	46.1	68 000	112	106 000	4 300	15	2.8	15.9
155	33.9	83 700	117	104 000	7 800	21	4.7	10.7
Gabbro 6	69.3	77 000	254	184 000	11 000	42	8.4	7.0
Gabbro 1	54.2	98 000	169	162 000	11 000	32	5.4	8.8
Air saturated water 10°C		4	25	29 000		9.4	1.3	

*) Based on the representative (Rep.) values given in Table 1, multiplied by the dry gas/condensed steam ratio.

†) Measured in the field during sample collection.

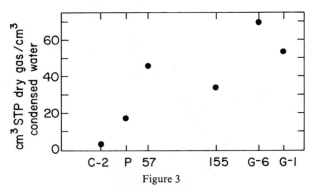

Figure 3
Relative amounts of dry gas in condensed steam along the traverse through the Larderello field (Table 2). C-2: Colombaia 2; P-Pineta; G-1: Gabbro 1 and G-6: Gabbro 6. Wells are spaced according to projected distance.

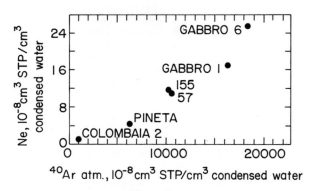

Figure 4
Ne and atmospheric Ar, expressed in parts per condensed steam. A remarkable correlation with geographic position is seen.

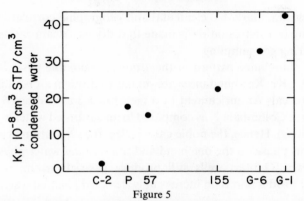

Figure 5

Krypton in wells along the traverse through Larderello expressed in parts per condensed steam. A gradual increase is observed towards the Gabbro wells.

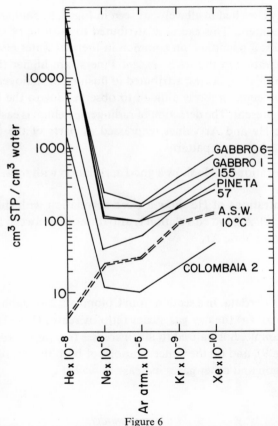

Figure 6

Noble gases in Larderello wells, as compared to air-saturated water at 10°C (A. S. W. 10°C), taken as representing the original recharging water. Colombaia 2 is fed by steam from gas depleted water, whereas towards Gabbro the wells are fed by increasingly gas-enriched steam. The large admixtures of radiogenic He are obvious.

second mode reveal remarkable elemental and geographic correlations (Figs. 4 and 5). Thus, a strong recommendation is made that dry gas/steam ratios should always be measured during gas sampling.

The relative abundance pattern of the atmospheric noble gases is shown in Fig. 6. The relative Ar;Kr:Xe abundances are similar to those in air equilibrated water at 10°C. Figure 6 reveals Ar enrichment by a factor of 6.3 at Gabbro 6 and a depleted value of only 0.4 at Colombaia 2, as compared to air saturated water at 10°C (assumed recharge conditions). Hence, the noble gases reflect the separation occurring between the steam and gas phase on the one hand and the residual water phase on the other. The relatively new Gabbro wells still produce gas-rich steam, whereas towards Colombaia 2 the production is on increasingly depleted residual water.

2.3. Radiogenic noble gases

He seems highly enriched in all wells, as seen in Fig. 6, as compared to the expectable atmospheric content. This excess is attributed to flushing of radiogenic helium from reservoir rocks, a phenomenon common in thermal water elsewhere.

The $^{40}Ar/^{36}Ar$ ratios in the wells, except Pineta, are higher than values in air (Table 1), indicating ^{40}Ar excesses, attributed to flushing of radiogenic ^{40}Ar from the aquifer rocks. Radiogenic argon is difficult to observe, due to the large amounts of atmospheric argon present. The detection of radiogenic helium is easier in this respect.

The radiogenic He and Ar values, expressed as parts of the condensed steam (Table 2) reveal the following patterns:

  (a) The concentrations of He show a good correlation with those of the *radiogenic* Ar (Fig. 7).
  (b) The concentrations of He show a clear correlation with the *atmospheric* Ar as well (Fig. 8). A good correlation is also observed between the radiogenic Ar and the atmospheric Ar.

3.3. Geographical trends

These are seen in the data. In a section from Colombaia 2 to Gabbro 1 the following changes are observed: (a) the dry gas/steam ratio increases (Fig. 3), (b) the ratio of $^{40}Ar/^{36}Ar$ in the sample, normalized to air, increases (i.e. more radiogenic Ar in the Gabbro wells) (Fig. 9), and (c) the concentrations of both the *radiogenic* helium and the *atmospheric* argon and other gases increase as well (Fig. 10).

3. A working model

The observations seem to indicate that water is always recharged into the deep part of the system so that the atmospheric gases brought with it get well mixed with the

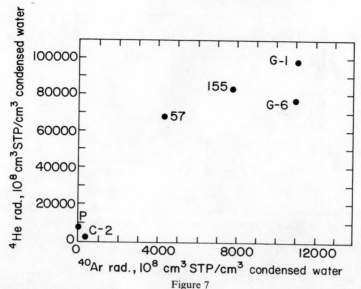

Figure 7
Radiogenic He versus radiogenic Ar. A positive correlation is observed.

radiogenic gases and CO_2 flushed from the heated rocks. Subsequent steam separation causes the observed depletions in the atmospheric and radiogenic noble gases and the reactive (mainly CO_2) gases. The good correlations of the atmospheric and deep seated components exclude shallow mixing of hot and cold water. In such a case no positive correlations between atmospheric and deep (radiogenic gases and CO_2) components could occur.

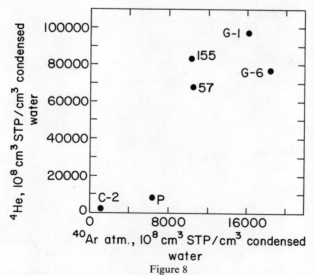

Figure 8
Radiogenic He versus atmospheric Ar. Again positive correlation is observed, indicating thorough mixing of atmospheric and radiogenic gases prior steam separation.

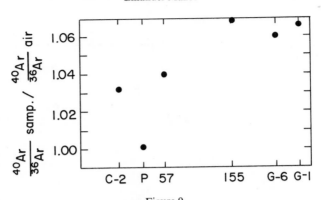

Figure 9

Relative amounts of radiogenic argon in a traverse through the Larderello system. Values increase from about 3% enrichment in Colombaia 2 to 7% in the Gabbro wells.

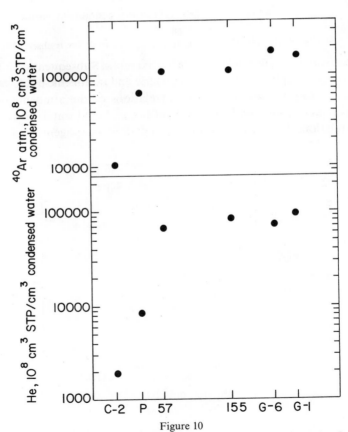

Figure 10

Atmospheric argon and radiogenic helium. Both are seen to increase along the section from Colombaia 2 to the Gabbro wells.

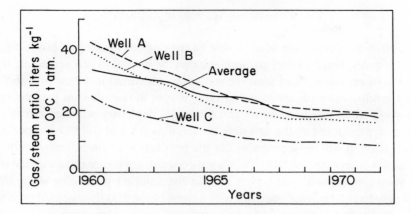

Figure 11
Trend of gas/steam ratio as a function of production (time) in three wells in the intensively drilled area of Castelnuovo (CELATI *et al.*, 1973).

The degree of gas depletion seems to be a result of production. CELATI *et al.* (1973) report such a decrease in the dry gas content in three wells near Castelnuovo, during production from 1960 to 1970 (Fig. 11). Data from FERRARA *et al.* (1963), plotted in Fig. 12 as a function of estimated production, reveal a similar decrease in the *relative* amount of radiogenic argon.

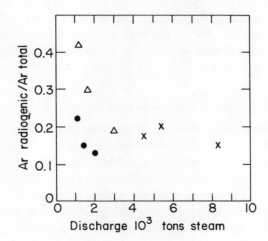

Figure 12
Change of radiogenic Ar/total Ar ratios (FERRARA, *et al.*, 1963) as a function of production (data from C. Panichi). An exponential decrease is seen. △ – BcF3, Serruzzano; ● – Campo Sportiva and × – Fabiani. The last two are from Larderello.

4. Possible implications to well spacing

It is often hard to decide what should be the optimal spacing of producing wells. The noble gases, besides other parameters, may be helpful in this respect. If the concentration of radiogenic and atmospheric gases, measured in a new prospection, or planned production well, are found to be as low as in the adjacent producing wells, one may conclude that production and noble gas depletion occurred already in the field section penetrated by the new well. Hence, if such a well will be equipped and put into operation, it will most probably cut the production of the adjacent wells.

If, however, in a new well high noble gas concentrations are observed, close to the initial values in the field and higher than in the adjacent producing wells, the conclusion may be reached that one deals with a niche of the field that was not produced so far. Hence, the new well should be equipped and put into production.

5. Neon excess – primordial or atmospheric?

5.1. Excess neon values

One of the most interesting outcomes of the noble gas study at Larderello is the detection of a significant excess of neon, as compared to the atmospheric argon (and krypton and xenon) contents in the assumed original recharge water. The relevant data are given in Table 3. It is seen that although the total noble gas contents vary in duplicate and triplicate samplings (footnote, Table 1), the $^{20}Ne/^{36}Ar$ ratios are well repeated in each well. The $^{20}Ne/^{36}Ar$ values in air-saturated water are 0.14 for equilibration at 5°C and 0.17 for equilibration at 25°C, hence a $^{20}Ne/^{36}Ar$ ratio of 0.15 seems well representative for the recharging waters of Larderello. The $^{20}Ne/^{36}Ar$ values in the studied Larderello wells are, however, 0.17 to 0.38 (Table 3), i.e. revealing Ne excesses of 12 to 250%.

5.2. Possible primodial origin of the excess neon

Neon excesses have been observed in igneous rocks. LORD RAYLEIGH (1939) reported Ne excesses (over Ar) in pumice from the Lipari Islands. This observation has been recently confirmed by BOCHSLER and MAZOR (1975) who found $^{20}Ne/^{36}Ar$ values up to 15.8. In a scorya sample from northern Israel a $^{20}Ne/^{36}Ar$ ratio of 23.7 was found (unpublished data). i.e. over two orders of magnitude above the atmospheric ratio and one order above the enrichments observed at Larderello. DYMOND and HOGAN (1973), FISHER (1974) and CRAIG and LUPTON (1976) reported $^{20}Ne/^{36}Ar$ higher than atmospheric in mid-oceanic basalts and discussed them in terms of primordial (solar and planetary) noble gases. The isotopic composition of the excess neon, in at least part of these rock samples, is similar to the atmospheric neon (BOCHSLER and MAZOR, 1975) and significantly different from meteoritic primordial

Table 3
Neon excesses, 10^{-8} cm^3 STP/cm^3 dry gas.

Well	Run no.	°C	^{20}Ne	^{36}Ar	^{20}Ne/^{36}Ar	Average ^{20}Ne/^{36}Ar
Colombaia 2	384	229	2.3	6.6	0.35	
	393		4.8	17.0	0.28	
	415		3.0	0.0	0.33	0.32 ± 0.03
Pineta	338	200	2.0	10.1	0.19	
	390		1.1	13.3	0.08	
	414		3.3	13.3	0.25	0.17 ± 0.07
57	389	241	1.9	6.2	0.31	
	391		1.2	5.6	0.21	
	412		3.5	11.5	0.31	0.28 ± 0.05
155	386	222	2.8	9.5	0.30	
	394		3.2	10.4	0.31	
	413		3.3	11.2	0.29	0.30 ± 0.01
Gabbro 6	387	242	3.4	7.6	0.44	
	395		2.6	6.5	0.40	
	416		4.0	12.8	0.31	0.38 ± 0.05
Gabbro 1	392	240	2.3	8.9	0.26	
	411		3.3	11.3	0.30	0.28 ± 0.02
Average Larderello (except Pineta)			3.0			0.31 ± 0.04
Air saturated water		5	19.8	145	0.14	
		15	18.1	118	0.15	
		25	16.7	98	0.17	

Average Ne excess: $3.0 \times 10^{-8} \times \dfrac{0.15}{0.31} = 1.4 \times 10^{-8}$ cm^3 STP/cm^3 dry gas.

Average He (except Pineta, Table 2): 1490×10^{-8} cm^3 STP/cm^3 dry gas.

Average He/Ne excess $\cong 100$.

neon, described by EBERHARDT (1974). The isotopic composition of the Larderello neon was atmospheric as well. One may consider that we observe indigenous, non-atmospheric neon but with an isotopic composition that is close, or equal, to the atmospheric composition. For the lack of a better term we will refer to it as 'crustal', indicating it is regarded as a left-over of the primitive earth, yet differing from extra-terrestrial primordial neon.

5.3. *Possible neon enrichment due to fractionation during boiling at high temperatures*

Partition coefficients of noble gases in gas and water phases have recently been obtained at temperatures up to the critical point (POTTER and CLYNNE, in preparation). The data are reproduced in Fig. 13. It is seen that $K_{Ne}/K_{Ar} \cong 3$ at the recharge temperature range but equals ~ 2 in the range of 200–300°C, hence a possible mechanism for Ne–Ar fractionation during boiling at high temperatures. The observed high Ne–Ar values reported in Table 3 could be formed by partial boiling

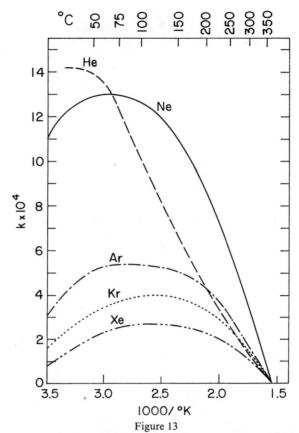

Figure 13

Henry's Law constants for solubilities in distilled water (POTTER and CLYNNE, in preparation).

at 250 ± 20°C (POTTER, MAZOR and CLYNNE, 1977), compared to measured temperatures of 222°–242°C. Hence, the observed Ne excesses (relative to Ar) may result from boiling at high temperatures (effected by the well operations).

Further work, including other geothermal systems, is needed to recognize excess 'crustal' neon from atmospheric neon concentrated by fractionation caused by boiling at high temperatures.

Acknowledgements

Valuable help in the sample collection and data discussion was given by Dr. A. H. Truesdell (U.S. Geological Survey, Menlo Park) and Dr. C. Panichi (International Institute for Geothermal Research, Pisa).

REFERENCES

BENSON, B. S. and KRAUSE, D., Jr. (1976), *Empirical laws for dilute aqueous solutions of nonpolar gases*, J. Chem. Phys. *64*, 689–709.

BOCHSLER, P. and MAZOR, E. (1975), *Excess of atmospheric neon in pumice from the island of Lipari*, Nature *257*, 474–475.

CELATI, R., NOTO, P., PANICHI, C., SQUARCI, P. and TAFFI, L. (1973), *Interactions between the steam reservoir and surrounding aquifers in the Larderello Geothermal field*, Geothermics *2*, 174–185.

CRAIG, H. and LUPTON, J. E. (1976), *Primordial neon, helium and hydrogen in oceanic basalts*, Earth and Planet Sci. Letters *31*, 369–385.

DOUGLAS, E. (1964), *Solubilities of oxygen, argon and nitrogen in distilled water*, J. Phys. Chem. *68*, 174–196.

DYMOND, J. and HOGAN, L. (1973), *Noble gas abundance patterns in deep-sea basalts – primordial gases from the mantle*, Earth Planet. Sci. Letters *20*, 131–139.

EBERHARDT, P. (1974), *A neon-E-rich phase in the original carbonaceous chondrites*, Earth Planetary Sci. Letters *24*, 182–187.

FERRARA, G., GONFIANTINI, R. and PISTOLA, P., *Isotopic composition of argon from steam jets of Tuscany*, in *Symp. on Nuclear Geology in Geothermal Areas* (Spoleto, 1963).

FISHER, D. E. (1974), *The planetary primordial component of rare gases in the deep earth*, Geophys. Res. Letters *1*, 161–164.

KÖNIG, VON, H. (1963), *Über die löslichkeit der Edelgase in Meerwasser*, Z. Naturforsh, *18a*, 363–367.

MAZOR, E. (1972), *Paleotemperatures and other hydrological parameters deduced from noble gases dissolved in groundwaters; Jordan Rift Valley*. Israel. Geoch. et Cosmochim. Acta *36*, 1321–1336.

MAZOR, E., *Geothermal tracing with atmospheric noble gases. Geothermics*, in, *Application of Nuclear Technics to Geothermal Studies* (I.A.E.A., Vienna, 1977), Special Issue of Geothermics, S.

MORRISON, T. J. and JOHNSON, N. B. (1954), *Solubilities of the inert gases in water*, J. Chem. Soc. *III*, 3441–3446.

POTTER II, R. W., MAZOR, E. and CLYNNE, M. A., *Noble gas partition coefficients applied to the conditions of geothermal steam formation*, Abstract, *Geochemistry of Geothermal Systems Symposium, Ann. Meeting G.S.A., Seattle* (1977).

RAYLEIGH, F. R. S., LORD (1939), *Nitrogen, argon and neon in the earth's crust with applications to cosmology*, Proc. Roy. Soc. *A170*, 451–464.

WEISS, R. F. (1971), *Solubility of helium and neon in water and sea water*, J. Chem. Eng. Data, *16*, 235–241.

(Received 10th November 1977, revised 14th March 1978)

Pageoph, Vol. 117 (1978/79), Birkhäuser Verlag, Basel

Gases and Water Isotopes in a Geochemical Section across the Larderello, Italy, Geothermal Field

By Alfred H. Truesdell and Nancy L. Nehring[1])

Abstract – Steam samples from six wells (Colombaia, Pineta, Larderello 57, Larderello 155, Gabbro 6, and Gabbro 1) in a south to north section across the Larderello geothermal field have been analyzed for inorganic and hydrocarbon gases and for oxygen-18 and deuterium of steam. The wells generally decrease in depth and increase in age toward the south. The steam samples are generally characterized by

(1) Total gas contents increasing south to north from 0.003 to 0.05 mole fraction;
(2) Constant CO_2 (95 ± 2 percent); near constant H_2S (1.6 ± 0.8), N_2 (1.2 ± 0.8), H_2 (2 ± 1), CH_4 (1.2 ± 1), and no O_2 in the dry gas;
(3) Presence of numerous, straight chain and branched C_2 to C_6 hydrocarbons plus benzene in amounts independent of CH_4 contents with highest concentrations in the deeper wells;
(4) Oxygen-18 contents of steam increasing south to north from −5.0‰ to −0.4‰ with little change in deuterium (−42 ± 2‰).

These observations are interpreted as showing:

(1) Decreasing gas contents with amount of production because the proportion of steam boiled from liquid water increases with production;
(2) Synthesis of CH_4 from H_2 and CO_2 with CO_2 and H_2 produced by thermal metamorphism and rock–water reactions;
(3) Extraction of C_2 to C_6 hydrocarbons from rock organic matter;
(4) Either oxygen isotope exchange followed by distillation of steam from the north toward the south (2 plates at ∼220°C) or mixture of deeper more-exchange waters from the north with shallow, less-exchanged recharging waters from the south.

Key words: H_2S, CO_2, N_2, H_2, Hydrocarbon gases; Oxygen isotopes; Deuterium content of steam; Isotope exchange; Thermal metamorphism; Rock–water reactions.

Introduction

The vapor-dominated geothermal fields of Larderello, Italy, and The Geysers, California, have been extensively developed for electric power production but their origin, reservoir fluid composition, and mechanism of steam production remain controversial (White, *et al.*, 1971; Truesdell and White, 1973). Recent data on the isotopic compositions of steam has suggested two modes of fluid movement in these systems, lateral recharge in the Larderello reservoir from the S.E. and S. (Celati, *et*

---

[1]) U.S. Geological Survey, 345 Middlefield Road, Menlo Park, California 94025, USA.

al., 1973; PANICHI *et al.*, 1974) and steam movement in the S.E. part of The Geysers by multistage distillation from the center toward the edge (TRUESDELL *et al.*, 1977). Analyses of gaseous constituents and water isotopes of steam samples from Larderello were undertaken to attempt to distinguish between these processes.

Steam from these systems contains inorganic and hydrocarbon gases in quantities that vary with time and with position in the field. Surveys of inorganic gases for these systems are in progress (FRANCO D'AMORE, personal communication, 1977) but we wish to report here detailed analyses of inorganic and hydrocarbon gases from a limited number of wells that form a cross section through the northern part of Larderello. The six wells sampled extend from the shallow, long produced area of Castelnuovo to the deeper, more recently drilled area of Gabbro. A physical description of the wells sampled is given in Table 1. Dr. Costanzo Panichi provided these data and selected and made available the wells for sampling.

Table 1

Physical data on six wells of the Larderello geothermal field sampled on September 5, 1975 for gas and isotope analyses. Wellhead temperature and pressure (WHT, WHP) are for flowing conditions. Gas/steam ratios are given in milliliters (STP) of gas per gram of condensate separated at ~30°C and do not include gas dissolved in the condensate. Data from CObottaNZO PANICHI *(personal communication, 1975).*

Well	Date drilled	Depth m	Production t/hr	In August 1975 WHT*) °C	WHP*) kg/cm^2	Measured Gas/Steam ml STP/gm
Colombaia	?	470	9.1	229	2.4	3.25
La Pineta	1942	316	11.1	200	4.3	17.5
Lard. 57	~1951	486	10.7	241	4.3	46.1
Lard. 155	~1961	844	15.3	222	4.9	33.9
Gabbro 6	1964	771	52.2	242	7.4	69.3
Gabbro 1	1962	853	44	240	7.4	54.2

*) Well head temperature and pressure during flow.

Collection and analysis

Gas and condensate samples were collected from a gas–water separator connected to the producing well through a stainless-steel, water-cooled condenser. The temperature of separation was maintained near 30°C to minimize fractionation of water isotopes by vapor separation and solution of gas in the condensate. The gas contents of steam given in Table 1 have been corrected to standard conditions (0°C and 1 atm). Gas contents of steam and analyses of major gases presented in Table 2 have been recalculated in Table 3 to include gas dissolved in the condensate. These corrections are significant for only one well. The loss of water vapor to the gas fraction at 30°C affects the isotope analyses of the condensate by less than 0.02‰ in $\delta^{18}O$ and 0.2‰ in δD in every case.

Table 2

Analyses of major gases separated from steam samples collected from six Larderello wells. Compositions are in mole percent of total dry gas separated from steam condensate at ~30°C and do not include gas dissolved in the condensate. Oxygen-18 and deuterium contents of steam condensate are also shown.

Well	CO_2	H_2S	N_2	Ar	O_2	H_2	He	CH_4	Total	$\delta^{18}O$, ‰ SMOW	δD, ‰ SMOW
Colombaia	93.3	1.8	0.77	0.01	n.d.	3.5	n.d.	0.58	100.0	−4.87	−42.0
Pineta	95.2	0.79	2.1	0.07	tr.	1.6	n.d.	0.27	100.0	−5.06	−44.6
Lard. 57	93.8	1.7	0.79	0.002	n.d.	1.8	0.004	1.7	99.8	−2.79	−43.1
Lard. 155	93.1	1.7	1.1	0.008	n.d.	2.0	0.006	2.2	100.1	−1.81	−41.8
Gabbro 6	96.7	0.92	0.45	0.007	n.d.	1.2	0.003	0.73	100.0	−0.47	−42.4
Gabbro 1	93.5	2.0	0.80	0.009	n.d.	2.4	0.005	1.2	99.9	−0.43	−39.8

Table 3

Analyses of major gases in mole percent recalculated to include gas dissolved in condensate. Gas solubility data are from the Handbook of Chemistry and Physics, v. 31, p. 1422–3.

Well	Gas/steam ml STP/gm	CO_2	H_2S	N_2	H_2	CH_4
Colombaia	3.88	93.5	2.40	0.65	2.90	0.50
Pineta	18.1	95.3	0.85	2.00	1.50	0.26
Lard. 57	46.6	93.8	1.75	0.78	1.78	1.68
Lard. 155	34.6	93.0	1.76	1.08	1.96	2.16
Gabbro 6	69.9	96.7	0.94	0.45	1.19	0.72
Gabbro 1	54.8	93.5	2.05	0.79	2.37	1.19

Gases were collected in evacuated 300-ml bottles containing 100 ml of 4N NaOH solution (GIGGENBACH, 1975). CO_2 and H_2S were analyzed by wet chemical methods on the solution, and residual gases were analyzed by gas chromatography. Major inorganic gases and methane were separated at room temperature on a 3 mm × 6 m stainless column packed with molecular sieve 5A and analyzed by a thermistor detector. For O_2, Ar, N_2, and CH_4, helium was used as a carrier gas; and for He and H_2, argon was used. The usual sample size was 0.5 ml STP. Hydrocarbon gases were analyzed with a temperature-programmed, flame-ionization gas chromatograph. Paired 3 mm × 3 m columns containing Porapak Q were programmed from 30°C to 150°C to obtain separation of $C_1 - C_6$ hydrocarbons. A typical chromatogram of hydrocarbon gases from the Gabbro 6 sample is shown in Fig. 1.

Inorganic and hydrocarbon gases were identified and their concentrations measured by comparison with known standards. The identification of hydrocarbon gases (Fig. 1 and Table 4) were confirmed for representative samples by GC-MS using a Perkin–Elmer 990 gas chromatograph coupled to a Bell and Howell 21-491 mass spectrometer. Special care was given to the identification of benzene because it appeared to be the only unsaturated hydrocarbon in these gases and occurred in relatively large amounts.

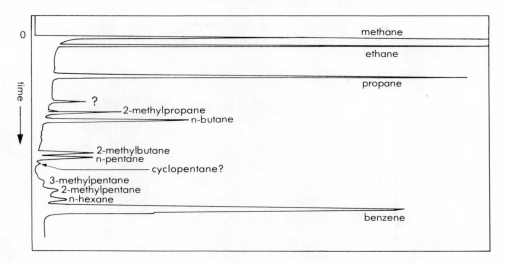

Figure 1

Chromatogram of hydrocarbon gases in steam from the Gabbro #6 well, Larderello geothermal field.
Cyclo-pentane was present in trace amounts in this gas but was not identified in other gases and has not been
included in Table 4.

Although three significant figures are given for analyses by wet chemistry and two
for those by gas chromatography, the reproducibility of these numbers depends on
the temperature of collection and other factors not entirely within our control and is
not as good as implied by the number of significant figures. Noble gases were analyzed
with a mass spectrometer by Emanuel Mazor of the Weizmann Institute, Israel, and
are described in a separate paper (MAZOR, 1978 this volume, p. 262). The contents of
all gases analyzed are presented graphically in Figure 2 without correction for gas
dissolved in condensate.

Oxygen-18 and deuterium analyses were made of condensed steam samples using
standard techniques (Table 2).

Interpretation

Interpretation of gas compositions of geothermal fluids is not so advanced as that
of water compositions. The clear magmatic source and inferred or demonstrated
equilibria which control the compositions of high temperature ($> 500°C$) volcanic
gases (ELLIS, 1957; HEALD et al., 1963; GIGGENBACH and LE GUERN, 1976) do not in
general apply to the lower temperature ($< 300°C$) gases of geothermal systems.
Suggestions have been made that all constituents (other than those from air con-
tamination) of geothermal gases are from a magmatic source (ALLEN and DAY, 1935,
p. 88) or from thermal metamorphism of sedimentary rocks (FACCA and TONANI,

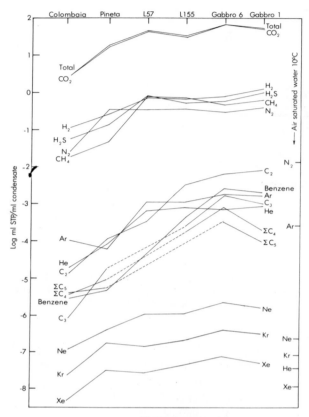

Figure 2

Major and minor gas contents in six wells in a south to north section from the edge to the center of the Larderello, Italy geothermal field. Compositions are expressed in cc gas at STP per cc of steam condensate. Data for noble gases are from Mazor (1978).

1964). Modern studies of geothermal systems do not entirely support either of these ideas. A definite determination of the ultimate sources of all gas constituents in geo-thermal systems is unlikely but isotope studies and equilibrium calculations described later have limited the possibilities from some species.

Whatever the ultimate sources, gases emanating from hot springs, fumaroles, and geothermal wells may be highly altered by reactions with other gases, rocks, liquid water, and dissolved substances. The composition of sampled gases may be affected by partitioning between subsurface water and steam existing naturally or produced during exploitation (ELLIS, 1962). In vapor-dominated geothermal systems such as Larderello a subsurface vapor phase exists in the natural state but most of the fluid mass is liquid water which boils in place to produce steam during exploitation (WHITE et al., 1971; TRUESDELL and WHITE, 1973). The steam produced is a mixture of original vapor (high in gases other than H_2O) and original liquid (low in other gases)

Table 4

Analyses of hydrocarbon gases separated from steam samples collected from six Larderello wells. Compositions are in moles per 10^6 moles of dry gas and do not include gas dissolved in condensate or in 4N NaOH–Na$_2$CO$_3$ solution.

Well	Methane C	Ethane C–C	Propane C–C–C	2 methyl propane C-C-C / C	n-butane C-C-C-C	2 methyl butane C-C-C-C / C	n-pentane C-C-C-C	3 methyl pentane C-C-C-C-C / C	2 methyl pentane C-C-C-C-C / C	n-hexane C-C-C-C-C-C	Benzene
Colombaia*)	5 800	4.0	0.3	tr	1.1	0.08	0.8	n.d.	n.d.	n.d.	0.9
La Pineta	2 700	6.6	1.1	0.07	0.5	0.2	0.1	tr	tr	0.05	0.3
Lard. 57*)	17 200	7.2	tr	n.d.	n.d.	n.d.	n.d.	n.d.	n.d.	n.d.	n.d.
Lard. 155*)	22 000	97	7.4	n.d.	5.1	n.d.	n.d.	n.d.	n.d.	tr.	13
Gabbro 6	7 300	98	27	4.2	7.6	2.6	2.6	tr	0.8	1.0	36
Gabbro 1	12 300	150	19	2.1	2.0	1.1	0.5	n.d.	n.d.	n.d.	38

*) Small sample;   tr: trace;   n.d.: not detected.

so the amount and composition of gases in the steam evolve in time towards decreasing total gas and increasing fractions of the more water soluble gases. This evolution can be studied only through repeated analyses of steam from the same source, not from the single sampling reported here. Gases in equilibrium in the unexploited reservoir may not remain in equilibrium during exploitation, or new equilibrium states may be established for the new conditions of vapor/liquid ratio, temperature, pressure, and composition. D'AMORE and NUTI (1977) suggest that equilibrium reactions generally control all gas compositions at Larderello. Their ideas are discussed in the following section.

Major gases

The major gases CO_2, CH_4, H_2, H_2S, and N_2 show a close grouping on Fig. 2 with CO_2 ranging from 93 to 97 percent of total gas and the other gases generally from 1 to 2 per cent of the total. These gases are found in all types of geothermal systems and except for N_2 appear to originate at least in part from rock–water reactions.

Carbon dioxide

Carbon dioxide is the major gas other than steam in most geothermal fluids and usually comprises more than 90 percent of the total dry gas (in our Larderello gases, 93 to 97 percent). Isotope studies have shown that the major sources of CO_2 are (1) thermal metamorphism of marine sedimentary carbonates which produces CO_2 with $\delta^{13}C$ values of -4 to $+4\%_0$ PDB, (2) oxidation of organic matter in sediments to produce $\delta^{13}C(CO_2)$ values of -30 to $-18\%_0$, and (3) magmatic or mantle sources with $\delta^{13}C(CO_2)$ near -6 to $-8\%_0$ (CRAIG, 1963). Unfortunately it is not always possible to isotopically distinguish magmatic CO_2 from mixed marine carbonate and organic CO_2. Analyzed samples (CRAIG, 1963) indicate a possible magmatic origin for CO_2 from Steamboat Springs and Lassen, predominantly marine

Table 5

Calculated aquifer gas pressures and equilibrium temperatures for the reaction $CO_2 + 4H_2 = CH_4 + 2H_2O$ compared with flowing wellhead temperatures and CO_2–CH_4 isotope temperatures interpolated from PANICHI et al. (1976). Gas pressures were calculated assuming H_2O pressure of saturated steam at the flowing wellhead temperature and perfect gas behavior. Equilibrium constants from ELLIS (1957) were used. Pressures are in atm. and temperatures in °C.

Well	t_{flow}, °C	P_{H_2O}	P_{CO_2}	P_{H_2S}	P_{N_2}	P_{H_2}	P_{CH_4}	t_{chem}	$t_{isotope}$
Colombaia	229	27.0	0.08	0.002	0.0005	0.002	0.004	153	<300
Pineta	200	15.3	0.21	0.002	0.004	0.003	0.0006	181	375
Lard. 57	241	33.5	1.1	0.02	0.009	0.009	0.02	225	>375
Lard. 155	222	23.8	0.60	0.01	0.007	0.01	0.01	206	325
Gabbro 6	242	34.1	1.8	0.02	0.008	0.02	0.01	235	300
Gabbro 1	240	32.9	1.7	0.04	0.01	0.04	0.02	251	<300

carbonate origin for CO_2 from Yellowstone, Wairakei and the Salton Sea, and a predominantly organic origin for CO_2 from The Geysers. A marine carbonate origin for CO_2 from the Salton Sea geothermal system was demonstrated petrographically by MUFFLER and WHITE (1968). Carbon dioxide from Larderello varies widely in carbon isotopes from -1 to $-7\%_0$ PDB but appears to originate predominantly from marine limestone (PANICHI and TONGIORGI, 1976).

Since CO_2 is the major gas, its amount and isotopic composition in the steady state is little affected by exchange with other forms of carbon, but other reservoirs of CO_2 (carbonate minerals and dissolved CO_2) may be important under transient conditions.

Methane

In geothermal gases, particularly from systems in sedimentary rocks such as Larderello, methane is often a major component (> 1 percent). Two origins have been suggested (1) the thermal breakdown of organic matter (GUNTER and MUSGRAVE, 1971) and (2) inorganic synthesis from CO_2 and H_2 (CRAIG, 1953; HULSTON and McCABE, 1962a). The weight of recent evidence is in favor of inorganic synthesis.

Methane may be generated by reaction of hydrogen with carbon dioxide

$$CO_2 + 4H_2 = CH_4 + 2H_2O$$

Equilibrium constants for this reaction have been calculated by ELLIS (1957) and fit the equation $\log K = 9089/T - 10.28$ where all gas pressures are in atmospheres and T is degrees Kelvin.

This reaction was tested for our Larderello analyses by calculating reservoir gas partial pressures and equilibrium temperatures (Table 5). In these calculations the pressure of water vapor was assumed to be controlled by coexisting steam and water at a reservoir temperature assumed to be equal to the measured flowing wellhead temperature. Ideal gas behavior was assumed and analyses corrected for gas dissolved in condensate (Table 3) were used.

The calculated temperatures (except for Colombaia) are within 20°C of those observed, suggesting that the contents of at least one of these gases is controlled by this reaction. Since H_2O and CO_2 are major gases and H_2 contents are probably controlled by reactions between water and minerals, it seems most likely that the CH_4 is produced or controlled by this reaction.

Inorganic synthesis of CH_4 is also suggested by ratios of CH_4 to other hydrocarbon gases (50 to > 1000) higher than those in natural gases resulting from breakdown of organic materials at elevated temperatures. Similar ratios may be produced by low temperature surficial bacterial decomposition, but there is no evidence for the existence of bacteria in high temperature geothermal environments.

Chemical equilibrium between CO_2, H_2, CH_4 and H_2O was suggested by HULSTON and McCABE (1962a) for Wairakei and by D'AMORE and NUTI (1977) for Larderello.

Isotopic equilibrium of CO_2 and CH_4 has been suggested for Wairakei and Broadlands (HULSTON and MCCABE, 1962b; LYON, 1974), but isotope temperatures are 120 to 160°C higher than those of the exploited reservoir and may occur only at greater depths in the system. Isotope temperatures at Larderello are also higher than those observed but they in general vary proportionally (PANICHI et al., 1976). These interpolated isotopic temperatures are also given in Table 5.

Hydrogen

Hydrogen is produced by high temperature reaction of water with ferrous oxides and silicates contained in reservoir rocks. Námafjall and other areas in central Iceland produce gases consisting of almost 50 percent hydrogen (ARNORSSON, 1974). Calculations by SEWARD (1974) indicate that the partial pressure of H_2 in equilibrium with H_2O, hematite (Fe_2O_3) and magnetite (Fe_3O_4) at 250°C should be about 0.1 atmosphere as observed in the hot-water systems of Broadlands and Wairakei, New Zealand and Hveragerdi, Iceland (ELLIS, 1967).

Calculated partial pressures of hydrogen in our Larderello gases (Table 5) range from 0.002 to 0.04 atmospheres and are generally lower than those of hot water systems possibly because much of the steam is produced by non-equilibrium vaporization of liquid water. Hydrogen pressure calculated by ELLIS (1967) for The Geysers is also low (0.007 atm). D'AMORE and NUTI (1977) suggest that the original hydrogen content of the gas has been depleted by the formation of methane and H_2S after it was no longer buffered by iron oxide–water reactions. This explanation appears less likely than ours because rock–water H_2 buffering should continue in the unexploited state even during CH_4 and H_2S formation.

Hydrogen sulfide

Hydrogen sulfide is usually present in moderate amounts (0.5 to 4 percent) in geothermal gases, but gases from some areas (Matsukawa, Tatun, Námafjall) contain much more. Sulfur isotope studies have indicated various possible origins; Jurassic seawater or magmatic sulfur highly altered by sulfide deposition have been suggested as the source at Wairakei ($\delta^{34}S(\Sigma S) = +16$; KUSAKABE, 1974; GIGGENBACH, 1977) and a magmatic origin is suggested at Yellowstone, the Salton Sea, and Cerro Prieto ($\delta^{34}S(H_2S) \sim 0$; SCHOEN and RYE, 1970; WHITE, 1968; R. O. RYE, unpublished data, 1978). Fractionation during boiling and steam separation and exchange of sulfur isotopes with dissolved sulfate and with sulfide minerals may occur and affect the isotopic composition of H_2S. Calculations by D'AMORE and NUTI (1977), suggest that H_2S, H_2, pyrite, and pyrrhotite are in equilibrium at Larderello, and calculations by SEWARD (1974) based on calculated pH values suggest that these species and SO_4^{2-} are in chemical equilibrium at Broadlands, New Zealand. Both studies conclude that the contents of H_2S in the natural state are controlled by mineral buffers. During exploitation increased subsurface boiling will produce gases relatively enriched in H_2S because H_2S is more soluble in liquid than other geothermal gases with the exception of NH_3. The evolution of the contents of CO_2, H_2S, and NH_3 in steam produced by partial boiling of Wairakei water was discussed by ELLIS (1962) and by GLOVER (1970).

Nitrogen, and the atmospheric noble gases

Nitrogen and most noble gases (other than He) are not contributed from magmas or thermal metamorphism of rocks and are not affected by rock–water interactions. In this volume, Emanuel Mazor discusses the analysis and interpretation of noble gases for this suite of Larderello samples (see also MAZOR, 1976) and little discussion of these gases will be given here. Our samples show a high ratio of N_2/Ar (average = 135) compared to that of air (83) or air-saturated water (~ 35). High N_2/Ar ratios have been interpreted as due to breakdown of NH_3 or organic matter (D'AMORE and NUTI, 1977), but these could equally well be caused by the much lower solubility of N_2 relative to Ar in water at high temperatures (POTTER *et al.*, 1977). During subsurface boiling, N_2 will partition preferentially into the steam phase and produce high N_2/Ar ratios during the early stages of boiling. As the liquid water is exhausted the ratio will decrease, and in the last stages an Ar-enriched steam will be produced. A choice could be made between these models by following the evolution of gas compositions with time.

Minor gases

Hydrocarbons other than methane and occasionally ethane are seldom reported for geothermal gases. Using a flame-ionization gas chromatograph, we have found that gases in geothermal systems contain a large number of C_2–C_6 trace hydrocarbons (NEHRING and TRUESDELL, 1978). Hydrocarbon patterns have been found to be characteristic of the temperature of the system and the type of reservoir rock. Relative to geothermal gases, volcanic gases (Kileaua, Augustine (Alaska), Mt. Hood) have low hydrocarbon contents with low $C_1/\Sigma C_n$ ratios, significant amounts of unsaturated hydrocarbons and little branching. Geothermal gases in volcanic rocks (Steamboat Springs, Yellowstone, Lassen) retain the predominance of unbranched hydrocarbons but have more total hydrocarbons, more methane and predominantly saturated hydrocarbons. Larderello hydrocarbon gases are very similar to those from other geothermal systems in sedimentary rocks such as The Geysers and Cerro Prieto. These gases have more total hydrocarbons, almost no unsaturated hydrocarbons, and more branched C_4–C_6 hydrocarbons. The ratio of $C_1/\Sigma C_n$ remains high. Figure 1 shows a typical pattern. The C_2–C_6 hydrocarbon contents of the northern 3 wells

(Larderello 155, Gabbro 6, and Gabbro 1) are about 20 times greater than those of the southeren wells, suggesting longer contact times between organic-bearing sediments and hot water and steam. This suggestion agrees with the much greater depth of the northern wells and with the model of lateral recharge of cold water from the south and east suggested by CELATI et al. (1973) and PANICHI et al. (1974).

Isotope measurements

An oxygen isotope gradient from more negative $\delta^{18}O$ values at the margins to more positive values at the center has been found at Larderello (CELATI et al., 1973; PANICHI et al., 1974) and at the S.E. Geysers (TRUESDELL et al., 1977). This gradient has been interpreted by the Italian workers to be the result of mixing of young marginal recharge water which has undergone little or no oxygen isotope shift with highly isotope-shifted deep water from the center of the field. This interpretation is supported by the existence of substantial amounts of tritium in steam from wells at the east and south margins of the field near outcrops of the reservoir rock. The isotope gradient also occurs, however, in the Serrazzano zone on the northwest edge of the field where no tritium is found in produced steam and no obvious possibility of recharge to the reservoir exists. At the southeast end of The Geysers an isotope gradient in both ^{18}O and D occurs with $D/^{18}O$ slope that suggests steam–water separation at 240°C, a temperature close to that found in the producing reservoir (247 $\pm$ 1°C). These observations lead to the multistage distillation model.

The isotopic compositions of the steam samples collected in this study (Table 2) agree very well for oxygen with earlier analyses (PANICHI et al., 1974) and little gradient appears to exist for deuterium. An oxygen isotope gradient without significant deuterium gradient may be attributed according to the recharge model to variable oxygen isotope shift of the same recharge water or according to the distillation model to steam-water separation at a temperature near 220°C where no deuterium fractionation occurs between liquid and vapor. The data are not sufficient to distinguish between the recharge-mixing model and the distillation model. If typical Larderello reservoir temperatures differed significantly from 220°C, a deuterium gradient would be produced by multistage distillation but not by mixing. Unfortunately, 220°C is a common temperature at Larderello. The average wellhead temperature of our samples is 229°C, and the average wellhead temperature from PANICHI et al. (1976) is 200°C.

Discussion

The partial pressures of H_2O, CO_2, CH_4, and H_2 reported here support an inorganic origin for methane and hydrogen buffering by rock reactions. Comparison with data from other geothermal systems suggests origin at Larderello of C_2–C_6 hydrocarbons by decomposition of organic matter in the sedimentary reservoir rock. We have no critical data to report concerning origin of other gases.

Unfortunately, we did not succeed in our object of distinguishing between the 'recharge' or 'distillation' models. The recharge model would require gases from the margin of the system to have a higher ratio of atmospheric gases to deep gases compared to those from the center. In these samples the amount and ratio of atmospheric gases is greatest in the center (Gabbro) and lowest at the margin (Colombaia). The distillation model would require a pumping of gas from the center toward the edge and steam from the margins would be expected to contain more total gas and a greater proportion of water–insoluble gas. However, the central (Gabbro) wells contain the most gas and the marginal wells the least with little systematic difference in gas compositions. Thus, neither model is supported by these analyses.

Apparently extensive production of steam from the Larderello reservoir has altered the original gas compositions which might have allowed a distinction between these models. Development of the field proceeded from the shallow steam zones at the east and south towards deeper zones in the northwest and the wells sampled for this paper decrease in age and increase in depth from south to north (Table 1). With production, the steam increasingly originates from evaporation of liquid water already impoverished in dissolved gases. Thus, the gradient in gas contents has apparently been produced by the pattern of development and does not represent original conditions.

The use of gas contents to indicate subsurface conditions due to production is more hopeful. Changes in gas compositions with flow rate may indicate the distance from the well to the evaporation front (gas in original reservoir steam differs significantly from that in steam vaporized from liquid water) and monitoring of gas compositions with time should indicate whether steam originates from subsurface boiling of dispersed liquid or from boiling of a single body of liquid below a deep water table. Repeated gas measurements may also indicate the degree of depletion of the liquid water if it behaves as a single well-mixed mass.

Acknowledgements

We thank Costanzo Panichi of the Instituto Internazionale per le Ricerche Geotermiche for selecting wells for sampling and providing physical data and we thank the technicians of the Ente Nazionale Energia Electtrica, Compartimento di Firenze for assistance in the collection. We also thank David DesMarais and Frank Church of the NASA–Ames laboratories for the GC-MS analyses and Lloyd D. White for the deuterium analyses. We have enjoyed discussions with Costanzo Panichi, Franco D'Amore, and other members of the IIRG on Larderello gas chemistry and with Emanuel Mazor of the Weizman Institute, Israel on the geochemistry of gases in general. We thank Emanuel Mazor for the use of his data to make Fig. 2 more complete.

RERERENCES

ALLEN, E. T. and DAY, A. L. (1935), *Hot Springs of the Yellowstone National Park*, Carnegie Inst. Washington Pub. *466*, 525 p.

ARNORSSON, S. (1974), *The Composition of Thermal Fluids in Iceland and Geological Features Related to the Thermal Activity*, in: *Geodynamics of Iceland and the North Atlantic Area*, Ed. Kristjansson, 307–323, D. Reidel Publishing Company, Dordrecht.

CELATI, R., NOTO, P., PANICHI, C., SQUARCI, P. and TAFFI, L. (1973), *Interactions Between the Steam Reservoir and Surrounding Aquifers in the Larderello Geothermal Field*, Geothermics, *2*, 174–185.

CRAIG, H. (1953), *The Geochemistry of the Stable Carbon Isotopes*: Geochim. Cosmochim. Acta, *3*, 53–92.

CRAIG, H. (1963), *The Isotopic Geochemistry of Water and Carbon in Geothermal Areas*, in *Nuclear Geology on Geothermal Areas, Spoleto*, Ed. E. Tongiorgi, 17–53, Consiglio Nazionale delle Ricerche Laboratorio di Geologia Nucleare, Pisa.

D'AMORE, F. and NUTI, S. (1977), *Notes on the Chemistry of Geothermal Gases*, Proc. 2nd. Int. Sym. on Water–Rock Interactions, sect. III, 61, CNRS, Strasbourg.

ELLIS, A. J. (1957), *Chemical Equilibrium in Magmatic Gases*, Am. J. Sci. *255*, 416–431.

ELLIS, A. J. (1962), *Interpretations of Gas Analyses from the Wairakei Hydrothermal Area*, N.Z. Jl. Sci. *5*, 434–452.

ELLIS, A. J. (1967), *The Chemistry of Some Explored Geothermal Systems*, in *Geochemistry of Hydrothermal Ore Deposits*, 465–514, Holt, Rinehart and Winston, Inc., New York.

FACCA, G. and TONANI, F. (1964), *Natural Steam Geology and Geochemistry*, in *Proc. U.N. Conference on New Sources of Energy, Rome, 1961*, *2*, 219–228, United Nations, New York.

GIGGENBACH, W. F. (1975), *A Simple Method for the Collection and Analysis of Volcanic Gas Samples*, Bull. Volcanol., *39*, 132–145.

GIGGENBACH, W. F. (1977), *The Isotopic Composition of Sulfur in Sedimentary Rocks Bordering the Taupo Volcanic Zone: Geochemistry*, N.Z. Dept. Sci. and Indus. Research Bull. *218*, 57–64.

GIGGENBACH, W. F. and LE GUERN, F. (1976), *The Chemistry of Magmatic Gases from Erta Ale, Ethiopia*, Geochim. Cosmochim. Acta, *40*, 25–30.

GLOVER, R. B. (1970), *Interpretation of Gas Compositions from the Wairakei Field over 10 Years*, in *Proc. U.N. Symp. on the Development and Utilization of Geothermal Resources, Pisa, 1970*, 1355–1366, Geothermics Spec. Issue 2, v. 2, pt. 2.

GUNTER, B. D. and MUSGRAVE, B. C. (1971), *New Evidence on the Origin of Methane in Hydrothermal Gases*, Geochim. Cosmochim. Acta, *35*, 113–118.

HEALD, E. F., NAUGHTON, J. J. and BARNES, I. L. (1963), *The Chemistry of Volcanic Gases 2 – The Use of Equilibrium Calculations in the Interpretation of Volcanic Gas Samples*, J. Geophys. Res. *68*, 545–557.

HULSTON, J. R. and McCABE, W. J. (1962a), *Mass Spectrometer Measurements in the Thermal Areas of New Zealand. Part 1. Carbon Dioxide and Residual Gas Analyses*, Geochim. Cosmochim. Acta, *26*, 383–397.

HULSTON, J. R. and McCABE, W. J. (1962b), *Mass Spectrometer Measurements in the Thermal Areas of New Zealand. Part 2. Carbon Isotopic Ratios*, Geochim. Cosmochim. Acta, *26*, p. 399–410.

KUSAKABE, M. (1974), *Sulphur Isotopic Variations in Nature*, N.Z. Jl. Sci., *17*, 183–191.

LYON, G. L. (1974), *Geothermal Gases*, in *Natural Gases in Marine Sediments*, Ed. I. R. Kaplan, 141–150, Plenum Press, New York.

MAZOR, E. (1976), *Atmospheric and Radiogenic Noble Gases in Thermal Waters: Their Potential Application to Prospecting and Steam Production Studies*, Proc. 2nd. U.N. Symp. on Geothermal Energy, San Francisco, 1975, 793–802.

MAZOR, E. (1978), *Noble Gases in a Section Across the Vapor Dominated Geothermal Field of Larderello, Italy*, Pure Appl. Geophys. *117*, 262–275.

MUFFLER, L. J. P. and WHITE, D. E. (1968), *Origin of CO_2 in the Salton Sea Geothermal System, Southeastern California, USA*, XXII Int. Geol. Cong., Prague, 1968, Proc. Symp. II, 185–194.

NEHRING, N. L. and TRUESDELL, A. H. (1978), *Some Hydrocarbon Gases in Geothermal and Volcanic Systems*, Geothermal Resource Council Trans. *2*, 483–486.

PANICHI, C., CELATI, R., NOTO, P., SQUARCI, P., TAFFI, L. and TONGIORGI, E. (1974), *Oxygen and Hydrogen Isotope Studies of the Larderello (Italy) Geothermal System*, in *Isotope Techniques in Ground–water Hydrology*, *2*, 3–28, International Atomic Energy Agency, Vienna, IAEA-SM-182/35.

PANICHI, C., FERRARA, G. C. and GONFIANTINI, R. (1976), *Isotope Thermometry in the Larderello Geothermal Field*, Geothermics, *5*, 81–88.

PANICHI, C. and TONGIORGI, E. (1976), *Carbon Isotope Composition of CO_2 from Springs, Fumaroles, Mofettes, and Travertines of Central and Southern Italy: A Preliminary Prospection Method of Geothermal Area*, Proc. 2nd. U.N. Symp. on Geothermal Energy, San Francisco, 1975, 815–825.

POTTER, R. W., MAZOR, E. and CLYNNE, M. A. (1977), *Noble Gas Partition Coefficients Applied to Conditions of Geothermal Steam Formation*, Geol. Soc. Am., Abstr. Programs, *9*, 1132–1133.

SCHOEN, R. and RYE, R. O. (1970), *Sulfur Isotope Distribution in Solfataras, Yellowstone National Park*, Science (AAAS), *170*, 1082–11084.

SEWARD, T. M. (1974), *Equilibrium and Oxidation Potential in Geothermal Waters at Broadlands, New Zealand*, Am. J. Sci., *274*, 190–192.

TRUESDELL, A. H., NEHRING, N. L. and FRYE, G. A. (1977), *Steam Production at The Geysers, California Comes From Liquid Water Near the Well Bottoms*, Geol. Soc. Am. Abstr. Programs, *9*, 1206.

TRUESDELL, A. H. and WHITE, D. E. (1973), *Production of Superheated Steam from Vapor-Dominated Geothermal Reservoirs*, Geothermics, *2*, 154–173.

WHITE, D. E. (1968), *Environments of Generation of Some Base-Metal Ore Deposits*, Economic Geology, *63*, 301–335.

WHITE, D. E., MUFFLER, L. J. P. and TRUESDELL, A. H. (1971), *Vapor-Dominated Hydrothermal Systems, Compared with Hot-Water Systems*, Econ. Geology, *66*, 75–97.

(Received 14th November 1977, revised 18th May 1978)

Pageoph, Vol. 117 (1978/79), Birkhäuser Verlag, Basel

Heat Extraction from Hot, Dry, Crustal Rock

By Morton C. Smith[1])

Abstract – Natural heat stored in the earth's interior represents an essentially inexhaustible energy supply which, at usefully high temperatures, is accessible at practical drilling depths from almost anywhere on the earth's land surface. The problems of extracting and using this heat are those of engineering and economics, and can be expected to vary with the local geology and value of thermal energy. The first major experimental system designed to investigate these problems in one common type of geologic environment has recently been completed in the crystalline rock underlying the Jemez Plateau of northern New Mexico. It consists principally of two boreholes connected at a depth of about 2.7 km by a system of hydraulic fractures produced in granitic rock at a temperature of approximately 185°C. Cool water injected through one hole is heated as it flows through the fractures, and is recovered through the second hole as pressurized, superheated water. In a surface heat-exchange system now being completed, this heat will be extracted and the cool water reinjected to maintain a continuous, closed, pressurized-water energy-extraction loop.

Key words: Heat extraction; Hot dry rock; Hydraulic fracture; Jemez Plateau, New Mexico.

1. Introduction

It is now generally appreciated that the earth's interior is a tremendous reservoir of heat, generated largely by the spontaneous decay of naturally occurring unstable isotopes of uranium, thorium, and potassium, and conserved because of the insulating value of the earth's upper crust. However, heat from this source has so far been used beneficially only where it has been extracted from the rock by the natural circulation of meteoric water and transported to, or nearly to, the earth's surface by the mass flow of hot water or steam. Elsewhere, in the absence either of water or of the permeability required for it to circulate, this vast energy resource remains largely unexplored and entirely unused.

In the Los Alamos geothermal energy program, we hope to enlarge the supply of useful geothermal heat by learning to extract it directly from the rock itself. This of course is not a new idea, and a variety of schemes for accomplishing it have been proposed over the years (Smith, 1975). To us it has seemed simplest to: (1) Identify a body of hot rock at an accessible depth, whose initial permeability is low enough that it will contain pressurized water; (2) Drill two holes downward into the hot rock; (3) Create a localized permeable zone between these holes within which water will be

[1]) Los Alamos Scientific Laboratory, The University of California, Los Alamos, New Mexico 87545, USA.

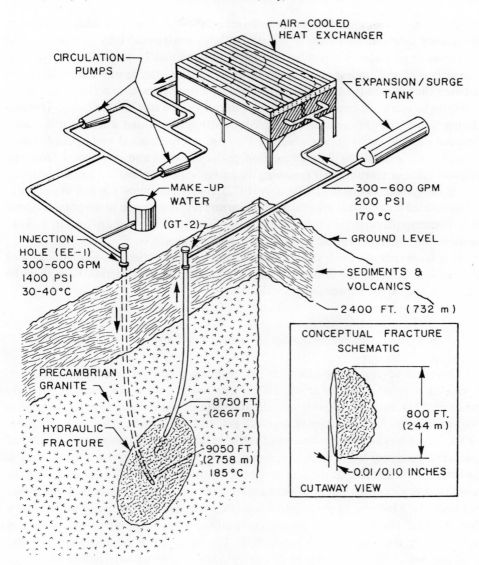

Figure 1
Schematic of prototype hot dry rock energy system at Fenton Hill in northern New Mexico, USA.

heated as it flows from one hole to the other; (4) Inject cool water through one hole, recover hot water from the other, extract its useful heat at the surface, and return it through the first hole to be reheated underground. Rate of heat extraction is then a function both of fluid-circulation rate and of the temperature rise that occurs during its flow through the underground loop (TESTER and SMITH, 1977).

One system of this general type is shown schematically in Fig. 1. In addition to being completely contained so that it should be both environmentally acceptable and conservative of water, it has other attractive possibilities. In particular, if the flow-impedance of the underground loop is not excessive, the temperature difference between the water in the two holes will create a thermal-siphoning action that may be sufficient to maintain circulation without pumping the water mechanically. Thermal contraction of rock in the region where cool water is introduced will almost certainly produce thermal-stress cracks which, if they are interconnected and opened widely enough, may extend the fluid-circulation paths outward and downward from the original fracture, continuously renewing the energy supply as heat is removed from the rest of the system (HARLOW and PRACHT, 1972). Finally, chemical control of the water being injected may control mineral dissolution and reprecipitation to the point not only of minimizing plugging, scaling, and corrosion problems but perhaps also of making possible the 'thermochemical mining' of mineral byproducts.

It is to investigate such possibilities as these and the thermal, mechanical, and geochemical behavior of a man-made, subterranean, heat-extraction loop that the present Los Alamos experiments are directed.

2. Creating the underground loop

There are several ways in which permeability might be created between two separated boreholes (SMITH et al., 1976). Our initial choice has been to use fluid pressure. This method, called 'hydraulic fracturing', is a familiar one for increasing fluid production from petroleum, natural-gas, and water wells (HOWARD and FAST, 1970). It involves isolating a section of the hole by the use of temporary seals called 'packers'; pumping water into this section until a circumferential tensile stress is developed sufficient to split the wall of the hole; and then continuing to pump until the resulting crack has been extended outward to almost any desired distance – which may be hundreds of meters. At the depth of a geothermal reservoir, the expected result is a thin, vertical crack, roughly circular in outline, whose azimuthal orientation is controlled by the local tectonic stress field. The operation is relatively inexpensive, produces no earth shock, leaves no debris, and is particularly suited to use in formations of low permeability – which minimize escape of the pressurizing fluid.

When such a fracture has been produced from one borehole, it is evidently well-connected to that hole and capable of accepting fluid at least at the rate and with the pressure drop used to create it. However, producing an equally good connection to the second hole is less straightforward. In the present state of the art, the dimensions, shape, and orientation of a hydraulic fracture cannot be predicted with confidence, and there are major uncertainties in mapping both the fracture and the boreholes and in guiding a drill bit precisely to a preselected intersection point. Our first attempt to do this was in granitic rock at a depth of about 2.9 km, and was unsuccessful. Having directionally-drilled a second hole to within about 10 m of a fracture produced from the first hole, we were then misled by inaccurate borehole surveys into turning the drill bit away from the fracture rather than toward it. By fracturing again from the second hole we did create a connection through which we could flow water from one hole to the other, but not at a rate sufficient for a good heat-extraction experiment. We therefore spent some months developing the instruments and techniques required to map the fracture and the boreholes more accurately. With a better understanding of the system geometry, we then sidetracked one hole at a depth of about 2.5 km and re-drilled it toward the fracture produced from the other hole. This re-drilling was successful in significantly improving the flow connection, but not to the desired degree – perhaps because the hole encountered the fracture at a point too near its upper edge. Accordingly we sidetracked the hole again, re-drilled it along a lower trajectory, and this time produced a satisfactory connection.

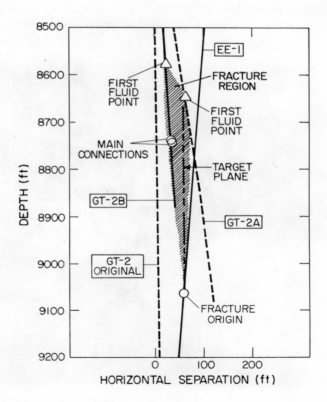

Figure 2
Horizontal projection (side view) of drilling trajectories at Fenton Hill.

3. System characteristics

As is illustrated by Fig. 1, the major fracture system through which the two boreholes are connected is centered in a biotite-granodiorite intrusive at a depth of about 2.7 km, where the rock temperature is approximately 185°C. The diameter of the main fracture system is of the order of 250 m, and its volume (uninflated) is about 35,000 l. (Apparently because of a mismatch of opposite fracture surfaces, the fractures do not close completely when fluid pressure in the system is reduced, and it has not been necessary to prop them open with injected particles.) Figure 2 is a horizontal projection (side view) of the two boreholes and of the additional hole segments produced by sidetracking one of them. Since the improved connection was produced at a point several meters short of that at which we expected to encounter a hydraulic fracture, it appears that either our mapping is still relatively inaccurate or that the fracture system is more complex than is suggested by Fig. 1. We believe that the latter is the case, at least in part because of the long series of pressurization experiments to which it has been subjected, and are continuing our efforts to understand it better.

In a preliminary experiment, we recently pumped water through the system for 20 hr at the rate of about 7 l/sec, maintaining an injection pressure of approximately 65

Figure 3
Superheated water recovered from hole GT-2 flashing to steam when discharged to the atmosphere.

bars in hole EE-1 and (to prevent boiling) a back pressure of about 3 bars in hole GT-2. At the end of this time the water leaving the system was at 130°C, and – as is shown in Fig. 3 – a fraction of it flashed to steam when it was released into the storage pond. Its temperature was still rising and, at this circulation rate, should eventually reach at least 150°C. After 20 hr, about 85 % of the injected water was flowing out of GT-2, and this should eventually increase to above 90 %. The recovered water was apparently free of both suspended solids and noncondensable gases. Its dissolved solids content was less than 2000 ppm, the major impurity being silica at about 240 ppm (corresponding to saturation with quartz at the reservoir temperature). A dye-injection experiment gave no evidence of short-circuiting of water from the injection to the recovery hole, confirmed the system volume stated above, and indicated that there was considerable mixing within the fracture system.

The underground circulation loop appears suitable for a long-term heat-extraction experiment, and the surface plumbing and heat-exchangers required for such an experiment are now being assembled.

4. Plans

This prototype system is a relatively small one. We hope and expect that, operated continuously as a pressurized-water loop, it initially will produce about 4 to 5 MW of thermal energy. This may increase briefly to some higher value, but within a few weeks – a conveniently short experimental period – it should begin to decrease rapidly, as the original fracture surfaces are cooled. However, we propose to operate it for several months to evaluate its physical and chemical behavior in detail, and to acquire the data that are needed to improve our mathematical models of such systems. If a reason appears to do so, we may attempt to improve it by acid treatment, explosive fracturing, chemical additions to the injected water, or other means. If its performance is generally satisfactory, we hope next year to begin construction of a larger, deeper, hotter system, capable of producing energy at a much higher rate for a much longer time, and perhaps eventually of supporting a small electrical-generating pilot plant. In the meantime we still have a great deal to learn about the details of this system, about more efficient ways to create the next one, and about other locations at which it would be useful to undertake generally similar experiments.

5. Conclusions

The prototype system and flow experiment described above represent the first demonstration of the technical feasibility of extracting energy at a useful rate and temperature from hot, dry, low-permeability rock in the earth's crust. The engineering problems and the economics of long-term operation of such a system remain to be determined, as do the system modifications that may be required in other geologic environments. We believe, however, that the problems which remain are at least as tractable as those that we have already solved, and we remain optimistic that in the foreseeable future hot dry rock can contribute substantially to the world's energy needs. It may, in ten to fifteen years, be possible to go almost anywhere that energy is needed, drill downward until a usefully high temperature is reached, and then – if steam or sufficiently hot water has not been discovered along the way – extract the necessary heat from the crustal rock itself. To this point we find no reason to suppose that this cannot or will not occur.

REFERENCES

HARLOW, F. H. and PRACHT, W. E. (1972), *A theoretical study of geothermal energy extraction*, J. Geophy. Res. *77*, 7038.

HOWARD, G. C. and FAST, C. R., *Hydraulic Fracturing* (Dallas, Soc. Petr. Eng. AIME, 1970).

SMITH, M. C., AAMODT, R. L., POTTER, R. M. and BROWN, D. W., *Man-made geothermal reservoirs*, in *Proc. Second U.N. Symp. on Dev. and Use of Geothermal Resources* (U.S. Govt. Printing Office, 1976), pp. 1781–1787.

SMITH, P. J. (1975), *Towards universal geothermal power*, Nature *257*, 10–11.

TESTER, J. W. and SMITH, M. C., *Energy extraction characteristics of hot dry rock geothermal systems*, in *Proc. 12th Intersoc. Energy Conversion Eng. Conf.* (LaGrange, Ill., American Nuclear Soc., 1977), pp. 816–823.

(Received 29th October 1977; revised 13th March 1978)

Pageoph, Vol. 117 (1978/79), Birkhäuser Verlag, Basel

Reservoir Lifetime and Heat Recovery Factor in Geothermal Aquifers used for Urban Heating

By Alain C. Gringarten[1])

Abstract – Simple models are discussed to evaluate reservoir lifetime and heat recovery factor in geothermal aquifers used for urban heating. By comparing various single well and doublet production schemes, it is shown that reinjection of heat depleted water greatly enhances heat recovery and reservoir lifetime, and can be optimized for maximum heat production. It is concluded that geothermal aquifer production should be unitized, as is already done in oil and gas reservoirs.

Key words: Reservoir lifetime; Heat recovery factor; Doublet; Geothermal aquifer; Reinjection.

Nomenclature

a	distance between doublets in multi-doublet patterns, meters	Δt	producing time, sec.
A	area of aquifer at base temperature, m^2	T	aquifer transmissivity, m^2/sec.
		v	stream-channel water velocity, m/sec.
	drainage area of individual doublets in multi-doublet patterns, m^2	$\Delta\theta$	actual temperature change, °C
		$\Delta\theta'$	theoretical temperature change, °C
D	distance between doublet wells, meters	θ	water temperature, °C
h	aquifer thickness, meters	λ	heat conductivity, W/m/°C
H	water head, meters	λ_r	rock heat conductivity, W/m/°C
Q	production rate, m^3/sec.	$\rho_a c_a$	aquifer heat capacity, $J/m^3/°C$
r_e	aquifer radius, meters	$\rho_a c_r$	rock heat capacity, $J/m^3/°C$
r_w	well radius, meters	$\rho_w c_w$	water heat capacity, $J/m^3/°C$
R_g	heat recovery factor, fraction	ϕ	aquifer porosity, fraction
s	water level drawdown, meters		

1. Introduction

One major question in the assessment of regional geothermal potential is the estimation of the quantity of the accessible resource base that can be extracted, and of the time during which such an extraction can be economically maintained (Muffler and Cataldi, 1977). These vary widely with the nature of the geothermal system, and various solutions have appeared in literature, mainly for hot water and steam geothermal reservoirs (Bodvarsson, 1974; Nathenson, 1975). The present paper deals more specifically with geothermal aquifers at intermediate temperatures that are used for urban heating.

[1]) Bureau de Recherches Géologiques et Minières, B.P. 6009, 45018 Orleans Cedex, France; now with Flopetrol, B.P. 592, 77005 Melun Cedex, France.

2. Production without reinjection

The lifetime of a geothermal reservoir of any type, depends primarily on how it is developed. There is no essential difference with normal groundwater aquifers, or oil and gas reservoirs, except that reservoir life may be shortened because of a lack of heat, in addition to a lack of pressure.

The simplest development scheme for a space heating project (and the one that requires the least investment) consists in producing the geothermal water without reinjecting it. In this case, the water will be produced at a constant temperature – equal to the initial reservoir temperature minus the temperature drop from the bottom to the top of the well – as long as the water supply lasts, i.e., as long as the reservoir pressure is sufficient for the production rate to be sustained. When no flashing can take place, the reservoir pressure is related to the depth of the liquid level in the wellbore, and production will continue until the level drops below the pump inlet. As submersible pump size and required power increase with depth, there is a practical limitation on the admissible well drawdown. Usually, it is of the order of 100 meters, although it is possible to obtain drawdowns as high as 200 meters.

For a certain period after it has been completed, a single well behaves as if it were isolated in an infinite system, but its behavior is ultimately affected by flow conditions at the boundaries of the drainage area. Typically, there can be a recharge (at the outcrop of the formation) or an impermeable barrier (due to sealing faults, or the other pumping wells in the vicinity). Although the geometry of the subsurface permeability is usually complex, the single well behavior with either boundary conditions can often be approximated by simple models leading to analytical solutions that can be used to estimate reservoir lifetime and heat recovery factors.

2.1. Single well in a constant pressure circle

The model in Fig. 1 represents a single well intersecting two circular reservoirs with constant pressure (zero drawdown) boundaries. Figure 1(A) is a plane view, showing the outer radius, and the 30°C and 60°C isotherms for the deeper aquifer. Figure 1(B) is a cross section indicating the limits of both aquifers.

If all production from one aquifer is supposed to take place from the single well, the model of Fig. 1 will give the production potential of that aquifer.

Assuming the reservoir to be horizontal, homogeneous, isotropic, of constant thickness and initially at uniform pressure, maximum pressure drawdown at the wellbore will be obtained under steady state conditions as (DIETZ, 1965):

$$s = \frac{Q}{2\pi T} \ln \frac{r_e}{r_w}, \tag{1}$$

where s = the water level drawdown, in meters
        Q = the constant production rate, in m^3/sec.
        T = the aquifer transmissivity, in m^2/sec.
  r_e, r_w = the aquifer and wellbore radius, respectively, in meters.

The maximum production rate that can be maintained indefinitely from such a reservoir is thus:

$$Q_{max} = \frac{2\pi T s_{max}}{\ln r_e/r_w}. \tag{2}$$

Where s_{max} is the maximum drawdown at the wellbore; with $s_{max} = 200$ m, $r_e = 300$ km, $r_w = 0.075$ m (6″ diameter hole) and the aquifer data shown in Table 1, one obtains:

$Q_{max} = 0.083$ m^3/sec (298 m^3/hr) for aquifer A,

$Q_{max} = 0.008$ m^3/sec (30 m^3/hr) for aquifer B.

It is important to emphasize that these Q_{max} values represent the *maximum total withdrawal rate from the aquifer*, under the specified boundary conditions, which is not dependent upon the number of wells. Greater values will only be obtained with higher T (aquifer of better transmissivity) or smaller r_e (recharge boundary closer to the

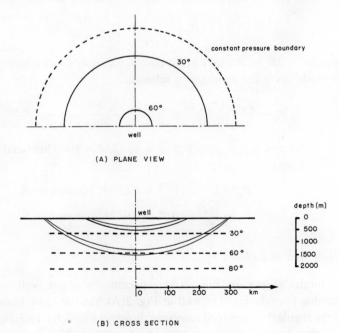

Figure 1
Schematic of a deep aquifer.

production well), but these usually imply a shallower aquifer whose temperature might not be adequate (cf. Fig. 1(B)).

Table 1
Example aquifer data

	Transmissivity (m²/s)	Storage coefficient fraction	Thickness (m)
Aquifer A	10^{-3}	10^{-4}	100
Aquifer B	10^{-4}	10^{-5}	15

The heat recovery factor can be defined as the ratio of extracted heat, $Q_{max} \Delta t \, \rho_w c_w \Delta\theta$, to the total theoretically recoverable heat in place, $\rho_a c_a A h \Delta\theta'$. Δt is the producing time, $\rho_w c_w$ and $\rho_a c_a$ are the heat capacity of the water and of the aquifer – $\rho_a c_a = \phi \rho_w c_w + (1 - \phi)\rho_r c_r$, where ϕ is the porosity and $\rho_r c_r$ the rock heat capacity – $\Delta\theta$ and $\Delta\theta'$ are the useful temperature drop that can be obtained with geothermal water (actual and theoretical). A should be taken as the area of the aquifer *corresponding to the appropriate water temperature*, which can be very different from the aquifer areal extent (Fig. 1(B)).

The heat recovery factor is thus equal to:

$$R_g = \frac{\rho_w c_w}{\rho_a c_a} \frac{Q_{max} \, \Delta t}{Ah} \frac{\Delta\theta}{\Delta\theta'}. \tag{3}$$

If we take $\Delta\theta = \Delta\theta'$ for the sake of simplicity, we are left with a heat recovery factor that only depends upon the production scheme:

$$R_g = \frac{\rho_w c_w}{\rho_a c_a} \frac{Q_{max} \, \Delta t}{Ah} = \frac{1}{\phi + (1 - \phi)(\rho_r c_r / \rho_w c_w)} \frac{Q_{max} \, \Delta t}{Ah}. \tag{4}$$

With $\Delta t = 30$ years, $\rho_r c_r / \rho_w c_w = 0.5$, $\phi = 15\%$, $A = 8000$ km² and the aquifer data of Table 1, one obtains:

$$R_g = 1.7 \times 10^{-4} = 0.017\% \text{ for aquifer A,}$$

$$R_g = 1.2 \times 10^{-4} = 0.012\% \text{ for aquifer B.}$$

2.2. Single well in a closed square

The model shown in Fig. 2(A) represents a single well in a reservoir with impermeable boundaries. The well in Fig. 2(A) has the same behavior as any of the wells in the regularly developed reservoir shown in Fig. 2(B). The model pictured in Fig. 2(A) can thus be used to obtain individual well potential, in addition to the aquifer potential obtained from the previous model.

(A) WELL IN A SQUARE WITH CLOSED BOUNDARIES

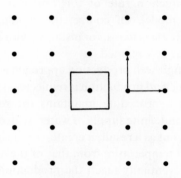

(B) REGULARLY DEVELOPED RESERVOIR

Figure 2
Single well in a closed square.

The long time well drawdown given by the model of Fig. 2(A) is equal to (MATTHEWS and RUSSELL, 1967):

$$s = \frac{Q\,\Delta t}{SA} + \frac{Q}{4\pi T}\ln\frac{A}{3.7r_w^2}, \tag{5}$$

where S is the aquifer storage coefficient and A is the well drainage area. Equation (5) indicates that no steady-state is reached with this model. Thus, if constant production is to be maintained throughout the total production period Δt, the maximum production rate must be less than or equal to:

$$Q_{max} = \frac{s_{max}}{\Delta t/SA + 1/4\pi T \ln A/3.7r_w^2}, \tag{6}$$

with $A = 1\ \mathrm{km}^2$, $\Delta t = 30$ years, $s_{max} = 200$ m and $r_w = 0.075$ m, Eqn. (6) yields:

$Q_{max} = 2\cdot10^{-5}\ \mathrm{m}^3/\mathrm{sec} = 0.076\ \mathrm{m}^3/\mathrm{hr}$ for aquifer A,

$Q_{max} = 2\cdot10^{-6}\ \mathrm{m}^3/\mathrm{sec} = 0.008\ \mathrm{m}^3/\mathrm{hr}$ for aquifer B.

The heat recovery factor is then:

$R_g = 3.5 \times 10^{-4} = 0.035\%$ for aquifer A,

$R_g = 2.4 \times 10^{-4} = 0.024\%$ for aquifer B.

3. Production with reinjection

It is clear that only very little production and very small recovery factors can be obtained with the type of development that was considered so far. Both models give results of the same order of magnitude; for a total useful area of 8000 km², the second model yields a total production rate of 0.167 m³/s for aquifer A, compared to 0.083 m³/s with the first model. Of course, this would only be valid in normal sedimentary aquifers. In karstic systems, for instance, with heavy recharge from below, very different results might be obtained.

An alternative to the single well production approach is the use of a doublet type of development (Fig. 3) in which all production is reinjected into the aquifer after the heat has been extracted. Such a procedure maintains the reservoir pressure, prevents subsidence, and insures an indefinite supply of water. It also permits the recovery of the heat contained in the rock, but as a result, it creates a zone of injected water around the injection well at a different temperature from that of the native water. That zone will grow with time, and will eventually reach the production well. After breakthrough occurs, the water temperature is no longer constant at the production well and this may reduce drastically the efficiency of the operation.

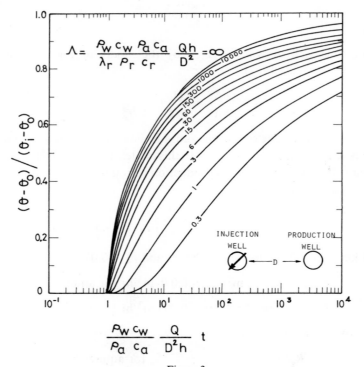

Figure 3
Temperature variation at the production well of a geothermal doublet.

In order to predict the temperature behavior of the aquifer during reinjection, it is necessary to utilize a heat/flow model. In such a model, the heat and flow equations, respectively:

$$\text{div}\,(K\,\nabla H) + Q = S\frac{dH}{dt},\tag{7}$$

$$\text{div}\,(\lambda\,\nabla T) - \text{div}\,(\rho_w c_w v T) = \rho_a c_a \frac{\partial T}{\partial t},\tag{8}$$

should be solved in a coupled manner, because the permeability K in Eqn. (7) includes the water viscosity and therefore depends upon the temperature, which in turn depends upon the potential field (velocity v in Eqn. (8)). This requires a rather complex computer program, the accuracy of which is usually not very satisfactory because of numerical dispersion. It is possible, however, to analytically approximate the solution, by using hypotheses that result in decoupling Eqns. (7) and (8). Such a simplified solution, published by GRINGARTEN and SAUTY (1975), will be used in this paper.

3.1. Single doublet

GRINGARTEN and SAUTY (1975), by neglecting horizontal thermal conduction in the aquifer and the confining rocks, have obtained the curves shown in Fig. 3 for an isolated doublet. In Fig. 3, the production well dimensionless temperature:

$$T_{wD} = \frac{\theta - \theta_0}{\theta_i - \theta_0}\tag{9}$$

is plotted versus a dimensionless time:

$$t_D = \frac{\rho_w c_w}{\rho_a c_a}\frac{Q\,\Delta t}{D^2 h}\tag{10}$$

for various values of a coefficient characteristic of heat exchange with the confining rocks:

$$\Lambda = \frac{\rho_w c_w \rho_a c_a}{\lambda_r \rho_r c_r}\frac{Qh}{D^2}\tag{11}$$

θ_0 and θ_i are the initial and injection temperature, λ_r is the confining rock vertical thermal conductivity, and D is the distance between the wells of the doublet.

It can be seen from Fig. 3 that the time of thermal breakthrough increases with decreasing Λ. It is minimum and equal to $\pi/3$ when $\Lambda = \infty$ (no heat loss to the confining rocks). In this paper, the theoretical thermal breakthrough time corresponding to $\Lambda = \infty$ will be defined as the *lifetime of the doublet*. It is also equal to the water breakthrough time, multiplied by $\rho_w c_w \phi / \rho_a c_a$. The following relationship is thus obtained:

$$\frac{\rho_w c_w}{\rho_a c_a} \frac{Q \Delta t}{D^2 h} = \frac{\pi}{3}.$$ (12)

Equation (12) yields the *minimum distance* D between the doublet wells for the production temperature to remain constant for a period Δt at a constant injection and production Q.

Another relationship between D, Q, and the steady state well drawdown is obtained from the potential theory:

$$s = \frac{Q}{2\pi T} \ln \frac{D}{r_w}.$$ (13)

Equation (13) yields the *maximum distance* D that will maintain the drawdown at the well above the maximum possible drawdown s_{max}.

It is then possible to combine Eqns. 12 and 13 in order to find the distance D that will provide the *maximum constant rate* Q corresponding to a lifetime Δt and to the maximum possible drawdown at the well, s_{max}. For instance, with $\Delta t = 30$ years, $s_{max} = 200$ meters, $r_w = 0.075$ m and the data of Table 1, one obtains:

$Q_{max} = 0.128$ m^3/sec (459 m^3/hr)   and $D = 1417$ m for Aquifer A,

$Q_{max} = 0.013$ m^3/sec   (47 m^3/hr)   and $D = 1170$ m for Aquifer B.

These figures are to be compared with those obtained previously with the model of a single well in a constant pressure circle. In this case, however, the *total production* from the entire aquifer could be much higher, because of the possibility of having several doublets in the same aquifer.

3.2. Multi doublet patterns

In order to evaluate the maximum aquifer production that can be reached with a doublet type of development, we will consider different doublet patterns, and calculate for each one the reservoir lifetime and heat recovery factor.

The reservoir lifetime is simply obtained by computing the time required for a water particle to travel from one input to one output well along the streamline of highest average velocity (MUSKAT, 1946) and multiplying it by $\rho_w c_w \phi / \rho_a c_a$.

The results of the calculations are presented in Figs. 4 and 5. Figure 4 shows the ratio of the reservoir lifetime with various *two doublet patterns* to that with a single doublet, as a function of a/D, where a is the 'distance' between the doublets (as indicated in Fig. 4). It appears that the lifetime ratio is always less than unity when the respective positions of injection and production wells are such that an impermeable (fictitious) boundary is created between the doublets (patterns (1) and (2) in Fig. 4). It increases with a/D and approaches unity within 1 % when $a/D = 2$. On the contrary, this ratio can be *greater than one* if the well positions are such that a constant pressure

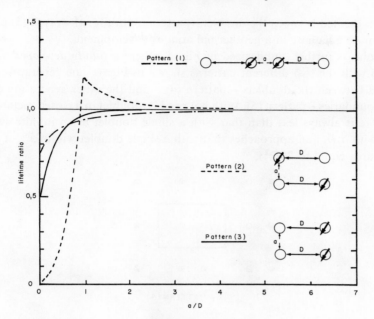

Figure 4
Comparison of reservoir lifetime for various two-doublet patterns.

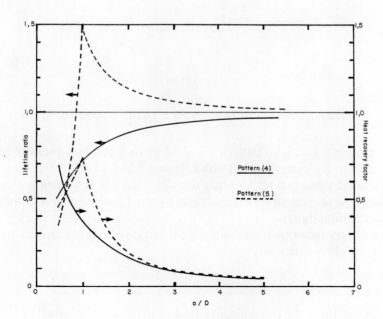

Figure 5
Comparison of reservoir lifetime and heat recovery factor for various patterns of doublet development.

boundary is created between the doublets (pattern (3) in Fig. 4). A type (3) pattern is therefore more adequate in a geothermal aquifer development.

This point is further emphasized in Fig. 5, where a *totally developed* aquifer is considered with the two different patterns shown in Fig. 6, one generating no flow boundaries between the doublets – pattern (4) – and the other generating constant pressure boundaries – pattern (5). With pattern (4) (doublet in a closed rectangle), the reservoir life is always less than that with a single doublet in an infinite system. It increases with a/D, and approaches that with a single doublet within 5% if $a/D \geq 4$ (drainage area equal to $20D^2$).

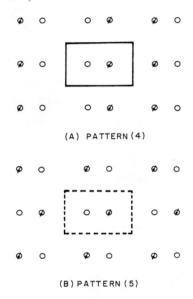

Figure 6
Various possible doublet arrangements in a totally developed reservoir.

On the other hand, with pattern (5) (doublet in a constant pressure rectangle), reservoir lifetime is greater than that with a single doublet for $a/D \geq 0.9$. It is maximum and equal to 1.5 times that with a single doublet when $a = D$ (drainage area equal to $2D^2$). The well arrangement corresponding to pattern (5) with $a = D$ is called a *five spot pattern* in the oil industry.

Heat recovery factors for both patterns (4) and (5) can be obtained from Eqn. (4), which can also be written as:

$$R_g = t_D \frac{D^2}{A},\tag{14}$$

where t_D (Eqn. 10) is the reservoir lifetime and A the drainage area:

$$A = a(D + a).\tag{15}$$

The results, shown in Fig. 5, indicate that heat recovery is much higher than with no reinjection and can reach 75 % with the most efficient pattern (pattern 5).

As an example, with $\Delta t = 30$ years, and the aquifer data of Table 1, and using pattern (5) with $a = D$, the *maximum constant* rate Q_{max} per doublet is:

$$Q_{max} = 0.134 \text{ m}^3/\text{sec } (481 \text{ m}^3/\text{hr}) \quad \text{and} \quad D = 900 \text{ m for aquifer A,}$$

$$Q_{max} = 0.014 \text{ m}^3/\text{sec } (49 \text{ m}^3/\text{hr}) \quad \text{and} \quad D = 750 \text{ m for aquifer B.}$$

The drainage area per doublet is thus 1.62 km^2 for aquifer A and 1.13 km^2 for aquifer B. Assuming a useful aquifer area of 8000 km^2, the total aquifer production rate would be:

$$662 \text{ m}^3/\text{sec } (2.4 \times 10^6 \text{ m}^3/\text{hr}) \text{ for aquifer A,}$$

$$96 \text{ m}^3/\text{sec } (3.5 \times 10^5 \text{ m}^3/\text{hr}) \text{ for aquifer B.}$$

If pattern (4) was used instead of pattern (5), but with the same well spacing, the *maximum rate Q_{max} per* doublet would be:

$$Q_{max} = 0.035 \text{ m}^3/\text{sec } (124.8 \text{ m}^3/\text{hr}) \text{ for aquifer A,}$$

$$Q_{max} = 0.005 \text{ m}^3/\text{sec } (18.7 \text{ m}^3/\text{hr}) \text{ for aquifer B.}$$

The total production rate for a total productive area of 800 km^3 would then be:

$$173 \text{ m}^3/\text{sec } (6.24 \times 10^5 \text{ m}^3/\text{hr}) \text{ for aquifer A,}$$

$$37 \text{ m}^3/\text{sec } (1.32 \times 10^5 \text{ m}^3/\text{hr}) \text{ for aquifer B.}$$

4. Conclusions

The results presented in this study can be summarized as follows:

(1) Reservoir lifetime and heat recovery factor in geothermal aquifers is very much dependent upon the development scheme.

(2) As a general rule, production with reinjection of heat depleted water increases the heat recovery factor and the aquifer production potential by several orders of magnitude.

(3) When using a doublet type of development (production with reinjection), it does not seem to be a good practice to try to isolate each doublet from the influence of the others. Greater reservoir lifetime and heat recovery factors are obtained by alternating injection and production wells. In other words, geothermal aquifer production should be *unitized*, as is already done in oil and gas reservoirs.

REFERENCES

BODVARSSON, G. (1974), *Geothermal resource energetics*, Geothermics *3*, 83–92.

DIETZ, D. N. (1965), *Determination of average pressure from build-up surveys*, J. Petrol. Tech., August, p. 955.

GRINGARTEN, A. C. and SAUTY, J. P. (1975), *A theoretical study of heat extraction from aquifers with uniform regional Flow*, J. Geophys. Res. *80* (35), 49–56.

MATTHEWS, C. S. and RUSSELL, D. G., *Pressure build-up and flow tests in wells*, *Monograph Series* (Society of Petroleum Engineers of AIME, Dallas (Tex.), USA, (1967) vol. 1.

MUFFLER, L. J. P. and CATALDI, R. (1977), *Methods for regional assessment of geothermal resources*, Larderello Workshop on Geothermal Resource Assessment and Reservoir Engineering, Sept. 12–16, Larderello (Italy).

MUSKAT (1946), *The Flow of Homogeneous Fluids Through Porous Media*, J. W. Edwards, Inc., Ann Arbor, Michigan, 507–617.

NATHENSON, M. (1975), *Physical factors determining the fraction of stored energy recoverable from hydrothermal convection systems and conduction dominated areas*, U.S. Geol. Survey Open-file Report, 75-525, 38 pp.

(Received 22nd November 1977, revised 9th March 1978)

Pageoph, Vol. 117 (1978/79), Birkhäuser Verlag, Basel

Flow in an Aquifer Charged with Hot Water from a Fault Zone

By J. W. Pritchett and S. K. Garg[1])

Abstract – Many geothermal anomalies are intersected by vertical fault zones (narrow zones of fractured material with large effective permeability). These conduits are probably responsible for much of the upwelling of hot water from depth. This paper considers a shallow aquifer intersected by a vertical fault. The fluid flow in the aquifer is numerically modeled as a two-dimensional problem. It is observed that the temperature distribution in the aquifer is governed primarily by lateral flow of hot water supplied from the intersecting vertical fault and only secondarily by conduction. The numerical results also provide a possible explanation for the local temperature maxima and inversions occasionally observed in borehole measurements. The present model is an alternative to that based on mushroom-shaped isotherm distributions found in high Rayleigh number large-scale circulation cell calculations.

Key words: Flow of hot water; Vertical fault; Finite difference method; Buoyant flow; Two-dimensional temperature field.

1. Introduction

To obtain useful estimates of heat and mass transport in geothermal systems, it is necessary to use physical models which are sufficiently representative of reality. In other words, the appropriate geological and geophysical data must be employed to provide realistic geometry (boundary conditions) and material properties (porosity, permeability, thermal conductivity, etc.). Several published studies on natural convection in geothermal reservoirs deal with large-scale convection of fluid in a porous matrix of constant porosity and permeability (see, e.g., Wooding [1957, 1963], Donaldson [1962], Elder [1967a, b], and Bories and Combarnous [1973]). A review of the geological literature (see, e.g., Grindley [1965], Kassoy and Zebib [1977]) shows that most geothermal anomalies are associated with tectonically active regions containing a complex network of vertical fault zones. The Wairakei geothermal field in New Zealand is traversed by several large northeast-striking faults and some smaller northwest cross-faults. At Larderello, Italy, the main faults strike northwest and the cross-faults northeast. The presence of faults has also been documented at many other geothermal fields (e.g., the Salton Sea and East Mesa anomalies in the Imperial Valley; the Geysers steam field).

---

[1]) Systems, Science and Software P.O. Box 1620, La Jolla, California 92038, USA.

According to GRINDLEY [1965], the most successful wells drilled at Wairakei and several other geothermal fields intersected one or more fault zones. This observation suggests that faults play a key role in geothermal systems. Some investigators have suggested that the fault zones are probably most responsible for the upwelling of hot water from depth. It is hypothesized that surface water gradually seeps down into permeable sediments or volcanic rocks where it comes close to a deep heat source. Since the heated water is buoyant relative to the cool recharge water, the heated water will rise through the faulted region which is characterized by a relatively high permeability [KASSOY and ZEBIB, 1977]. If the fault intersects a horizontal aquifer of high permeability, then the hot water rising through the fault will charge the aquifer. The temperature distribution in the aquifer will then be governed primarily by lateral movement of hot water from the fault.

The general equations expressing the balance of mass, momentum and energy for single-phase flow through porous media have previously been developed by several authors. In the following discussion, we will employ the terminology of BROWNELL, *et al.* [1977]. Assuming (1) that the rock matrix is rigid such that porosity $\phi(\mathbf{x})$ is constant in time, (2) that the fluid and the rock matrix are in local thermal equilibrium ($T_r = T$), and (3) that the fluid motion is governed by Darcy's law, the equations governing mass, momentum and energy can be written as follows:

Mass

Fluid:

$$\phi \frac{\partial \rho}{\partial t} - \nabla \cdot \left\{ \frac{k\rho}{\mu} \left(\nabla P - \rho \mathbf{g} \right) \right\} = 0. \tag{1}$$

Momentum

Fluid:

$$\mathbf{v} = -\frac{k}{\phi\mu} \left(\nabla P - \rho \mathbf{g} \right). \tag{2}$$

Energy

Rock/fluid mixture:

$$\frac{\partial}{\partial t} \left[(1 - \phi)\rho_r C_r T + \phi \rho E \right] - \nabla \cdot \left[\frac{k\rho}{\mu} \left(\nabla P - \rho \mathbf{g} \right) E \right] = \nabla \cdot (\kappa_m \nabla T) \tag{3}$$

where C_r = rock grain heat capacity
$\quad\quad E$ = fluid internal energy per unit mass
$\quad\quad \mathbf{g}$ = acceleration due to gravity
$\quad\quad k$ = rock permeability
$\quad\quad P$ = fluid pressure
$\quad\quad t$ = time
$\quad\quad T$ = temperature
$\quad\quad \mathbf{v}$ = fluid velocity (vector)
$\quad\quad \kappa_m$ = mixture (solid/fluid) thermal conductivity
$\quad\quad \mu$ = fluid viscosity
$\quad\quad \rho_r$ = rock grain mass density
$\quad\quad \rho$ = fluid density
$\quad\quad \phi$ = porosity.

The balance equations for fluid mass (1) and total energy (3) constitute a set of two equations in two unknowns (ρ, E). The equations also contain fluid viscosity, pressure and temperature. Furthermore, the specification of mixture conductivity requires knowledge of the thermal conductivity of the fluid. It is convenient to utilize a tabular representation for the equation-of-state for water. Given fluid density ρ and internal energy E, the equation-of-state sub-routines used in the simulation code yield pressure P, temperature T, viscosity μ, and thermal conductivity κ (see also BROWNELL, et al. [1977]).

For the rock matrix, the following constitutive functions are required:

$\kappa_r(T)$ – dependence of rock grain thermal conductivity on temperature
$\quad k(\phi)$ – dependence of permeability on porosity ϕ.

In the present work, both κ_r and ϕ will be regarded as simple constants. Furthermore, the mixture thermal conductivity κ_m will be approximated by Budiansky's formula [BROWNELL, et al., 1977].

Since the times of interest in geothermal reservoirs are long and the governing equations are nonlinear, it is necessary to employ an implicit iterative finite-difference procedure. A convenient numerical technique is the so-called Alternative Direction Implicit (ADI) method; ADI allows one to reduce a multi-dimensional (2-D or 3-D) problem to a series of one-dimensional problems. To illustrate the ADI method, consider the linear heat conduction equation in two dimensions (planar geometry):

$$\frac{\partial T}{\partial t} = \kappa \left(\frac{\partial^2 T}{\partial x^2} + \frac{\partial^2 T}{\partial y^2} \right). \tag{4}$$

The first-order implicit (time) finite-difference analogue of equation (4) is:

$$\frac{T^{n+1} - T^n}{\Delta t} = \kappa \left(\frac{\partial^2}{\partial x^2} T^{n+1} + \frac{\partial^2}{\partial y^2} T^{n+1} \right) \tag{5}$$

where the superscript n denotes the value of the superscripted variable at $t = n \, \Delta t$. The ADI technique replaces equation (5) by two one-dimensional problems:

$$\frac{^{j+1}T^{n+1} - T^n}{\Delta t} = \kappa \left(\frac{\partial^2}{\partial x^2} \, ^{j+1}T^{n+1} + \frac{\partial^2}{\partial y^2} \, ^{j}T^{n+1} \right) \tag{6a}$$

$$\frac{^{j+2}T^{n+1} - T^n}{\Delta t} = \kappa \left(\frac{\partial^2}{\partial x^2} \, ^{j+1}T^{n+1} + \frac{\partial^2}{\partial y^2} \, ^{j+2}T^{n+1} \right) \tag{6b}$$

where the superscript j denotes the jth iteration. Given T^n and $^{j}T^{n+1}$, one first solves for $^{j+1}T^{n+1}$ from equation (6a); T^n and $^{j+1}T^{n+1}$ are then employed in equation (6b), to solve for $^{j+2}T^{n+1}$. Note that $^{j}T^{n+1}$ in equation (6a), and $^{j+1}T^{n+1}$ in equation (6b) appear merely as source terms. This iterative procedure is repeated until some preset convergence criterion is met. The application of ADI to solve the present system of balance equations involves essentially (apart from quite involved algebra) the steps outlined above. The advection terms for both mass and energy are treated in a first-order fashion (often called 'upwind' or 'donor-cell' flux differencing) to avoid antidiffusive computational instability.

The finite difference scheme outlined above has been incorporated into the MUSHRM computer code [GARG et al., 1977]. The MUSHRM numerical simulator is capable of treating multi-dimensional (2-D or 3-D) problem geometries, heterogeneous rock formations, and all practical boundary conditions. It is also capable of treating multiphase (steam/water) fluid flow; the present problem is, however, single-phase liquid. Boundary condition options include (1) impermeable, insulated, (2) impermeable, prescribed heat flux or temperature, (3) prescribed mass flux, insulated, (4) prescribed mass flux, prescribed heat flux or temperature, and (5) prescribed pressure/heat content. A more extensive discussion of the MUSHRM simulator may be found in GARG et al. [1977].

2. Physical considerations

We consider a shallow aquifer intersected by a vertical fault (Fig. 1). The fault is a narrow zone of highly fractured material with large effective permeability, and provides a conduit for the upwelling hot water. In the following discussion it will be assumed that the fault is sufficiently tectonically active that continuous mechanical fracturing will counteract the tendency of the fault zone to be blocked by solid salts precipitated from the rising and cooling column of saline water [KASSOY and ZEBIB, 1977]. We shall not, however, try to characterize the fault zone in detail since we are mainly interested in the charging of the hot water aquifer. The fluid flow in the aquifer will be numerically modeled in vertical section as a two-dimensional problem (see Fig. 2). Note that owing to symmetry only the right (or left) half of the aquifer need be considered. The left boundary ($x = 0$) represents the fault and has specified

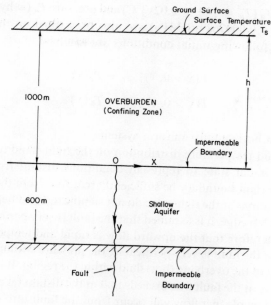

Figure 1
Shallow aquifer intersected by a vertical fault.

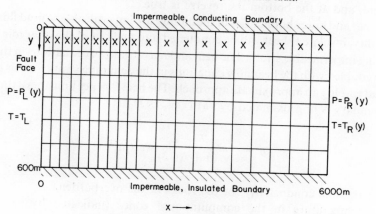

Figure 2
Two-dimensional numerical model. The reservoir cross-section is approximated by a 20 × 6 mesh (120 zones; 6 divisions in y-direction, $\Delta y = 100$ m; 20 divisions in x-direction, first $10\,\Delta x = 200$ m, second $10\,\Delta x = 400$ m). Note the difference in length scales in the x and y directions.

temperature T_L (=temperature of the upwelling hot water = 200°C) and pressure P_L (=hydrostatic pressure, in MPa, corresponding to T_L):

$$P_L(y) = (10.3823 + 0.0087y) \text{ MPa}$$

where the depth y is in meters. It is assumed, somewhat arbitrarily, that the pressure at $y = 0$ is 10.3823 MPa; the second term represents the pressure gradient due to gravity.

Fluid temperature T_R ($T_R = [40 + 0.02y]°C$) and pressure P_R ($=$ hydrostatic pressure at $T_R = [10.0116 + 0.0097689y]$ MPa) are also prescribed on the right boundary ($x = 6000$ m). The following initial conditions are assumed:

$$P(x > 0, y)|_{t=0} = P_R(y)$$

$$T(x > 0, y)|_{t=0} = T_R(y)$$

corresponding to a horizontally uniform system.

The pressure and temperature distribution on the right-hand (outer) boundary is maintained at its initial state to represent conditions distant from the fault. It is important that the right boundary be sufficiently removed from the fault face so that the boundary conditions at the right edge do not significantly influence the flow field in the aquifer. At the left edge, it is assumed that the fault is very permeable compared to the aquifer, and therefore that the upward flow is rapid and temperature is constant. Furthermore, since the fault is very permeable, the pressure within it may be taken as equal to the weight of the overlying (hot) fluid within the region. It is worth noting that the average pressure at the fault face exceeds that at the distant face by 0.05 MPa. This means that, on the whole, net flow will occur from the fault into the aquifer. On the other hand, at the top of the system, the pressure within the fault is greater than that in the far field, and at the bottom the reverse is true.

The upper and lower boundaries of the aquifer are impermeable to fluid flow but are not, in reality, insulated; heat will be conducted into and out of the reservoir from the surrounding impermeable strata. The heat and mass transport equations in the aquifer should be coupled with the heat flow equations in the confining strata. In this study, we will instead employ an approximate approach. The heat flux per unit area q through the upper boundary is approximated by [MILLER, 1977]:

$$q = -\kappa \frac{T - T_s}{(h + \Delta y/2)} \text{ watt/m}^2$$

where $\kappa =$ thermal conductivity for the rocks of the overburden,

$T =$ temperature in the computational zones (indicated by $\times$ in Fig. 2) immediately adjoining the overburden,

$T_s =$ surface temperature,

$h =$ overburden thickness,

$\Delta y =$ thickness of top layer of computational zones (indicated in Fig. 2 by $\times$).

In the numerical simulator, the effect of heat loss through the overburden is represented by a volumetric energy sink $\dot{e}$ added to the computational zones adjoining the upper boundary of the aquifer such that

$$\dot{e} = q/\Delta y.$$

Taking $\kappa = 4.2$ watt/m°C (representative of liquid-saturated natural sandstones), $h = 1000$ m, $\Delta y = 100$ m and $T_s = 20$°C, we obtain:

$$\dot{e} = -4 \times 10^{-5} (T - 20) \text{ watt/m}^3$$

where T is in °C.

The problem of heat flux through the bottom boundary is much more complex. In some previous work (e.g., NORTON [1977]), the bottom surface is assumed to be a constant temperature boundary, but this is unrealistic. Initially, the heat flux will be into the reservoir due to the natural thermal gradient in the crust. At intermediate times, as the hot fluid flows into the aquifer from the fault zone, the direction of the heat flux will be reversed since the reservoir is now at a higher temperature than the surrounding formations. For very late times, the reservoir will be in thermal equilibrium with the formations lying below it. Since we are primarily interested in late time solutions, we take the bottom boundary as thermally insulated.

To complete the problem description it is necessary to specify certain material properties for the reservoir rock. The following values, typical of natural sandstones, will be assumed for these material properties:

ρ_r (rock grain density) $= 2.65 \cdot 10^3$ kg/m^3
κ_r (rock grain thermal conductivity) $= 5.25$ watt/m°C
C_r (rock grain heat capacity) $= 10^3$ joule/kg°C
ϕ (rock porosity) $= 0.20$
k (rock permeability) $= 10^{-13}$ m^2.

Because of the small pressure variations in the present problem, the rock porosity ϕ and permeability k are assumed to be constants independent of fluid pressure and temperature.

3. Numerical results

The aquifer geometry, boundary conditions and the finite-difference grid are shown in Fig. 2. It should be noted that vertical dimensions in Figs. 2 and 3 are exaggerated by a factor of five. In order to study the evolution of the flow field with time, the numerical simulator was run in a time-marching mode. The simulator chooses its own time-step; and the results of the computations are printed out at specified (usually multiples of 100) computational cycles. Thus, the value of time t at which results are displayed are dictated principally by the computer rather than the user.

The calculation was carried out to a point in time corresponding to 8341 years. Temperature distributions in aquifer at $t = 1137, 4739$ and 8341 years are illustrated in Figs. 3(a–c). The flow pattern near the fault face at 8341 years is illustrated in Fig. 4. Figure 5 shows more quantitatively the horizontal component of velocity (u) as a function of distance from the fault along the top, bottom and mid-plane of the aquifer

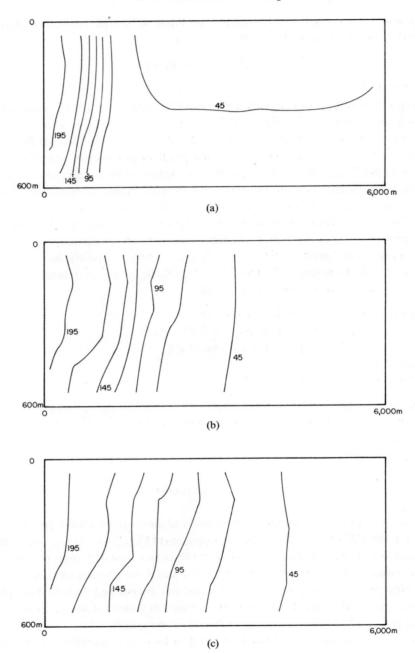

Figure 3
(a) Temperature (°C) contours at $t = 1137$ years. Vertical dimensions exaggerated five-fold. (b) Temperature (°C) contours at $t = 4739$ years. Vertical dimensions exaggerated five-fold. (c) Temperature (°C) contours at $t = 8341$ years. Vertical dimensions exaggerated five-fold.

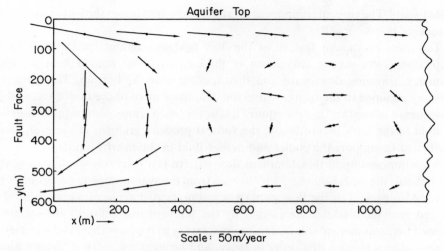

Figure 4
Velocity vectors near the fault face at $t = 8341$ years.

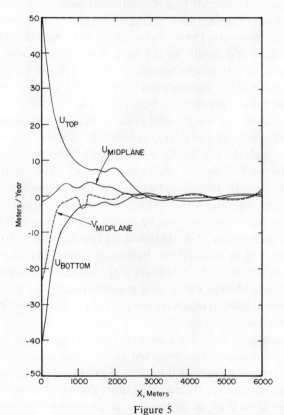

Figure 5
Horizontal and vertical velocities along the top, bottom and mid-plane of the aquifer at $t = 8341$ years.

at 8341 years. The vertical component (v) at the mid-plane is also shown; at the top and bottom, of course, $v = 0$.

The most prominent feature of the flow field throughout the duration of the calculation is the strong circulation of fluid entering the aquifer near its upper boundary, traveling downward and then leaving near the bottom. This pattern is generally confined to the region within one kilometer or so of the fault zone, and is a consequence of the fact that the aquifer has not reached temperature equilibrium with the fluid in the fault. Accordingly, the vertical pressure gradient at the fault face is insufficient to support the cooler and denser fluid in the nearby aquifer.

Superimposed upon this dominant flow pattern is a relatively weak net outward flow towards the outer boundary which arises from the average pressure drop described above. Also present in the flow pattern is a complicated array of weak eddies. It is believed that these eddies are caused by the buoyant instability of the system as opposed to the numerical solution technique, owing to the facts that (1) the numerical discretization scheme is first order in space, and hence stable to such computational instabilities, (2) the characteristic wavelength of the eddies is comparable to the size of the system as a whole as opposed to computational zone dimensions, as would be expected were the eddies computational artifacts, and (3) such a system of eddies is not surprising on physical grounds. The average net outward flow slowly drives the cold fluid out of the aquifer and results in the slow shift of the isotherms to the right (compare Fig. 3(a–c)). Along the right boundary, the fluid flow is generally outwards in the top part and inwards in the bottom portion of the grid. In Fig. 6 we show the total energy change in the computational grid with respect to the initial state. As might be expected, the energy change rate is greatest at early times; it then decreases with time. Basically, the energy change is due to (1) the hot fluid charging the aquifer (positive) and (2) heat loss through the top boundary (negative). With the passage of time, the heat loss through the top becomes more and more important due to the shift of 100°C isotherm to the right; it may, therefore, be expected that eventually the reservoir will reach a state of dynamic equilibrium such that the heat loss balances the heat gain. The dynamic equilibrium will, however, only be attained at times much greater than the maximum time (8341 years) of this calculation. (This also helps to underscore the approximations involved in treating the bottom boundary as thermally insulated. Indeed, the correct treatment of this boundary may significantly alter the time required to achieve dynamic equilibrium.) It is thus possible that many existing geothermal reservoirs are dynamical systems that have not attained a steady-state (see also NORTON [1977] in this connection).

The numerical results also provide a possible explanation for the local temperature maxima and inversions sometimes observed in borehole measurements (see, e.g., GRINDLEY [1965]). Temperature inversions with depth can clearly be observed in Figs. 3(a–c). The Rayleigh number for the aquifer system discussed herein is of the order of 50. The present model is thus an alternative to that based on mushroom-shaped isotherm distributions found in high Rayleigh number large-scale circulation cell

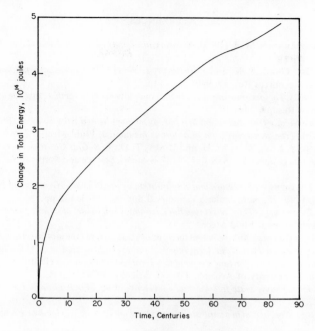

Figure 6
Change in total energy per meter aquifer thickness as a function of time.

calculations in constant property isotropic porous media (c.f., ELDER [1967a], WOODING [1957, 1963]).

4. Concluding remarks

A typical geothermal system consists of a flowing convective fluid heated at depth and rising towards the surface as a result of the reduced density. The system is not only non-isothermal but also a dynamic system as a consequence of buoyant flow. The three-dimensional temperature field is profoundly affected by the heterogeneity of the reservoir (e.g., rock types, geologic structure, faults). This paper presents a model of the role of faults in charging a hot water aquifer. This can be improved by (1) including a more detailed treatment of the thermal boundary conditions and (2) by incorporating actual fault geometries and heterogeneous rock formations.

Acknowledgments

This work was sponsored in part by NSF/RANN grant GI-44212 to the University of Colorado and Systems, Science and Software and also by NSF/RANN grant ENV 75-14492-A01 to Systems, Science and Software.

REFERENCES

BORIES, S. A. and COMBARNOUS, M. A. (1973), *Natural convection in a sloping porous layer*, J. Fluid Mech. *57*, 63.

BROWNELL, D. H., JR., GARG, S. K. and PRITCHETT, J. W. (1977), *Governing equations for geothermal reservoirs*, Water Resources Res. *13*, 929.

DONALDSON, I. G. (1962), *Temperature gradients in the upper layers of the earth's crust due to convective water flows*, J. Geophys. Res. *67*, 3449.

ELDER, J. W. (1967a), *Steady free convection in a porous medium heated from below*, J. Fluid Mech. *27*, 29.

ELDER, J. W. (1967b), *Transient convection in a porous medium*, J. Fluid Mech. *27*, 609.

GARG, S. K., PRITCHETT, J. W., RICE, M. H. and RINEY, T. D., *U.S. Gulf Coast Geopressured Geothermal Reservoir Simulation* (Report No. SSS-R-77-3147, Systems, Science and Software, La Jolla, California, 1977), 116 pp.

GRINDLEY, G. W., *The Geology, Structure and Exploitation of the Wairakei Geothermal Field, Taupo, New Zealand* (Bulletin No. 75, New Zealand Geological Survey, 1965), 132 pp.

KASSOY, D. R. and ZEBIB, A. (1977), *Convection fluid dynamics in a model of a fault zone in the earth's crust*, manuscript submitted to J. Fluid Mech.

MILLER, R. E. (1977), *A Galerkin, finite-element analysis of steady-state flow and heat transport in the shallow hydrothermal system in the East Mesa Area, Imperial Valley, California*, J. Res. U.S. Geol. Survey *5*, 497.

NORTON, D., *Exploration Criteria for Low Permeability Geothermal Resources* (Report Under Contract No. EY-76-S-02-2763, University of Arizona, Tucson, Arizona, 1977), 186 pp.

WOODING, R. A. (1957), *Steady state free thermal convection of liquid in a saturated permeable medium*, J. Fluid. Mech. *2*, 273.

WOODING, R. A. (1963), *Convection in a saturated porous medium at large Rayleigh or Peclet number*, J. Fluid Mech. *15*, 527.

(Received 25th October 1977, revised 3rd April 1978)

Pageoph, Vol. 117 (1978/79), Birkhäuser Verlag, Basel

Flow Visualization Studies of Two-Phase Thermal Convection in a Porous Layer

By C. H. Sondergeld[1,2]) and D. L. Turcotte[1])

Summary – Two-phase thermal convection has been studied in a porous layer heated from below. A water saturated porous layer was heated so that boiling occurred on the lower boundary. In order to observe flow patterns one lateral dimension of the apparatus was made small. At moderate heat fluxes a water zone overlay a two-phase, steam-water zone. Water velocities and streamlines were obtained as well as the location of the two-phase zone for several heat fluxes. Within the water zone heat transfer took place due to both conduction and convection. In the two-phase zone heat transfer took place due to counterpercolation of steam and water.

Key words: Convective heat transfer; Conductive heat transfer; Water velocity; Nusselt number; Water-steam boundary.

1. Introduction

Two-phase thermal convection in porous media occurs in nuclear reactors and in vapor-dominated geothermal reservoirs. The basic processes associated with this convection is poorly understood. Schubert and Straus (1977) have studied the stability of two-phase thermal convection in a porous medium. They modeled the system as a homogeneous fluid with a single Darcy equation to describe the flow. Convective instabilities were shown to be more sensitive to perturbations caused by the phase change than to the buoyancy forces, due to thermal expansion. They found that flows and temperature variations were concentrated near the two-phase zone at the base of the porous layer.

In order to understand the basic heat transfer processes in two-phase thermal convection we have carried out a series of experiments involving the convection of steam and water in a porous layer. Initial experiments (Sondergeld and Turcotte, 1977) were carried out in a 'sand box' with dimensions of 0.92 by 0.99 m. The 'sand box' was filled with glass beads (mean diameter 5.3×10^{-4} m) to depths between 0.10 and 0.20 m. The saturated porous layer was heated from below and cooled from above. Thermocouple grids were used to obtain temperature distributions. At low and

[1]) Department of Geological Sciences, Cornell University, Ithaca, New York 14850, USA.
[2]) Present address: CIRES, University of Colorado, Boulder, Colorado 80302, USA.

moderate Nusselt numbers a water zone was found to overlie a two-phase (water-steam) zone. The density difference between the two zones generated cellular convection. In order to explain the transport of heat in the isothermal, two-phase zone a counterpercolation model was developed. In this model steam is formed on the base plate, percolates upward through the porous matrix, rises vertically, and condenses at the upper boundary of the two-phase zone. Condensation cannot occur within the two-phase zone since it is isothermal. Water percolates downward to provide the steam at the lower boundary.

In order to better understand the processes associated with two-phase thermal convection, a flow visualization facility has been built. A series of experiments carried out in this facility are reported in this paper.

2. Experimental Facility

Two-phase thermal convection occurs in a porous layer heated to the boiling point from below and cooled from above. In order to observe the convection processes a thin container with glass side walls was built; the length, width and height were 0.457 m, 0.019 m and 0.305 m, respectively. The apparatus is shown schematically in Fig. 1 (end view). The front and rear walls were made of two sheets of double thickness glass panels separated by a 3.2×10^{-3} m air gap. The thermopane construction minimized lateral heat loss. A 2.54×10^{-2} m thick anodized aluminium plate served as the base plate. The vertical glass walls were glued into slots milled into the base plate. The bottom of the base plate was painted flat black to promote absorption of the radiant heat from 2000 watt resistance heater. The heater box was enclosed with transite panels covered with reflective metal sheets.

The interior of the tank was partially filled with deaerated distilled water to which was added glass beads. A 8.0×10^{-3} m thick aluminium bar covered the matrix layer (see Fig. 1). Holes in the bar prevented the potentially dangerous steam accumulation and provided a permeable upper boundary. A cooling coil was placed above the top plate in a layer of water nominally 0.05 m deep. Tap water was circulated through the helically wound cooling coil at a controlled and measured flow rate. An evaporation barrier, a 0.05 m thick styrofoam layer, was placed above the water layer.

The temperature of the water layer and base plate were continuously recorded. The temperature change of the cooling water was also monitored and used with the flow rate to determine the measured heat flux, Q_m (SONDERGELD and TURCOTTE, 1977). The thermocouples, type T used in the temperature measurements were calibrated to 0.01°C.

Transmitted light photography of dye, $KMnO_4$, movements was used to determine fluid velocities and streamlines. Velocities could be determined to within $\pm 1.2 \times 10^{-7}$ m/sec.

All results reported in this paper were obtained using glass beads with a mean diameter of 530 μm, and a layer depth $H = 0.14$ m. The measured porosity, ϕ, was 34.7% and the measured permeability, k, was 7.05×10^{-11} m^2. The thermal conductivity of the glass from which the glass beads were fabricated was 1.05 watts/m°C.

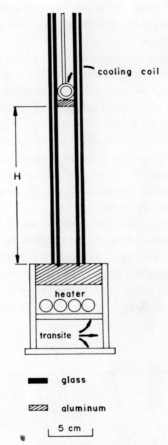

Figure 1
Cross-section of the flow visualization facility.

3. Results

A series of experiments were carried out to determine the fluid velocities, streamlines and the location of the two-phase zone as a function of heat flux. The heat flux Q_m was nondimensionalized by using the calculated conductive heat flux Q_c, for the layer with the boundary at 100°C but no steam present

$$\text{Nu} = Q_m/Q_c \tag{1}$$

where Nu is the Nusselt number. As the Nusselt number is increased above unity a two-phase (steam-water) zone develops. Up to a Nusselt number of about 30 (for the conditions studied) the lower boundary remains at the boiling temperature (100°C) and the additional heat is transported by boiling at the lower boundary and convection of steam through the two-phase zone. At larger Nusselt numbers a zone of dry steam appeared (burnout). The extent of the steam zone could be easily visualized because of the appearance of steam bubbles.

Without boiling no convection occurred. The presence of the two-phase zone induced convection in the overlying water zone. The streamlines and the velocities associated with this convection were obtained from numerous photographs of dye tracks. The reproductions of these photographs are given in Figs. 2 through 5 for Nusselt numbers of 3.1, 10.0, 18.4 and 30.0 respectively. The horizontal scale x is referenced to the center of the chamber and the vertical scale z to the top of the matrix layer. The plots have a vertical exaggeration of approximately 1.8. The dashed lines represent the interfaces between the water and two-phase zones and the solid lines indicate streamlines. The number adjacent to arrows give fluid velocities in m/sec $\times 10^{-5}$; the arrows give direction of the motion.

Observations during heating indicated that the steam zone formed through the coalescing of numerous two-phase (steam-water) fingers. As the boiling temperature of the fluid was reached, small nearly vertical fingers evolved, penetrated, and partially displaced water from the pores of the matrix. The fingers spread laterally as they grew vertically, eventually coalescing into a continuous two-phase zone. At low heat fluxes, slightly above the conductive flux, fingering was favored (SCHEIDEGGER, 1974).

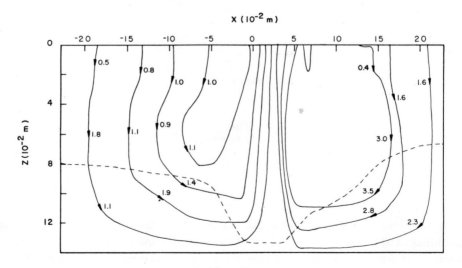

Figure 2
Flow streamlines (solid lines) at a Nusselt number of 3.1; the numbers give velocities in m/sec $\times 10^{-5}$ and the dashed line is the boundary between the water and two-phase (steam-water) zones.

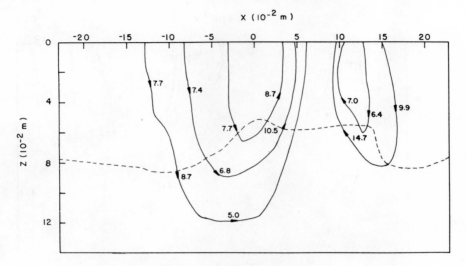

Figure 3
Flow streamlines (solid lines) at a Nusselt number of 10.0; the numbers give velocities in m/sec $\times$ 10^{-5} and the dashed line is the boundary between the water and two-phase (steam-water) zones.

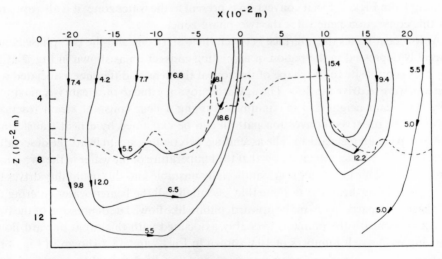

Figure 4
Flow streamlines (solid lines) at a Nusselt number of 18.4; the numbers give velocities in m/sec $\times$ 10^{-5} and the dashed line is the boundary between the water and two-phase (steam-water) zones.

However, at Nusselt numbers above about 1.8 the two-phase zone was essentially a continuous region with the pore fraction occupied with coexisting steam and water.

No perceptible dye movement was observed until boiling occurred. This is strong evidence that the two-phase zone was responsible for the hydrothermal convection. It is

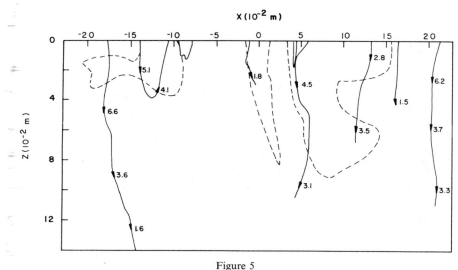

Figure 5

Flow streamlines (solid lines) at a Nusselt number of 30.0; the numbers give velocities in m/sec $\times$ 10^{-5} and the dashed line is the boundary between the water and two-phase (steam-water) zones.

apparent from Figs. 2–5 that convection is present in the water zone; it is also apparent that this convection penetrates the two-phase zone.

At low heat fluxes the interface between the water zone and the two-phase zone is noticeably depressed in the region of ascending convection as shown in Fig. 2 at Nu = 3. This is opposite to the shape of the zone if the density difference associated with the steam were to drive the flow. The upward buoyancy due to the steam is a maximum where the steam height is a maximum (assuming a near constant steam fraction). However, the observed convection pattern can be explained by a heat transfer and thermal expansion mechanism. The ascending, percolating steam heats the descending water to 100°C. Measurements show that the temperature of the water in the two-phase zone is 100°C. This hot water is gravitationally unstable and this instability drives the rapid upward flow near $x = 0$. Note that essentially all the heated water entering the two-phase zone participates in the upward, plume-like flow. The depression of the two-phase zone beneath the plume is probably associated with this rapid upward flow.

At higher Nusselt numbers of 10.0 shown in Fig. 3 and 18.4 shown in Fig. 4 the thickness of the two-phase zone increased and the flow velocities increased considerably. In these cases the two-phase zone is elevated in the region of ascending flow. It is likely that the density difference between the water and two-phase zones drives the flow generating the large velocities. At a Nusselt number of 30 the two-phase zone occupies almost the entire porous layer as shown in Fig. 5. Although it is difficult to trace streamlines because of the vigorous counterpercolation the observations do indicate a reduction in the convective velocities. Since the water zone is virtually nonexistent, the density difference is ineffective in generating convective flows. The two-phase counterpercolation mechanism is primarily responsible for the transport of heat across the nearly isothermal layer.

4. Discussion

We have carried out an experimental study of heat transport across a porous layer saturated with water when boiling occurs at the lower boundary. We find that a water zone overlies a two-phase (water-steam) zone. The boundary between the two zones is distorted by convective processes. Within the water zone, heat transport takes place due to both thermal conduction and convection. Since no convection occurs prior to boiling, we conclude that the convection in the water zone is associated with production of steam on the lower boundary and the resultant two-phase zone.

However, neither thermal conduction nor convection of water alone can result in heat transport in the two-phase zone since it is isothermal (at 100°C). Heat transport across the two-phase zone requires a counterflow of steam and water. Steam is formed on the base plate by boiling, percolates upward, and condenses on the upper boundary of the two-phase zone. Water percolates downward to provide the steam at the lower boundary. There is convective heat transport due to the difference in enthalpy between the steam and water, i.e., the latent heat of vaporization.

In order to model the transport of heat through the two-phase zone we consider the one-dimensional, steady counterpercolation of water and steam through a porous medium (SONDERGELD and TURCOTTE, 1977). The heat flux Q_m is related to the mass flux of steam $\rho_v V_v$ upward from the lower boundary by

$$Q_m = L\rho_v V_v, \tag{1}$$

where L is the latent heat of vaporization. A mass balance between water and steam at the lower boundary requires

$$\rho_v V_v = -\rho_w V_w \tag{2}$$

where the velocities V_v and V_w are Darcian velocities. We assume that the flow of both steam and water satisfy Darcy equations of the form

$$V_v = -\frac{k(1-S)}{\mu_w}\left[\frac{dp}{dz} + \rho_v g\right] \tag{3}$$

$$V_w = -\frac{kS}{\mu_\omega}\left[\frac{dp}{dz} + \rho_w g\right] \tag{4}$$

where k is the permeability, μ is the viscosity, and S is the saturation (fraction of the porosity filled with water).

Combining (1) to (4) we find that

$$S = \frac{1}{2}\left[1 - \frac{Q_m}{kgL(\rho_w - \rho_v)}\left(\frac{\mu_v}{\rho_v} - \frac{\mu_w}{\rho_w}\right)\right]$$

$$+ \frac{1}{2}\left\{\left[1 - \frac{Q_m}{kgL(\rho_w - \rho_v)}\left(\frac{\mu_v}{\rho_v} - \frac{\mu_w}{\rho_w}\right)\right]^2 - \frac{4Q_m\mu_w}{kgL(\rho_w - \rho_v)}\right\}^{1/2} \quad (5)$$

This relation gives the saturation; $1 - S$ is the steam fraction required to transport the heat flux Q_m across the two-phase zone. Taking the properties of water at 60°C and the properties of steam at atmospheric pressure and 100°C the saturations determined from (5) for Nusselt numbers of 3.1, 10.0, 18.4, and 30.0 are 0.980, 0.936, 0.881, and 0.806 respectively. According to this model at a Nusselt number of 30, about 20% of the porosity is filled with steam; this volume fraction is sufficient to provide the required heat flux. We have previously shown (SONDERGELD and TURCOTTE, 1977) that the steam volumes predicted by this model are in reasonable agreement with observations.

If the only heat transport mechanism in the water zone was conduction, then the height of the two-phase zone h_s is easily determined (SONDERGELD and TURCOTTE, 1977)

$$h_s = H\left(1 - \frac{1}{\text{Nu}}\right) \quad (6)$$

This predicted height is compared with the mean of the observed two-phase heights in Fig. 5. The observations indicate that the thickness of the water zone, $H - h_s$, is

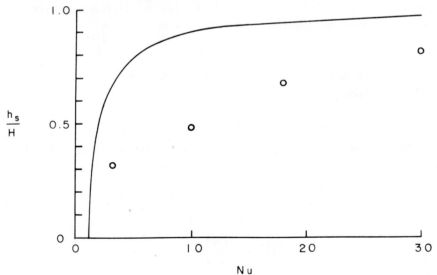

Figure 6
Dependence of the height of the steam zone on Nusselt number; the circles are the mean observed heights and the solid line is the predicted height from (6).

considerably larger than would be expected. This is evidence that a significant fraction of the heat transport across water zone is by convection, since the Nusselt number for the water zone is significantly greater than one.

In the absence of the two-phase zone the stability of the saturated porous layer heated from below is determined by an appropriately defined Rayleigh number (WOODING, 1957)

$$Ra = \frac{kgH\rho_w \alpha_w \, \Delta T}{(\lambda_m/\lambda_w)k_w \mu_w} \tag{7}$$

For a permeable upper boundary and an isothermal lower boundary, thermal convection will occur if this Rayleigh number exceeds 27.1. The calculated Rayleigh number with the lower boundary at 100°C but no steam present is 32, slightly above the critical value for the onset of convection. The fact that the layer is slightly supercritical is probably the reason that the additional heat transfer associated with small concentrations of steam at low Nusselt numbers is able to induce thermal convection as shown, for example, in Fig. 2 for Nu = 3.1.

The appropriate stability analysis with a two-phase, counterpercolating zone has not been carried out. In order to approximate the effect of the two-phase zone we define a Rayleigh number, Ra_s, based on the density difference between the two zones. In order to define this Rayleigh number we replace the density difference due to thermal expansion $\rho_w \alpha_w \, \Delta T$ in (7) with the density difference between the water and two-phase mixture $(1 - S)(\rho_w - \rho_v)$ with the result

$$Ra = \frac{kgH(1 - S)(\rho_w - \rho_v)}{(\lambda_m/\lambda_w)k_w \mu_w} \tag{8}$$

Substituting the values of S given above we find that $Ra_s = 16.2, 51.9, 96.5,$ and 157.3 for the Nu = 3.1, 10.0, 18.4, and 30.0. Although this calculation is only approximately valid, it should indicate the relative importance of the two-phase zone in driving thermal convection.

At a Nusselt number of 3.1 the phase change is only about one-half as effective as thermal expansion in driving convection. This appears to be consistent with our observation that the phase change boundary is depressed in the region of ascending convection as shown in Fig. 2. At Nusselt numbers of 10 and 18.4 the density difference between the zones is several times larger than the density difference due to thermal expansion. In these cases the two-phase zone is elevated in the region of ascending convection so that the density difference does in fact drive the flow as shown in Figs. 3 and 4. At a Nusselt number of 30 the water zone has virtually disappeared so that the density difference is ineffective in driving convection. Since the two-phase zone is isothermal and the steam fraction is nearly constant, there are no density differences to drive thermal convection. This explains the low observed flow velocities associated with this high heat flux.

5. Acknowledgements

This research has been supported by the Division of Engineering of the National Science Foundation under grant ENG 76-11821 and by the Division of Earth Sciences of the National Science Foundation under grant EAR 72-01522. One of the authors (C.H.S.) held a National Science Foundation Energy Related Traineeship (Geothermal). This paper is contribution number 607 of the Department of Geological Sciences of Cornell University.

REFERENCES

SCHEIDEGGER, A. E., *The Physics of Flow in Porous Media* (University of Toronto Press, Toronto 1974), p. 313.

SCHUBERT, G. and STRAUS, J. M. (1977), *Two-phase convection in a porous medium*, J. Geophys. Res. *82*, 3411–3421.

SONDERGELD, C. H. and TURCOTTE, D. L. (1977), *An experimental study of two-phase convection in a porous medium with applications to geological problems*, J. Geophys. Res. *82*, 2045–2053.

WOODING, R. A. (1957), *Steady state free convection of liquid in a saturated permeable medium*, J. Fluid Mech. *2*, 273–285.

(Received 29th November 1977, revised 1st March 1978)

Pageoph, Vol. 117 (1978/79), Birkhäuser Verlag, Basel

Thermal Structure and Energy of the Hakone Volcano, Japan

By Jun Iriyama[1]) and Yasue Ōki[2])

Abstract – The thermal energy balance and the temperature profile of the Hakone volcano are considered quantitatively. Across the Hakone volcano and its surroundings the heat flow values vary from 10^{-1} to 10^3 mW/m², due to thermal conduction and mass flow involving volcanic steam and hot spring discharge. An area with extremely low heat flow is observed in the western side of the caldera showing the presence of percolating meteoric water. The hydrothermal activity is intense in the eastern half of the caldera.

The total heat discharge from the high temperature zone (discharge area) of the Hakone volcano amounts to 11.0×10^7 W. The magmatic steam energy discharge is 95.0×10^6 W. The thermal energy by redistribution of the terrestrial heat flow by the lateral deep ground water flow is calculated to be 9.00×10^6 W. For the model having the vertical vent in the volcano's central part up to 1 km depth below the ground surface from a magma reservoir the computed temperature distribution is consistent with the observed values. The depth of the magma reservoir is 7 km below the ground surface and the diameter is 5 km.

Key words: Thermal energy redistribution; Magmatic steam discharge; Heat flow variations; Deep ground water flow; Magma reservoir.

1. Introduction

Hakone is a strato volcano with overlapping calderas and seven post-caldera cones, belonging to the Fuji volcanic zone on the northern extension of the Izu-Mariana Island Arc. The volcanic eruption began about 400,000 years ago (Suzuki, 1970). The final eruption, which occurred at the central cone Kamiyama, was a violent steam explosion followed by the appearance of a lava dome at 2900 years ago. The hydrothermal activity is still intense discharging about 11×10^7 W as volcanic steam, hot springs, and conductive heat flow.

In the present paper we will discuss the geothermal structure, heat flow, and the size and depth of the magma chamber which would fit well to the present thermal budget of the Hakone volcano, based on a steady-state model.

[1]) Chubu Institute of Technology, Department of Natural Sciences, Kasugai-shi, Aichi 487, Japan.
[2]) Hot Spring Research Institute, Yumoto 997, Hakone, Kanagawa 250-03, Japan.

2. *Heat flow distribution*

Ōki and Hirano (1970, 1974) described two isothermal maps of the Hakone volcano at sea-level and −400 m below sea-level. The temperature gradient is obtained from the temperature difference between 0 m and −400 m below sea-level. The thermal conductivity is assumed to be 2.1 W/m·deg for the whole area; mostly composed of hydrothermally metamorphosed submarine pyroclastic compact rocks, called Yugashima formation (Ōki, 1971). We calculated the areal distribution of heat flows using the data. Across the Hakone and its surroundings the heat flow values vary from 10^{-1} to 10^3 mW/m². Figure 1 shows the heat flow, along an east-west cross section, and the geological setting of the Hakone volcano. Figure 1 represents the distribution of the heat flow by means of the thermal conduction and mass flow involving volcanic steam and hot spring discharge.

The high heat flow area overlies roughly the high temperature area. An area with extremely low heat flow is observed in the western side of the caldera suggesting the presence of percolating meteoric water down to a depth of 1 km below sea-level. The hydrothermal activity is intense in the eastern half of the caldera. The heat flow in the central part of the volcano is extremely high.

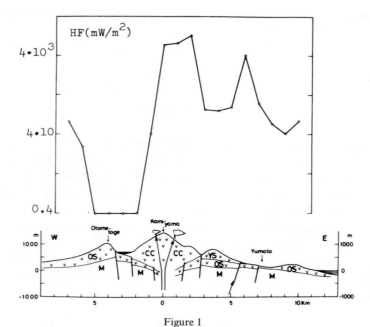

Figure 1
Heat flow in the Hakone volcano (thermal conduction and mass flow involving volcanic steam and hot spring discharge) and the geological setting, along an east–west cross section. CC: central cone lavas, M: Yugashima group M (older Miocene), OS: old somma lavas, YS: young somma lavas.

3. Energy balance

Explanation of the positive and negative heat flow anomalies as compared to the standard crustal heat flow value of 84 mW/m² (UYEDA, 1972; WATANABE, 1972) in this territory will be given by the presence of a hydrothermal system (YUHARA, 1972, 1973). Analysis of the energy balance is useful to consider the thermal structure of the volcano (IRIYAMA, 1977). Table 1 shows the calculated thermal energy balance of the Hakone volcano based on the observed values of the thermal discharge as steam and hot water, and of the calculated conductive heat flow.

Table 1

Thermal energy balance of the Hakone volcano

1. High temperature zone (discharge area; 59 km²)

   Volcanic thermal discharge as steam and hot water (14 km²)

V_k (Kamiyama)	33.6×10^6 W
V_s (Soun-zan)	43.5×10^6 W
V_y (Yumoto)	17.1×10^6 W
V_{ko} (Komagatake)	8.40×10^6 W

   Conductive heat flow (45 km²)

   $$C_f = 45 \text{ km}^2 \times 4.0 \times 41.8 \text{ mW/m}^2$$
   $$= 7.52 \times 10^6 \text{ W}$$

   Total heat discharge

   $$T_H = V_k + V_s + V_y + V_{ko} + C_f = 1.10 \times 10^8 \text{ W}$$

   Standard conductive heat flow (59 km²)

   $$S_{ch} = 59 \text{ km}^2 \times 2.0 \times 41.8 \text{ mW/m}^2$$
   $$= 4.93 \times 10^6 \text{ W.}$$

2. Low heat flow zone (water recharge area 140 km²)

   Conductive heat flow of 0.42 mW/m² zone (100 km²)

   $$C_{h1} = 100 \text{ km}^2 \times 0.01 \times 41.8 \text{ mW/m}^2$$
   $$= 0.04 \times 10^6 \text{ W}$$

   Heat flow subtracted by migration of water

   $$W_{h1} = 100 \text{ km}^2 \times (2.00 - 0.01) \times 41.8 \text{ mW/m}^2$$
   $$= 8.32 \times 10^6 \text{ W}$$

   Conductive heat flow of 42 mW/m² zone (40 km²)

   $$C_{h2} = 40 \text{ km}^2 \times 1.0 \times 41.8 \text{ mW/m}^2$$
   $$= 1.67 \times 10^6 \text{ W}$$

   Heat flow subtracted by migration of water

   $$W_{h2} = 40 \text{ km}^2 \times (2.0 - 1.0) \times 41.8 \text{ mW/m}^2$$
   $$= 1.67 \times 10^6 \text{ W}$$

3. Energy balance

   $$W_{h1} + W_{h2} + S_{ch} + M_{sf} = T_H = 1.10 \times 10^8 \text{ W}$$

   Magmatic steam energy flow $M_{sf} = 9.50 \times 10^7$ W

In the discharge area (59 km²; high temperature zone) the steam and hot water energy flow amounts to 10.3×10^7 W. The thermal discharge as conductive heat flow (45 km²) amounts to 7.52×10^6 W. Thus, the total heat discharge in the discharge area, T_H, becomes to 11.0×10^7 W. While, the heat loss by standard crustal heat flow of this area, S_{ch}, is 4.93×10^6 W. The calculating procedures and the parameters used are given in Table 1.

In the recharge area (140 km²; low heat flow area) the conductive heat flow value is divided into two classes. The energy released by conductive heat flow of the

0.42 mW/m² area (100 km²), C_{h1}, is 0.04×10^6 W. After the procedure of YUHARA (1973), the heat energy subtracted by migration of underground water, W_{h1}, is $10^8 \times (83.6 - 0.4) = 8.32 \times 10^6$ W. In the 42 mW/m² area (40 km²) the energy released by the conductive heat flow, C_{h2}, is 1.67×10^6 W. The heat energy subtracted by water migration, W_{h2}, becomes 1.67×10^6 W. Thus a considerable amount of the standard crustal heat flow of this area is subtracted by the lateral flow of deep ground water; the thermal energy so removed amounts to 9.99×10^6 W.

The energy balance at the discharge area of the Hakone volcano is given by

$$W_{h1} + W_{h2} + S_{ch} + M_{sf} = T_H = 11.0 \times 10^7 \text{ W}$$

where M_{sf} denotes the magmatic steam energy flow. From this the magmatic steam energy flow becomes 9.50×10^7 W. Figure 2 shows a schematic illustration of the energy balance and the redistribution of the terrestrial heat flow by the hydrothermal system. The major amount of the thermal discharge is derived from the magma reservoir.

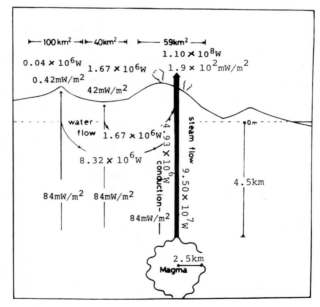

Figure 2
Schematic illustration of the thermal energy balance of the Hakone volcano and the redistribution of terrestrial heat flow by the hydrothermal system.

4. Thermal structure

In this section the thermal structure of the volcano is considered quantitatively. Disturbances of the temperature field in the region associated with the size, shape and depth of magma reservoir are discussed. For the calculation of the temperature field, we improved upon the method of McCRAKEN (1967). Figure 3 shows a steady state temperature distribution in the model, assuming a vertical vent in the central part up to 1 km depth below the surface from magma reservoir. The diameter of the vertical vent is assumed as 200 m. The depth of spherical magma reservoir is 7 km and the diameter is 5 km. The temperature of the magma is supposed to be 1000°C. In Fig. 3 letter A denotes the surface temperature. Letter H denotes 500°C. In this model the temperatures in the central part are high which is almost consistent with the observed values.

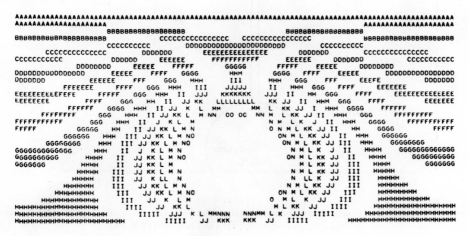

Figure 3

Steady-state temperature profile of the volcano having a vertical cent in the central part up to 1 km depth below the surface from magma reservoir. The depth of spherical magma reservoir is 7 km and its diameter is 5 km. A: 10°C, C: 150°C, E: 300°C, H: 500°C, N: 900°C, O: 1000°C.

The size and depth of magma reservoir are determined from geological evidence, i.e., the width of the caldera (KUNO et al., 1970) and the distribution of epicenters of the volcanic earthquakes (MINAKAMI, 1960; MINAKAMI et al., 1969; HIRAGA, 1972). Most micro-earthquakes that occurred in the last 20 years are concentrated in the highest temperature area at depths of foci between 1 and 5 km below the surface.

5. Summary and conclusions

The computed temperature field of the Hakone volcano is shown in Fig. 4. The temperature at the surface layer is inferred from the data of deep drillings by ŌKI and HIRANO (1970, 1974). The shape of the isotherms > 500°C is strongly affected by the geometry of the magma reservoir. The near-surface temperatures, i.e., the 50–100°C isotherms are affected by the ground water flow. The existence of the asymmetric distribution of the isotherms at shallow depth less than 1 km below the

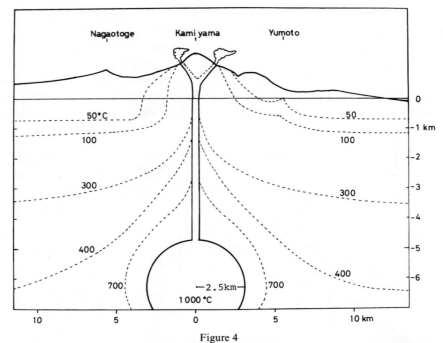

Figure 4

Computed temperature profile of the Hakone volcano. The near-surface temperatures are inferred from the data of deep drillings by ŌKI and HIRANO (1970, 1974).

surface is caused by the existence of hydrothermal system. The magma reservoir of the Hakone volcano is 7 km below the surface and its diameter is 5 km. The rising hot magmatic steam flow exists, in the central part, through a vertical vent from the magma reservoir. The diameter of the vertical vent is inferred about 200 m from the numerical calculations.

The life time of the magma reservoir, t_m, is given by $t_m = M_i/M_{sf}$ where M_i is the internal energy of the magma. The formula for M_i can be written as (cf. HAYAKAWA, 1967)

$$M_i = (4/3)\pi r^3 C\rho\, \Delta T$$

where C = specific heat of magma = 1.1×10^3 J/kg·deg, ρ = density of magma = 2.5×10^3 kg/m^3 (after NAKAMURA, 1965), and ΔT = difference of the temperature = $1000 - 400 = 600°$C (from the temperature profile of Fig. 4). Thus, the magmatic internal energy, M_i, is 1.1×10^{20} J. The life time of the magma reservoir, t_m, becomes about 4×10^4 years.

The total heat discharge from the high temperature zone (discharge area) of the Hakone amounts to 11.0×10^7 W. The magmatic steam energy discharge is 9.50×10^7 W. The thermal energy by redistribution of the terrestrial heat flow is $9.99 \times$

10^6 W, by the presence of a hydrothermal system. The thermal energy released by standard crustal heat flow is 4.93×10^6 W in the discharge area.

Acknowledgements

The numerical calculations of the thermal structure were carried out while one of us (J.I.) was staying at the Earth Physics Branch of EMR, Canada. One of us (J.I.) thanks Dr. Alan M. Jessop for his guidance.

Further, we would like to thank Dr. L. Rybach for his comments. A part of our travel expenses to IASPEI/IAVCEI meeting at Durham 1977 was defrayed by the fund of Nomura Gakugei Foundation, to whom we are grateful.

REFERENCES

HAYAKAWA, M. (1967), *Geophysical study at Matsukawa geothermal area*, Jour. J.G.E.A., *4*, 35–51 (in Japanese).

HIRAGA, S. (1972), *Earthquakes swarms of geothermal fields in Japan*, Jour. J.G.E.A., *9*, 30–39 (in Japanese).

IRIYAMA, J. (1977), *Energy balance in the earth's interior*, Tectonophysics, *41*, 243–249.

KUNO, H., ŌKI, Y., OGINO, K. and HIROTA, S. (1970), *Structure of Hakone caldera as revealed by drilling*, Bull. Volcanologique, *34*, 713–725.

McCRAKEN, D. D., *Fortran with Engineering Applications* (John Wiley & Sons, New York, 1967).

MINAKAMI, T. (1960), *Fundamental research for predicting volcanic eruptions*, Part I, Bull. Earthq. Res. Inst., *38*, 497–544.

MINAKAMI, T., HIRAGA, S., MIYAZAKI, T. and UTIBORI, S. (1969), *Fundamental research for predicting volcanic eruptions*, Part II, Bull. Earthq. Res. Inst., *47*, 893–949.

NAKAMURA, K. (1965), *Energies dissipated with volcanic activities – classification and evaluation*, Kazan, *10*, 81–90 (in Japanese).

ŌKI, Y. (1971), *The hot springs of the Hakone volcano*, in *The Hakone Volcano*, pp. 139–176, Volcan. Soc. Japan (in Japanese).

ŌKI, Y. and HIRANO, T. (1970), *The geothermal system of Hakone volcano*, Geothermics, Spec. Issue 2, Vol. 2, pp. 1157–1166.

ŌKI, Y. and HIRANO, T. (1974), *Hydrothermal system and seismic activity of Hakone volcano*, in *The Utilization of Volcanic Energy* (Proc. U.S.-Japan Coop. Sci. Sem. at Hilo), pp. 13–40.

SUZUKI, M. (1970), *Fission track dating and Uranium contents of obsidian (II)*, Quat. Res., *9*, 1–6.

UYEDA, S. (1972), *Heat flow; Crust and upper mantle of the Japanese area*, Part I Geophysics, Japan. Commit. for UMP, pp. 97–105.

WATANABE, T. (1972), *Heat flow through the ocean floor*, in *Physics of the ocean floor*, Tokai Univ. Press, Tokyo, pp. 1–107 (in Japanese).

YUHARA, K. (1972), *Geophusical aspects of the hydrothermal system*, Jour. J.G.E.A., *9*, 3–14 (in Japanese).

YUHARA, K. (1973), *Effects of the hydrothermal system on the terrestrial heat flow*, Kazan, *18*, 129–141 (in Japanese).

(Received 25th October 1977, revised 2nd May 1978)

Subject Index

A

Applied volcanology, 242
Ascent of magma, 3
Atmospheric Ar, Xe, Kr, 262
Aquifer
- charged with hot water, 309
- flow in, 309
- in volcanic rocks, 242
- in warm water, 196

B

Basaltic magmatism, 34
Basin and Range Province, 34
Bouguer anomalies, 15
Brawley geothermal area, USA, 51
Brazil, 180
British Columbia, Canada, 172
Bulk density, 75
Buoyant flow, 309
Burial history, 83

C

California, 51
Canada, 172
Carpathians, 104
Cation packing index, 75
Central Europe, 109
Coal reflectivity, 83
Coalification rank, 83
Coast Plutonic Complex, Canada, 172
Convection, two-phase, 321
Convective heat transfer, 34, 92
Continental
- crust, 3
- mantle heat flow, 65
Coso Hot Springs geothermal area, USA, 51
CO_2, 276
Crete, 150
Crimea peninsula, USSR, 104
Crustal
- geotherm, 109
- low velocity layer, 75
- melting, 15
- rocks, 75
- spreading, 51
- spreading center, 51
- temperature, 92

- thickening, 150
- thickness, 124
- thinning, 12
Cyprus, 150

D

Deep groundwater flow, 331
Denmark, 205
Density, bulk, 75
Deuterium content of steam, 276
Dniepr-Donetz depression, USSR, 104
Doublet, 297

E

Earthquake swarms, 51
Eastern Mediterranean, 150
East European Platform, 104
Economics of geothermal energy utilisation, 205
Egypt, 150, 213
Europe, 92
Extensional strain rate, 34

F

Fault plane solutions, 51
Fault zone, 309
Finite difference method, 309
Flow
- buoyant, 309
- of hot water, 309
- of producing wells, 253
- visualization, 321
Fluid transit time, 253
Fracture zone, 196

G

Gas content of steam, 262
Geochemical section, 276
Geopressurized system, 160
Geotherm, 109
Geothermal
- aquifer, 297
- exploration, 242
- fluid, 253
- gradient, 180
- potential, 242
Geothermal areas
- Brawley, 51
- Coso Hot Springs, 51

Geothermal resources
- assessment, 160
- Brazil, 180
- Canada, 172
- Denmark, 205
- Egypt, 213
- Sweden, 196
Geothermometer
- paleo, 109
- silica, 213
- sodium-potassium-calcium, 213
Gravity anomalies, 15, 124
Greenland, 15
Groundwater
- circulation, 227
- silica content, 227
Gulf of Suez, 150, 213

H
H_2, 276
Hakone Volcano, Japan, 331
Heat extraction, 290
Heat generation, 75, 104, 172
Heat flow, 34, 83, 124, 135, 150, 172, 213
Heat flow
- age relationship, 65
- corrections, 135
- crustal thickness relationship, 109
- heat generation relationship, 109
- provinces, 65
- reduced, 34
- regional, 227
- transition zones, 109
- variations, 331
Heat flow map
- of Europe, 92
- European part USSR, 104
- USA, 227
Heat production, 75, 104, 172
Heat recovery factor, 297
Heat transfer
- conductive, 321
- convective, 321
Hot dry rock, 290
Hot igneous system, 160
H_2 S, 276
Hyaloclastite, 242
Hydraulic fracture, 290
Hydrocarbon gases, 276
Hydrothermal convection system, 160

IJ
Iceland, 242
Igneous rocks, 75
Isotope exchange, 276
Israel, 150

Italy, 135, 253, 262, 276
Japan, 331
Jemez Plateau, New Mexico USA, 290

L
Larderello, Italy, 253, 262, 276
Lava pile, 15
Levantine Sea, 150
Lithosphere
- continental, 150
- cooling, 65
- oceanic, 150
- thermal models, 34
Lithostatic equilibrium, 3
Low enthalpy
- geothermal resource, 205
- geothermal system, 180

M
Magma
- ascent of, 3, 15
- density of, 3, 15
- reservoir, 3, 331
- traps, 3, 15
Magmatic steam discharge, 331
Magmatism, basaltic, 34
Mantle heat flow, 92, 109
Mantle heat flow, continental, 65
Mediterranean plate tectonics, 135
Metamorphism of organic matter, 83
Metamorphic rocks, 75
Moho temperature, 109

N
N_2, 276
New Zealand, 15
Noble Gases, 262
Nusselt number, 321

O
Oceanic crust, 3
Organic matter
- metamorphism of, 83
- in sedimentary rocks, 83
Oxygen isotopes, 276

PQ
Paleo-geothermometer, 83
Parana sedimentary basin, Brazil, 180
Partition coefficients of noble gases, 262
Pillow lava, chemical composition of, 242
Plate tectonics, 172
- Mediterranean, 135
Poland, 109
Porosity of volcanic rocks, 242
Porous layer, 321

Precambrian
- granite, 196
- rocks, 213
- shield, 180
Production well
- flow rate, 253
- spacing, 262
Quaternary volcanism, 172

R
Radioactive heat production, 65, 75, 104, 172
Radiogenic ^{4}He, ^{40}Ar, 262
Radon-222
- concentration of steam, 253
- emanating power of rocks, 253
Recharging water, 262
Red Sea, 213
Reduced heat flow, 34
Regional heat flow, 227
Reinjection, 297
Reservoir lifetime, 297
Resource
- assessment, 160
- base, 160
Rhine Graben, Germany, 83
Rock-water reactions, 276
Romania, 124

S
Sedimentary basin, 205
Seismic velocity, 75
Silica geothermometer, 213, 227
Silica content of groundwater, 227
Sodium-potassium-calcium geothermometer, 227
Steam
- deuterium content of, 276
- deuterium content of, 276
- gas content of, 262
- magmatic discharge, 331
- reservoir, 253
Strain rate, extensional, 34
Stress directions, 51
Strike-slip faults, 51
Sweden, 196

T
Tectonic extension, 34
Thermal
- conductivity, 83, 124
-convection, two-phase, 321
-energy redistribution, 331
-history, 83
-metamorphism, 276
-models of lithosphere, 34, 65
Temperature-depth profiles, 104, 124
Temperature field, two-dismensional, 309
Temperature gradient, 213

U
United States, 160, 227, 290
Urban heating, 297

V
Vapor dominated field, 262
Vertical fault, 309
Volcanology, applied, 242
Volcanism, quaternary, 172

W
Water isotopes, 276
Water-steam boundary, 321
Water velocity, 321

Date Due